Renewable Energy and Landscape Quality

Renewable Energy and Landscape Quality

Michael Roth
Sebastian Eiter
Sina Röhner
Alexandra Kruse
Serge Schmitz
Bohumil Frantál
Csaba Centeri
Marina Frolova
Matthias Buchecker
Dina Stober
Isidora Karan
Dan van der Horst
(eds.)

Content

PREFACE

Climate change was considered the biggest potential threat to the global economy in a survey of 750 experts at the World Economic Forum in 2016 (http://reports.weforum.org/global-risks-2016/). This risk is linked to other global risks such as social instability and large-scale involuntary migration (ibid.) which shows the interrelation between environmental, economic, and socio-cultural aspects at the global scale. Both the problems of climate change mitigation/renewable energy production and the loss of landscape/environmental quality have to be addressed at various scales from global policy down to local action.

On a regional and local level, Nuertingen-Geislingen University (NGU) as a university of applied sciences intensively pursues inter- and transdisciplinary research and teaching of economic, ecological, and societal aspects of sustainable development. The German name of the University—Hochschule für Wirtschaft und Umwelt or University for Economy and Environment—underlines this integrative approach. Also, on the research map of the German Rectors' Conference, NGU is included with two research priorities related to environment/landscape and energy/economy:

- Applied agricultural research, landscape development, environmental planning and nature conservation
- Sustainable management in the energy, automotive, and real estate industries

Against this background, the COST Action TU1401 'Renewable Energy and Landscape Quality' has been fully within the research scope of our university and contributed significantly to the international visibility of NGU as a research institution with a strong focus on transfer and application. Leading an international research network of this size with more than 200 participants from 37 countries in Europe and beyond would not be possible without both institutional support of the university as well as personal dedication and devotion of the faculty and staff involved.

With 97 contributing authors, the book *Renewable Energy and Landscape Quality* as a main product of the four-year COST Action shows the potential of international and interdisciplinary collaboration. I hope that this book finds a responsive audience, so that future policies, political decisions, and planning documents can contribute to optimise trade-offs between renewable energy systems and landscape quality protection by promoting an effective and efficient renewable energy policy without jeopardising the assigned values and inherent qualities of European landscapes.

Nuertingen, April 2018

Prof. Dr. Carola Pekrun
Vice-Rector for Research and Transfer at
Nuertingen-Geislingen University

INTRODUCTION

Michael Roth & Sebastian Eiter

In response to climate change, limited fossil fuels, and rising energy demand and prices, renewable energy is being heavily promoted throughout Europe. While objectives to boost renewable energy and trans-European energy networks are ambitious, it is increasingly understood that public acceptance becomes a constraining factor, and general support for green energy does not always translate into local support for specific projects. Perceived landscape change and loss of landscape quality have featured heavily in opposition campaigns in many countries, even though renewable energy can facilitate sustainable development, especially in disadvantaged regions rich in wind, water, biomass, geothermal, or solar energy.

Climate change mitigation and adaptation is a major societal challenge, and renewable energy is a core element in the transition to a low-carbon society. This will reshape our landscapes. It is unlikely that existing landscape management mechanisms will be effective in adapting to climate change and facilitating renewable energy development. New deliberative, interdisciplinary, and integrated approaches are needed to inform and guide the transformation process and to create a vision and coalition for reconciling renewable energy systems and landscape quality across public, stakeholders, and sectoral, administrative, and geographical boundaries.

Against this background, COST Action TU1401 'Renewable Energy and Landscape Quality (RELY)', running from 16 October 2014 to 15 October 2018 investigated the interrelationships between renewable energy production and landscape quality, and the role of public participation for the acceptance of renewable energy systems. Starting as a relatively small network with around 20 academics from 18 institutions in 13 European countries and

Canada at the proposal stage, the partnership grew rapidly over the lifetime of the Action: more and more countries joined the Action and individual attention was raised through networking tools and events like training schools, special sessions and co-organisation of scientific conferences, and a traveling exhibition. In the final phase of the Action, the research network consisted of more than 200 individual members from nearly 100 institutions (academic, governmental, and non-governmental) in 35 European countries, Canada, and Israel. The disciplinary backgrounds of the members involved include social sciences, engineering, political sciences, and interdisciplinary fields like geography, landscape planning, and landscape architecture. With this wide coverage in terms of geographical scope and disciplinary background, the Action network formed an ideal basis to overcome fragmented national and sectoral research, language, and cultural barriers. Moreover, the Action consolidated existing research networks across the natural science/social science/engineering divide, thereby creating a network of networks:

- EEEL: Emerging Energies, Emerging Landscapes
- PECSRL: The Permanent European Conference for the Study of the Rural Landscape
- EUCALAND: European Culture expressed in Agricultural Landscapes
- RESERP: Spanish Renewable Energy and Landscape Network
- IALE-Europe: International Association for Landscape Ecology—European Chapter
- NLRN: Nordic Landscape Research Network
- NIES: Nordic Network for Interdisciplinary Environmental Studies

This book presents the results of almost four years of collaboration. The large network of the Action has made it possible to produce a pan-European synopsis of 32 contributing countries regarding their national situations concerning renewable energy and landscape quality (section 1).

The Action was organised in four working groups (WGs): WG 1 reviewed specific renewable energy production systems and their impacts on landscape character and quality in Europe from a past, present, and future perspective and produced a systematic review of the nexus between renewable energy systems and Europe's landscapes' qualities (section 2). WG 2 assessed landscape functions and qualities and their sensitivity to and potential for specific renewable energy production systems. These analyses were used to produce: (i) a typology of best practices of sustainable, landscape-compatible renewable energy production systems, (ii) guidance for assessing the potential of areas for specific renewable energy systems in terms of effects on landscape quality or character, (iii) a catalogue of relevant criteria, indicators, and respective GIS-available proxy-data for assessing the suitability of landscapes for renewable energy systems (section 3). WG 3 investigated socio-cultural aspects of sustainable renewable energy production and proposed modes and means of integrating specific aspects of renewable energy in participatory toolkits to increase public acceptance of renewable energy projects (section 4). WG 4 focused on the synthesis of findings, the dissemination of results towards different target groups, and the facilitation of collaboration across working groups by providing a multi-lingual glossary of terms (section 5).

COST stresses cooperation in science and technology by addressing academics, public and private (research) institutions, as well as non-governmental organisations (NGOs), in order to increase research impact on policy-makers, regulatory bodies, and national decision-makers as well as on the private sector. That emphasis is also reflected in this book: to supplement existing communication channels like scientific articles, conference presentations, and the Action's website (http://www.cost-rely.eu/), a book format and layout were chosen, which is intended to motivate potential readers to explore the multi-facetted aspects of renewable energy landscapes. At the same time, the book addresses policy-makers at EU and national levels as well as decision-makers in public agencies and business to encourage internationally accepted best practice. Following the general principle of the European Landscape Convention, and general provision of the Aarhus Convention on Access to Information, Public Participation in Decision-Making and Access to Justice in Environmental Matters, that the public is entitled to environmental information, this book can also inform and empower citizens and NGOs to build on solid research results in participation and decision-making processes.

It is with great appreciation that we acknowledge the funding provided by the COST Association over the past four years as part of the EU Framework Programme Horizon 2020. Without that specific funding scheme that allows both cooperation and exchange, targeting a wide geographical scope across Europe and beyond, leveraging national research investments and building capacity by connecting high-quality scientific communities in Europe and worldwide, this book would not have been possible.

0.2

COST RELY FACTSHEET: A SUCCESS STORY

Sina Röhner & Alexandra Kruse

Participating countries in 2018	37
ITC countries	18
Individual participants	200
MC meetings, always combined with WG meetings	7
WG meetings	14
Core group meetings	4
Thematic meetings	2
Training schools	2 with 44 participants
Co-organisers of conferences	3
STSM (Short term scientific missions)	17
Exhibitions of the COST RELY Travelling Exhibition	15
Translation of the COST RELY Flyer	10
Publications (from 2014-2018 in April, around 10 to come)	37
Surveys	2
Case Studies collected	WG 2: 51 in 20 countries, WG 3: 25 in 12 countries
COST RELY Glossary	48 terms translated into 28 languages including Esperanto
Photo Database with special RE types, RE landscapes and RE and landscape quality	> 100 photos
Photo competitions	3

Table 0.2.1
COST RELY in figures

The main objective of the Action was to develop a better understanding of how European landscape quality and renewable energy deployment can be reconciled to make socio-environmental contributions to the sustainable transformation of energy systems. Four Working Groups put their focus on different aspects during the four-year lifetime of the Action:

1. Renewable energy production systems and impacts on landscape quality
2. Landscape sensitivity and potentials in terms of renewable energy production
3. Socio-cultural aspects of sustainable renewable energy production
4. Synthesis of findings and dissemination

The Core Group of the Action consisted of the Action Chair Michael Roth from Germany and Action Vice-Chair Sebastian Eiter from Norway, the working group chairs and vice-chairs as listed in Figure 0.2.1, and the STSM Coordinator, Serge Schmitz from Belgium. The position of WG4 vice-chair was transferred during the Action from Malgorzata Lachowska (Poland) to Isidora Karan.

In addition the activities of the working groups, the Action was quite active in dissemination activities and events. Figure 0.2.2 shows the time table of the work done during the four years of the Action.

The Action was submitted by academics from 13 European countries plus Canada. At the kick-off meeting in October 2014, the Action had already grown to members from 27 European countries plus Canada, and it kept growing to 200 participants from

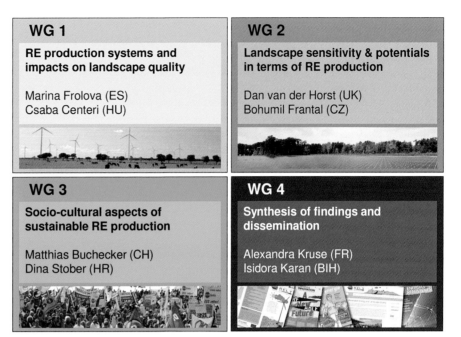

Figure 0.2.1
The four working groups of the COST Action RELY and their topics

Milestones and timetable

Duration: October 2014 - October 2018

Activity	Year			
	1	2	3	4
Kick-off phase	■			
Working Group 1: Systematic review, meta-analysis				
Working Group 2: Strategic case studies				
Working Group 3: Multidimensional scenarios				
Working Group 4: Synthesis, dissemination				
Milestones				
Meeting, incl. kick off meeting	X X	X X	X X	X X
Annual progress report	X	X	X	
Action conferences	X	X		X
Training Schools with special focus on ECI		X	X	
Publication of a comprehensive Action book				X
Final Action report				X

Figure 0.2.2
Timetable of RELY

35 European countries as well as from Canada and Israel until the final conference in September 2018.

Besides Cyprus and Luxembourg **inclusiveness target countries** (ITC) were well represented in the Action. Almost 50 % of the participating countries and almost 40 % of participants belong to ITCs, as well as all four WG vice-chairs. The share of participants from IT countries at meetings was between 40 % (Lisbon, Portugal, 2015) and up to 67 % (Brno, Czech Republic, 2018). Nearly 65 % of the STSMs between 2015 and 2018 were carried out by members from ITCs and nearly half of the participants of the two training schools also came from ITCs. Meetings in ITCs were held in Bosnia & Herzegovina, Czech Republic, Croatia, Hungary, Portugal, and Slovenia.

Regarding **gender balance** COST RELY was doing fine: 47 % of all participants were female. In the Core Group, 50 % of the members were female. The share of female participants at meetings was between 37 % (kick-off Meeting) and 53 % (Lisbon, Portugal, 2015). Half of the STSMs were carried out by female participants and 59 % of the training school participants were also female.

The action chair and three out of four WG vice-chairs are **early career investigators** (ECI), who were also well represented within the whole Action.

Figure 0.2.3
Countries participating in
the COST Action RELY.
Author: Sina Röhner.

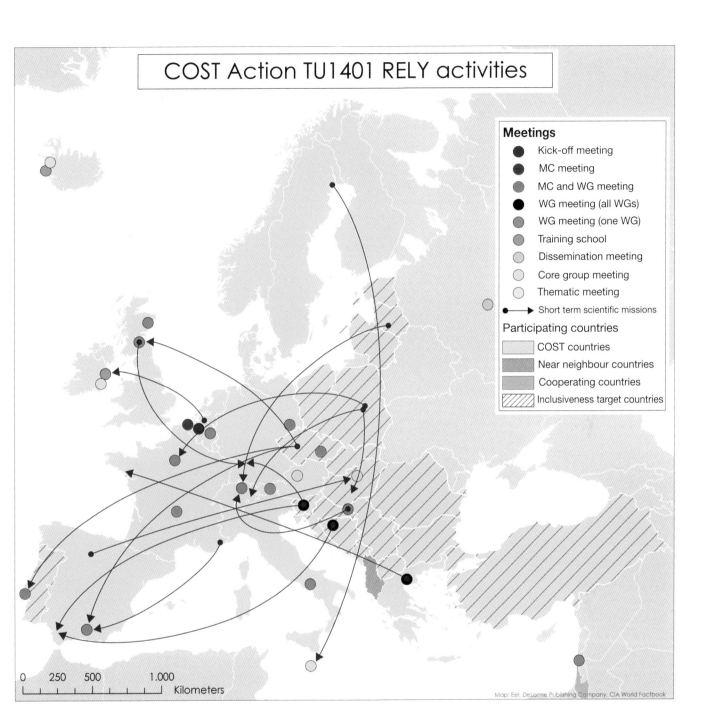

Figure 0.2.4
Meetings and STSMs of
the COST Action RELY.
Author: Tadej Bevk.

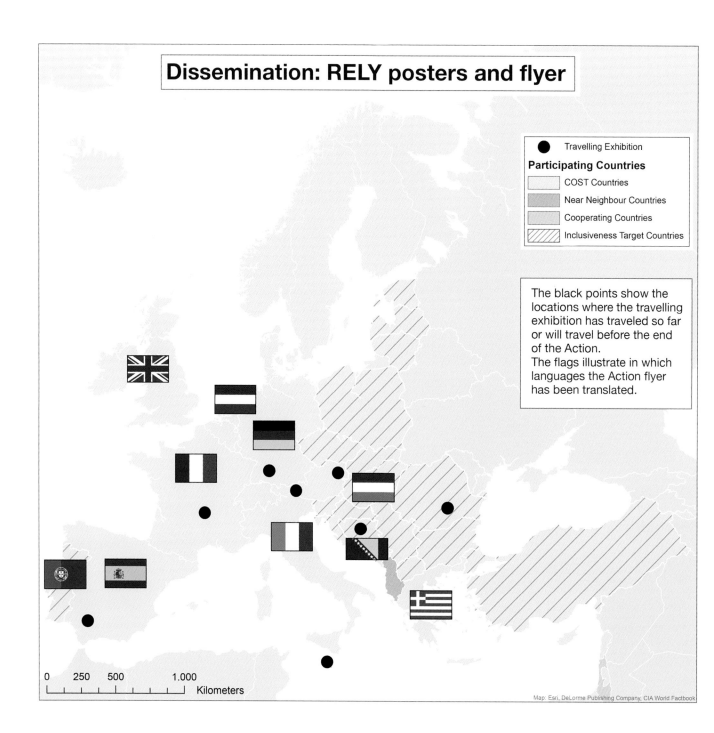

Figure 0.2.5
Travelling exhibition
and flyer translations.
Author: Sina Röhner.

1 NATIONAL OVERVIEWS

Renewable energy (RE) and landscape quality in the different countries of the COST Action are characterised by numerous similarities and differences, according to the presence of natural resources and their use histories across Europe. Similarities and differences are related to, for example, types of energy production, when production started, capacity installed, amounts of energy produced, procedures for landscape character assessment, rules and practices of landscape protection, and the role of landscape in planning processes with respect to RE production. To illustrate the diversity and commonality of characteristics in as meaningful a manner as possible, Action members from participating countries were invited to contribute with brief overviews of the national situations of renewable energy, data available on landscape quality, and the interactions between renewable energy and landscape quality.
Authors from 33 countries provided contributions. Countries are presented in alphabetical order, basically according to the official COST Member States, nomenclature (www. cost.eu/about_cost/cost_member_states), and all texts follow the same structure. Differences in content may, for example, reflect the different professional backgrounds of the authors. National experience in landscape management and data availability relating to landscape quality differs. For several countries the most reliable data were on renewable energy consumption (e.g. from Eurostat) rather than production. However, as countries can cover demands for energy through importing electricity or, in case of renewable electricity, through purchasing certificates from producers in other countries, energy consumption is not necessarily reflected in the national landscape. Landscape quality impacts happen first and foremost where renewable energy is produced, independent of whether the energy is consumed domestically or exported. On the basis of Eurostat data, Figure 1.0.1 presents a first insight of RE primary production and gross inland consumption across Europe.

Sebastian Eiter & Serge Schmitz

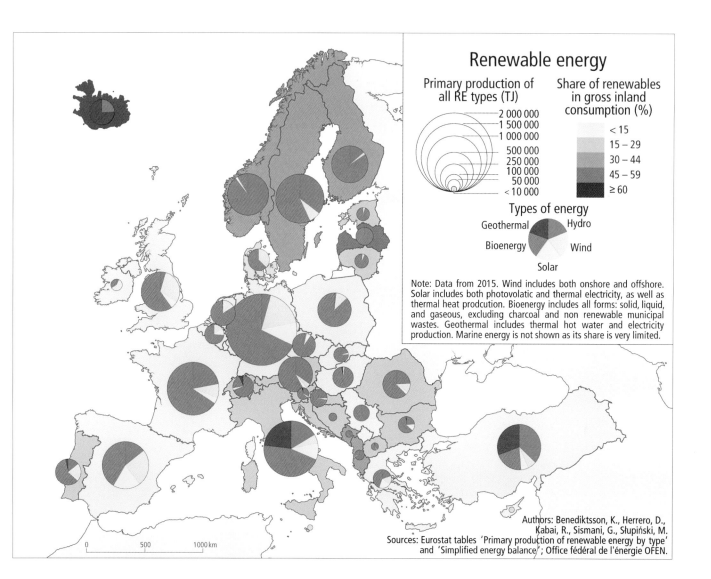

The map legend reads:

Renewable energy

Primary production of all RE types (TJ)

2 000 000
1 500 000
1 000 000
500 000
250 000
100 000
50 000
< 10 000

Share of renewables in gross inland consumption (%)

< 15
15 – 29
30 – 44
45 – 59
≥ 60

Types of energy

Geothermal
Hydro
Bioenergy
Wind
Solar

Note: Data from 2015. Wind includes both onshore and offshore. Solar includes both photovolatic and thermal electricity, as well as thermal heat prodcution. Bioenergy includes all forms: solid, liquid, and gaseous, excluding charcoal and non renewable municipal wastes. Geothermal includes thermal hot water and electricity production. Marine energy is not shown as its share is very limited.

0   500   1000 km

Authors: Benediktsson, K., Herrero, D., Kabai, R., Sismani, G., Słupiński, M.
Sources: Eurostat tables 'Primary production of renewable energy by type' and 'Simplified energy balance'; Office fédéral de l'énergie OFEN.

Figure 1.0.1
RE primary production and gross inland consumption across Europe. Authors: Karl Benediktsson & Daniel Herrero-Luque.

1.1

ALBANIA

Sokol Dervishi & Artan Hysa

Figure 1.1.1
Albanian RE and landscape
resource use diversification
(Original source: AKPT)

Situation of Renewable Energy

Albania relies on hydropower for about 90 % of its electricity production. Historically, domestic energy production primarily comes from the use of big and medium hydropower plants which implies a relatively high risk and uncertainty regarding the security of supply and vulnerability to climatic variations. Climate variability can considerably affect energy production, and climate change is creating further challenges (Ebinger 2010).

The Drin river provides about 90 % of domestic electricity generation. It delivers power for local industry and households. Hydropower production can vary between almost 6 TWh in very wet years to less than half of that amount in very dry years. Limited energy interconnections and inefficiencies in domestic energy supply, demand, and water use constrain the ability to manage energy challenges (IFC 2010). In addition, the needs of the agricultural sector for irrigation is a great concern, especially during the dry summer period. Efficient use of water is vital for irrigation systems and for adopting mechanisms that support the energy and agricultural sectors.

A National Energy Strategy aims to improve energy security. It describes plans to extend capacity, diversify the energy system, and encourage the development of alternative RE generation (Ministry of Energy and Industry 2015). Figure 1 illustrates the diversified RE and landscape resource use potential in Albania. Alternative RE installations might decrease the dependence on imports and improve the security of the energy supply and even

the macroeconomic and political security of the country through decreasing the domestic budget deficit. Albania is characterised by a high potential for renewable energy estimated at around 25 GW capacity of small hydropower, wind, solar, and geothermal energy. The Adriatic coastline has unexploited potential for wind power. Zones have been identified (Ebinger 2010) which are characterised by an average annual wind speed of 6 to 8 m/s and an energy density of 250 to 600 W/m^2. In addition, agricultural waste provides considerable potential for biomass, assessed to about 2.3 TWh/year.

Data on Landscape Quality

Albania has various types of natural landscapes, with terrain comprising 28 % mountains, 47 % hills, and 25 % plains up to 300 m above sea level. The average altitude is 708 m, twice as high as the European average. Although it occupies no more than 0.3 % of European territory, more than 30 % of Europe's flora and fauna are present in Albania (Mullaj et al. 2017).

Albania also includes a variety of cultural landscapes. These can be categorised into three main groups: (i) traditional, (ii) socialist, (iii) and post-socialist. Traditional landscapes consist of historical sites originating from the Illyrian, Hellenic, Roman, and Ottoman periods, some of which are listed World Heritage sites (UNESCO 2015). The four socialist legacies in the Albanian landscape (Rugg 1994) are 1) reclamation of the Myzeqe Plain, 2) new urban centres, 3) use of the Drin River for

Labels in image: wind energy, citrus, tourism, aquaculture, oil/olive, F VAJI, vineyards, solar energy, hydrocarbone, thermocentral, hydropower, agricultural products, mountai..., livest..., agro

hydroelectricity, and 4) the creation of socialist Tirana. The first two post-socialist decades were dominated by a process of 'anarchical' urban and peri-urban expansion due to free domestic migration (Strazzari 2009). During the second half of the 1990s, Tirana expanded in area by 37 %, and its suburbs by 400 % (Aliaj & Lulo 2003).

In recent years, efforts have been made to consider landscape quality as a crucial objective of the national development strategy and aspirations of EU membership. Institutions which are responsible for territorial development have been created, such as the National Agency of Territorial Planning, the National Authority of Geospatial Information, the National Environmental Agency, the National Agency of Protected Areas, and the National Inspectorate on Environment, Forests and Waters. The Territorial and Administrative reform in 2014 was an attempt to reduce the fragmentation of decision-making through the amalgamation of 374 administrative units into 61 municipal districts (Ndreu 2016). A further example is the law which froze all construction permits at a national level between 2014 and 2016.

According to the State of Environment Report prepared by the National Environmental Agency, in recent years a main priority of the government has been the designation of new Protected Areas. This has resulted in a progressive expansion, starting from 6.4 % of the national area in 2005 up to 16 % in 2013 (NEA 2015). The protected areas in Albania consist of 2 strict/scientific nature reserves, 14 national parks, 750 natural monuments, 22 managed nature reserves, 5 protected landscapes, and 4 RAMSAR areas (Haska 2010). Besides state institutions, there are non-governmental organisations such as the Institute of Nature Conservation, which aim to increase citizen participation.

Albania is a collaborating member of the European Environment Agency (EEA). Consequently, there are a variety of periodic assessment reports on environmental quality and biodiversity. The only source of data on landscape assessment is that of CORINE Land Cover. Albania remains one out of seven countries of the 47 member states of the Council of Europe that has not signed the European Landscape Convention. Landscape character assessment, landscape quality assessment, visual landscape assessment, and landscape impact assessment are research fields to be developed in the country.

Interaction between Renewable Energy and Landscape Quality

Over recent decades, unique natural sites have come under threat. Apart from deforestation due to uncontrolled harvesting, and degraded natural monuments due to lack of maintenance, the natural landscapes face mega-sized infrastructure projects. Projected hydropower plants in the cascade of the Vjosa River and in the valley of Valbona are considered threats to the most significant natural and cultural landscapes of Albania, generating a lot of controversy.

AUSTRIA

Gerald M. Leindecker

Figure 1.2.1
First alpine photo-
voltaic plant at the
Mountain Loser
(Photo: Gerald
Leindecker)

Situation of Renewable Energy

Hydropower has been an important source of energy in Austria for a considerable time. More recently, development of photovoltaic plants and wind turbines has increased. The opportunities for their uptake improved due to the development of taller wind turbines and better price-to-performance ratio of photovoltaic panels. The first wind turbine was built in 1994 in St. Pölten with a rating of 0.11 MW. By the end of 2017 there were 1,260 wind turbines with a total capacity of 2,844 MW. Sixty-eight new wind turbines, equivalent to a generating capacity of 210 MW, are planned for 2018. In Munderfing five 140 m towers are operational with a total capacity of 15 MW.

The use of small-scale solar thermal panels at a household level started in the late 1990s, with significant uptake in areas such as Upper Austria which currently has 1,309,000 m² of panels installed, providing a capacity of 920 MW. The production of electricity from photovoltaics contributes 2 % of the total demand, with the plan to increase this to 15 % by 2030 and to 27 % by 2050. This should require a surface of 174 km² of south-facing roof space. The current, existing potential south facing roof and elevation space covers 230 km² (Fechner et al. 2016). The first alpine photovoltaic plant was installed in 1988 on the mountain

Loser at an altitude of 1,600 m. Panels covering an areas of 263 m² have a capacity of approximately 30 kW, the output from which is also used for the ski lift. The loss of performance after 30 years is only 10 %. In the mountains, the peak performance is in June and July, compared to April and May in the lowlands. This is because of the high solar input together with a comparatively low temperature in the cells. The problem of snow in winter is neutralised by fewer days with fog and dust compared to regions at lower altitudes.

The idea of an Alpine photovoltaic plant to support the energy demand of ski lifts was put into practice in 2010 on the Wildkogel, at an altitude of 2,100 m. This plant provides 75 % of the current electricity demand for the nearby ski lift. In this high-altitude site, the brightness of the sky is enough to produce electricity even without direct sunlight. One of the largest photovoltaic parks in Austria is a 1 MW plant in Eberstalzell consisting of 8,000 m² of panels, i.e. an area the size of two soccer fields.

Data on Landscape Quality

Austria is organised into nine federal states. These state governments are responsible for legislation on regional planning which

Figure 1.2.2
Photovoltaic park
in Eberstalzell,
1 MW 2010
(Photo: Gerald
Leindecker)

includes landscape assessment and the management of ecologically sensitive areas such as lakes, rivers, and mountains. Additionally there are some special zones such as world heritage sites, national nature parks and European wild corridors designated under Natura 2000. In these areas landscape quality is protected and landscape assessment requires a full analysis of all local parameters such as zoning, population development, economic parameters, wildlife, heritage buildings, and landscape scenery. Changes in the built environment (including installation of photovoltaics) need the permission of the local building department as well as consent of the regional nature protection department on the protection of landscape scenery.

Interaction between Renewable Energy and Landscape Quality

For the location of wind turbines, relevant Austrian legislation is at the federal level. In the federal state of Lower Austria the minimum distance between wind turbines has to be 1,200 m, whereas in Upper Austria the equivalent was changed in 2017 from 800 m to 1,000 m for wind turbines of greater than 0.5 MW. There is also a federal masterplan for potential wind turbines sites and for areas in which they are forbidden (Land OÖ 2017). For photovoltaic plants in the non-built environment, freestanding plants greater than 500 m^2 require nature protection permission or in sensitive areas a landscape protection permission (e.g. within 500 m from lake shores or 200 m from rivers).

Landscape quality is highly significant with respect to securing public acceptance of RE projects. In the built environment this becomes a matter of design. Since the predicted growth in RE production to 2030 needed to meet the energy plan is based on photovoltaics, the full integration of panels into the built envelope becomes of great significance. New design solutions utilising new technology panels means that photovoltaic panels can function as well as solar thermal panels in an integrated manner (Leindecker & Krstic-Furundic 2017). Such an integrated approach needs to be manifested into new building codes.

With respect to wind turbines, there is a need for a communication plan to ensure that the advantages are understood. For example, electricity generation from one wind turbine equals an area of six soccer fields of photovoltaic cells. For photovoltaics, the above-mentioned need of full integration in the building infrastructure, and the construction of combined systems in the Alps for tourism (e.g. ski lifts) and renewable energy production, need to be reflected in future regulations.

1.3

BELGIUM

Symi Nyns, Anneloes van Noordt & Serge Schmitz

Situation of Renewable Energy

A share of RE in gross final energy consumption of 7.9 % in 2015 is one of the lowest in Europe, although it has increased from 2.6 % in 2006. Looking closer at the share of RE, it comprised 15.4 % in electricity, 7.3 % in heating and cooling, and 4.8 % in transport (Eurostat 2017b). Biofuels and waste were the main renewable sources, providing 83 % of total primary energy supply. Wind power accounted for 10 % and solar energy for 7 % in 2014 (IEA 2016). A favourable policy environment means that these two sources of renewable energy were mainly responsible for the increase in RE production over the last decade.

The development of RE is overseen by regional authorities and the federal authority, in line with their sustainable energy policies. Regional authorities deal with regulations and set targets, while the federal authority manages the overall development of offshore wind energy. Belgium follows the EU 20-20-20 strategy with a goal of 13 % RE in gross final energy consumption by 2020. Diverse measures to develop green electricity production have been implemented such as green certificates (SPF Economie 2010).

Regarding the shift to a low-carbon economy, Belgium faces three main challenges. The first is the high level of suburbanisation, resulting in a dominance of car transport and reducing the opportunities to site wind turbines. The second is that a significant majority of the building stock is old and inefficient to heat.

The third is the ongoing debate and uncertainty about nuclear energy, resulting in an unfavourable investment climate for RE. The following can be said for each RE source (APERe 2017):

- Wind energy: 938 wind turbines in 2016 (756 onshore and 182 offshore) with a total installed capacity of 2386 MW. Belgium's offshore wind power capacity is the fourth highest in Europe. The development onshore lost momentum over the last five years because of a lack of clarity in regulations and the increase in opposition of residents against new projects. However, an increase in the amount of installed capacity has occurred over the last year.
- Solar energy: photovoltaics had a collective installed capacity of 3,423 MW in 2016. Three quarters of the facilities are in Flanders. Because the Flemish government stopped subsidising solar panels in 2013 and the Walloon Region changed the rules in 2014, Belgium faced a temporary decrease in installed capacity. Recent years show an increase again.
- Hydroenergy is produced by 153 hydroelectric power plants in 2016, 136 of which are situated in Wallonia with a total capacity of 106.5 MW. The major role of these plants is the storage of electrical energy through power plants of pump-turbines for meeting peaks in consumption.
- Geothermal energy is mainly localised in Mons. There is potential in Flanders where tests are ongoing.

1.3.1
Biogas production on farm, new landmarks in the countryside (Photo: Serge Schmitz)

• Biomass energy provided 96 % of the renewable heat consumption and 38 % of renewable electricity production in 2014. In Flanders, 190 installations produced 2.6 TWh of electricity and 1.7 TWh of heat in 2016. In addition to its considerable forest resources that produce approximately 3.5 TWh and two first-generation biofuel plants (114 GWh) (Quadu 2013), Wallonia has 49 biomethanisation units which produce 188 GWh of electricity and 195 GWh of heat (Val Biom 2017).

Data on Landscape Quality

Belgian landscapes are diverse with a pronounced regional character and identity. Belgian agriculture specialises in livestock breeding and is characterised by a continuous increase in efficiency and productivity. It configures the landscape with intensive farming and small farms in Flanders, and with extensive agriculture and big farms in Wallonia. Moreover, the Belgian landscape is marked by industrialisation and urbanisation, and in particular suburbanisation.

In accordance with the ELC, landscape types are defined and grouped into landscape units. In Flanders, there is an atlas of traditional landscapes with a description, cartography, and an inventory of cultural and historical characteristics; while in Wallonia, there is a classification of landscape areas documented by several sub-regional atlases in the form of monographs that detail the scenery and landscape dynamics (Van Hecke et al. 2010).

Interaction between Renewable Energy and Landscape Quality

RE projects must respect environmental and urban regulations. Rules can be different in Flanders and Wallonia, due to differences in population density. Sometimes specific processes are prescribed for technologies according to the size of installations. Criteria for authorisation are represented in regional codes of territorial development. Moreover, landscape uses are integrated into regional strategic plans; town and country planning should also integrate landscape perspectives. The number of wind farms increased over the last decade, but faced strong opposition from residents. Now, the focus is on offshore wind turbines and, due to legislative restrictions, on onshore installations in harbours and industrial areas or along motorways. Photovoltaic panels are mostly placed on rooftops, the preferred location from a planning perspective. Large solar fields on the ground are rare. Anarchic development created financial issues, problems for the electricity grid, and inappropriate visual impacts in traditional neighbourhoods. Local companies have developed 'invisible' solar panels but they are not yet in use.

BOSNIA AND HERZEGOVINA

Isidora Karan & Igor Kuvac

Situation of Renewable Energy

Bosnia and Herzegovina has a GDP of 14,975 M euros, an area of 51,197 km², and a population of 3.77 million. Energy issues and landscape protection are under the jurisdiction of two entities and one district with different legal frameworks. Gross production of electricity is 15,629 GWh of which 36 % comes from RE sources. Installed capacity of RE plants in 2014 was 2.1 GW, of which 2 GW was from hydropower. The total number of RE plants is 169.

The country has a great potential and a long history of hydropower production. The first plant was constructed in 1899 on the Pliva River, which was the largest in Europe at that time, with a capacity of 7 MW (B&H CPNM 2017). There is potential for energy production from other RE types. There is potential to exploit geothermal resources in the northern part of the country (Pannonia Basin). The southern part of the country belongs to the Mediterranean region with a great capacity for solar and on-shore wind energy production. As forests occupy approximately 53 % of the country's biomass is also considered as an important energy resource (Gvero 2017).

Data on Landscape Quality

There is no national or regional landscape assessment, except for protected areas or as part of planning documents. Landscape quality is mainly regulated by a law on national parks and nature parks. There are three national parks, three nature parks and four protected natural areas (RIPCHNH RS 2016).

Interaction between Renewable Energy and Landscape Quality

For RE projects that could have a significant impact on the landscape in terms of their nature, size, or location, an environmental impact assessment must be carried out according to the national laws that regulate landscape protection and energy production.

Awareness of landscape quality and planning processes for RE is increasing, and becoming the subject of community resistance. The level of public involvement is low, seen as being imposed by international agreements, and not as a substantial and important part of planning processes. The public relates neither to the system nor with institutions they do not trust (Transparency International B&H 2014). Public criticism is usually not clearly

Figure 1.4.1
Delibašino selo, the first
small hydropower plant
constructed in Banja Luka
in 1899.
(Photo: Sebastian Eiter 2017)

articulated and public participation tools are underdeveloped. However, there are some positive examples of RE projects involving stakeholders and the public, such as construction of the hydropower plant Čajdraš in Zenica Canton. The most valuable lesson learned from this project was the responsible management of natural resources through respecting all legislative procedures and including active public participation.

Table 1.4.1
RE production
in Bosnia and
Herzegovina

RE type	First plant constructed	Number of plants (2017)
Small and micro hydropower	1899	56
Large hydropower	1899	16
Solar PV power	2012	96
Biomass	under construction	
Biogas	under construction	
Wind onshore energy	under construction	27 planned
Geothermal power	under construction	

1.5

BULGARIA

Vania Kachova

Structure of RE production in Bulgaria 2016

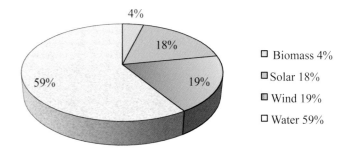

- Biomass 4%
- Solar 18%
- Wind 19%
- Water 59%

Figure 1.5.1
Structure of RE in
Bulgaria (2016)

Situation of Renewable Energy

The development of RE production is a strategic priority for Bulgaria because it reduces the country's dependence on external energy sources and supports the transition to a sustainable energy system. The strategic documents for the development of RE aims at a 16 % share of RE in the final consumption by 2020. However, as early as 2014, the share was 18 % (of which 5.3 % was transport), and by 2016 the share had reached 18.2 %.

Resource	Installed capacity	Production	
	(MW)	(GWh)	%
Water	3327.0	4 438	59
Solar	1027.4	1 381	18
Wind	698.7	1 425	19
Biomass	51.6	19	4

Table 1.5.1
Installed capacity and
electricity production
from RE sources in
2016 (Hydroenergy
Association 2017)

Bulgaria has a law relating to the production of renewable energy and a directive for the Promotion of the Use of Energy from Renewable Sources.

In 2009 Bulgaria was ranked 59th in the world in terms of energy production, with a total of 40 TWh. However, the energy efficiency in Bulgaria is four times lower than the European average. In addition, Bulgaria is dependent on energy imports, for example, in 2004 70 % of the final energy consumption was from imported oil, gas, and nuclear sources.

The first power plant using a renewable production source, a hydroelectric plant, was opened in 1900 in Pancharevo, 18 km from Sofia on the Iskar River. It was the first such plant on the Balkan Peninsula and had a capacity of 1.6 MW. It marked the beginning of a public electricity supply in the country, and the capital Sofia became one of the first electrified cities in Europe. Now, the main potential for development of RE in Bulgaria is from wind, sun, and water. Geothermal sources, tidal energy, and biomass offer lower potential return.

Data on Landscape Quality

Bulgaria has ratified the ELC, and landscape studies are taught at universities. The country has landscape types based on different features. The Bulgarian landscape classification consists of 5 zones, 17 areas, and many sub-areas of which large-scale landscape maps have been produced (Batakliev 1934, Petrov 1997). Typological, large-scale landscape maps provide a good basis for identifying physical-geographical units. The application of the method provides good opportunities to set different prerequisites for the use of their resources and to protect the environment.

Figure 1.5.2
Wind turbines in
Bulgaria (Photo:
Vania Kachova)

The landscape typological mapping by the administrative-territorial authorities provides a basis for the proper use of territorial resources for the development of settlements and RE production systems.

Landscape development plans are an integral part of site management and planning processes, but they are developed only for landscapes with exceptional features. For example, in the management plans of the Pirin and Vitosha Nature Parks (2014) comprehensive landscape assessments of the territories and landscape maps are presented, and measures for minimising the impact on landscape qualities are outlined. The overall landscape planning and activities in Bulgaria are coordinated by the Ministry of Regional Development and the Ministry of Ecology and Water, as well as by the local administrative authorities. The development of the spatial planning plans in the municipalities, preparation of town-planning plans, and selection of sites for the different functional orientations are also connected with the protection and management of the landscape. Landscape planning has an ecological focus and takes into account socio-economic, political, and aesthetic requirements for producing the best balance in the structure of the territory of the given municipality. Local communities are involved in discussions about RE developments. The degree of vulnerability of natural landscapes, the preservation and maintenance of landscape components, and landscape types are important when investment proposals are implemented in a protected area. Protected areas cover 4.5 % of the area of Bulgaria, (11.1 million ha). This ranks the country amongst the highest for proportions of the country protected, after Finland and Norway. There are six categories of protected landscapes: strict nature reserves; national parks, natural landmarks, maintained reserves, nature parks, and protected areas. Activities within maintained reserves are allowed only if their natural character is preserved; in nature parks they are only permitted if they do not pollute the environment, and in protected areas only if they do not threaten typical and remarkable landscapes and preserve all other landscape components. Natura 2000 sites in Bulgaria occupy 34.3 % of the area in which activities including renewable energy generation may be permitted, but only after if there is an individual assessment of the potential impacts on the environment, and on the landscape in particular.

Interaction between Renewable Energy and Landscape Quality

The construction of RE power plants passes all the required procedures set out in ecological legislation. The investment proposals of projects for which an environmental impact assessment (EIA), ecological assessment (EA), or assessment of the compatibility with the type and aims of protected areas conservation (ACSA-PAC) are required have to be approved by the environmental authorities, taking into account recommendations that have been made. A landscape assessment is carried out for each investment proposal requiring an ecological assessment. Such landscape assessments include a characterisation of the landscape, an assessment of impacts and of the sustainability and vulnerability of the natural features, and recommendations of mitigation measures. The public acceptability of such proposals is particularly important with a recognised need to raise public awareness of the benefits of such RE projects.

1.6
CROATIA

Nela Jantol, Matea Kalčiček, Dina Stober & Zlata Dolaček-Alduk

Figure 1.6.1
Biogas plant, eastern
Croatia. (Photo: Romulic &
Stojcic multimedia studio)

Situation of Renewable Energy

The Energy Development Strategy of the Republic of Croatia (OG 130/09) has three energy objectives: (1) security of energy supply, (2) competitiveness of the energy system, and (3) sustainability of energy development. Croatia has the goal of a 35 % share of electricity generation from renewable energy sources, including large hydropower plants, in overall electricity consumption by 2020. In 2013, the Government adopted a National Action Plan for Renewable Energy Sources by 2020 which revised targets based upon the technology available, and in line with market changes and changes in energy consumption.

In 2014, 27.9 % of energy production in Croatia came from renewable sources (European Commission 2017). Total electricity production was 13,553.8 GWh, 74.2 % of which was produced from renewable energy, of which 67.3 % was from large hydropower plants (9124 GWh) and 6.9 % from other renewable sources such as small hydropower plants, wind energy, solar energy, biomass, biogas, and photovoltaic (Energy in Croatia 2014). There are 18 large hydropower plants, 11 of which are reservoir plants and 7 of which are run-off river plants, with a combined installed capacity of 2,158 MW. Excluding large hydropower plants, by 31 December 2014, there were 1,324 renewable energy plants with a total installed capacity of 411,901 kW. The renewable energy systems comprised 1,068 solar power plants,

16 wind farms with 184 turbines, 18 biogas plants, 28 small hydropower plants, and 10 biomass plants (extrapolation of data from Eurostat, HROTE 2015, Croenergo 2015). The majority of the installed capacity comes from the wind power plants with 339.25 MW which generated 729.97 TWh, followed by solar power plants with an installed capacity of 33.52 MW and 35.17 TWh generated (Ministry of Economy 2015).

Data on Landscape Quality

A basic landscape inventory of Croatia was developed in 1997, within the framework of the Physical Planning Strategy, in which 16 basic landscape units were defined based upon their natural values. The most recent document relating to landscape quality, the Physical Planning Development Strategy, was adopted in 2017 (OG 106/17), setting the preservation of identity of spaces as one of its priorities, and one activity is the creation of a landscape atlas. Inventory, typological classification, and landscape assessments were carried out in certain areas at local and regional levels, and currently there are several innovative practices and efforts to include landscape values in spatial plans (e.g. UN-DP-GEF/Project COAST, the City of Zagreb, the City of Sveta Nedelja, Zagreb County, Dubrovačko-neretvanska County).

There are three main frameworks for nature protection in which landscape is directly or indirectly protected: the national

Figure 1.6.2
Photovoltaic plant, Adriatic Croatia (Photo: Romulic & Stojcic multimedia studio, permission)

ecological network (Natura 2000), Nature Protection Act (OG 78/15), and tge Protection of Cultural Heritage Act (OG 44/17) which includes natural heritage and cultural landscapes. Currently, Natura 2000 covers 47 % of the national territory while 85 landscapes are protected as significant landscapes according to the Nature Protection Act. Twelve locations are registered as areas of national cultural heritage landscapes, and one location is registered on the UNESCO World Heritage List.

Interaction between RE and Landscape Quality

Croatia has set the goal of increasing the number of renewable energy projects and at the same time preserving its valuable landscapes. These two goals are often in conflict with one another. Hydro and wind energy projects are planned, mostly in areas of great landscape value such as prominent mountain areas and river areas. Most of the wind plants are situated in the Adriatic part of Croatia, with impacts upon residents and tourism, which is of great importance for this region. In such situations EIA and SEA are formal and legally regulated instruments that protect the most valuable areas from damage. The era of largescale hydropower plants ocurred during the second half of the 20th century, and now it is small scale hydropower plants which are on the Croatian energy agenda. These types of interventions in the landscape provoke the most public protest, for example, the hydropower

plants planned for the Mura, Ombla, and Dobra Rivers. There are a few bioenergy plant projects, an aim of which is to be multipurpose and to align with rural development (Domac et al. 2015). Examples of good practice in strategic landscape planning integration with renewable energy projects are still rare, and most of these projects are pilot or pioneering projects (Aničić 2013). The plan for use of renewable energy sources in Dubrovacko-neretvanska County (Botinčan et al. 2015) integrated a model of landscape vulnerability within a multi-criteria methodology. Spatial analysis was used for the selection of the most appropriate sites for siting wind and solar power plants.

Recently, Croatia has made considerable efforts to provide a digital database of spatial plans at all levels. To fulfil EU (Infrastructure for Spatial Information in Europe, INSPIRE) and international (United Nations Initiative on Global Geospatial Information Management, UN-GGIM) obligations, the National Spatial Data Infrastructure and geospatial services have been developed and made available to the public. The experience in spatial analysis and the use of contemporary tools for the integration of renewable energy into landscapes is still limited; however, considerable expectations are raised by the new generation of spatial plans.

1.7

CZECH REPUBLIC

Bohumil Frantál, Stanislav Martinát & Dan van der Horst

Figure 1.7.1
Czech energy landscape: city of Kladno with a coal-fired power plant, Pchery wind farm, rooftop PV installations, and the 'national' hill of Říp in the background. (Photo: Bohumil Frantál)

Situation of Renewable Energy

The Czech Republic is one of the most energy-intensive economies in the EU. Current energy policy is highly dependent on fossil fuels and nuclear energy. Overall electricity generation is predominantly thermal power plants (51 %), most of which are fired by domestic coal, and nuclear power plants (37 %), with a mix of renewable energy sources (12 %) (Energostat 2017). Over the last decade, the Czech Republic has been among the biggest net exporters of electricity, exporting approximately a quarter of its total production. While the Czechs are among the EU leaders in the production of solar energy and biogas (particularly in agricultural anaerobic digestion plants), they are comparatively slow in the implementation of wind energy (Table 1.7.1).

The recently approved Update to the State Energy Policy (USEP) to 2040 deals with the progressive decline in energy from coal, an increase in production from nuclear power plants, and development of RE, particularly from biomass, wind, and small-scale solar installations. The construction of new large hydropower plants is almost impossible due to the depleted capacity of river flow, and plans for the construction of additional pumped-storage plants face strong opposition from regional authorities and local residents.

The Czech Republic still has one of the highest levels of public support for nuclear energy and distrustful and partly utilitarian attitudes towards renewable energies (Frantál 2015, Frantál & Prousek 2016). The public image of, and political attitudes towards, renewables has been adversely affected by: 1) a boom of 'solar business', driven by cheaper technology and over-generous support schemes, which has seen the installed capacity of PVs increase from 3 MW in 2008 to 2,000 MW in 2011, with most plants installed on agricultural land, and 2) a few examples of 'bad-practice' of wind farms and biogas plants, which are often presented in the media as typical rather than exceptions. As a result, the government made retrospective changes through a tax on solar developments in 2014 which then destabilised the business environment, practically cutting off support for new solar, biogas, and wind installations.

With regard to global trends, obligations to EU directives, and the fact that the potential for renewables (particularly wind, biomass, and on-roof solar installations) is far from being fully utilised, further development can be expected over the coming few years. The question is to what extent this development will be effective, environmentally and economically sustainable, and reduce prevailing socio-political and land use conflicts.

Data on Landscape Quality

The Czech Republic is a relatively small-scale country with a wide variety of natural conditions and landscape structures. A monumental *Landscape Atlas* (Hrnčiarová et al. 2009) presents a

Renewable energy type	Installed capacity (MW) in 2016	Number of plants in 2016	The first plant installed (year, location, capacity)
Biofuel	No data for the installed capacity. The annual production of biodiesel was circa 220,000 tons; the production of bioethanol was circa 130,000 m³.		
Biogas	393	> 500	1974, Třeboň, 1 MW
Biomass	No data for the installed capacity. The annual electricity production from biomass was estimated to be circa 2,000 GWh.		
Geothermal	No operational power plant. Thermal energy has been used for heating of buildings or swimming pools in few cities in Northern Bohemia.		
Hydropower small (< 10 MW)	335	> 100	1912, Čeňkova pila, 100 kW
Hydropower large (> 10 MW)	753	12	1933, Vranov, 19 MW
Pumped-storage hydroelectric	1,172	3	1930, Černé jezero, 1.5 MW
Photovoltaic (on-land and on-roof)	2,073	> 28,000	1997, Mravenečník, 10 kW
Solar thermal	No available data		
Solar thermoelectric	No operational power plant		
Wind onshore	283	183	1990, Kuželov, 150 kW
Wind offshore	As a landlocked country CZ has no offshore wind energy potential		

Sources of data: CZ Biom (2017), CZSO (2016), Energostat (2017), ERO (2016)

Table 1.7.1
Basic data about RE development in the Czech Republic

thematically arranged collection of 906 maps showing the history, the contemporary natural and socio-economic conditions, the environmental problems, and the heritage of Czech landscapes, including 'Landscape character types', 'Natural and cultural importance of landscape', and 'The limits and potentials of landscape'.

The protection of landscape character has been enshrined in the Act on the Conservation of Nature and Landscape. A number of methodologies and expert approaches have been developed for the assessment of landscape quality and landscape character being based either on biogeographical or architectural approaches (e.g. Vorel et al. 2006).

Interaction between Renewable Energy and Landscape Quality

Potential landscape impacts are major limiting factors for the further development of renewable energy. In 2009, the Ministry of the Environment issued a Methodological Guide to the Assessment of the Location of Wind and Photovoltaic Power Plants in Terms of the Protection of Nature and Landscape. This set out a procedure for the preparation of studies to identify the interests, value, and significance of nature and landscape protection at the regional scale and determine the inappropriateness or potential suitability of the construction of power plants in a particular territory.

However, most regions commissioned their own methodological studies as a basis for regional territorial planning. The general approach is use of a geographic information system (GIS) for 'sieve mapping', in which the interests which might be harmed by energy developments are mapped (see Van der Horst 2009). This approach can be considered a type of spatial multi-criteria analysis, which takes only negative criteria into account but does not provide a clear answer of where to direct development. Differences in attitudes of regional authorities towards RE have caused significant spatial differences in the rate of implementation of projects, particularly wind farms (Frantál & Kunc 2010). Since 2011, an ambitious project called Regional Sustainable Energy Policy based on the Interactive Map of Sources, supported by the European Commission and the Ministry of the Environment and led by the Czech University of Life Sciences, introduced a new comprehensive method for urban management and regional planning for proposing and assessing renewable energy projects. The result is an interactive web-based map of GIS layers (https://restep.vumop.cz/), which define the potential for development and contexts of all renewable energy sources in the selected territory. This takes into account the landscape character and technical capabilities of the transmission grid, and the current needs of municipalities and regions.

1.8
ESTONIA

Elis Vollmer, Monika Suškevičs, Ain Kull, Mart Külvik & Hannes Palang

Situation of Renewable Energy

The RE sector in Estonia has been growing steadily for the last three decades. In the 1990s, development was dominated by the conversion of conventional fuel (oil, coal, oil shale), based district heating boiler houses, to biomass (primarily wood chips, and to a lesser extent peat). In the following decade, the share of wood-based heat energy production continued to increase and a rapid development of onshore wind farms started. In the current decade, the most active development has been in wind and solar energy (ETEK 2015). In particular, electricity production from renewable sources is increasing, reaching 13.2 % in 2014. This comprised a total of 1,356 GWh from renewable sources out of 11,013 GWh of electricity, which was 18 % more than in 2013. Most of this production (753 GWh) was from biomass, biogas, and organic waste. Wind farms produced 576 GWh, which represents 42 % of the RE production. However, 83 % of the total electricity production in 2014 was still based on oil shale. The latest development in the RE sector is the planning of several offshore windfarms which will affect coastal landscapes and seascapes. The sector which has the lowest use of RE is that of transport (0.3 %).

There is a constant increase in the use of biomass in the heating and cogeneration sector (district heating accounts for heating 60 % of Estonian households), as solid biomass is cheaper than natural gas or oil. In 2014, 45 % (circa 4,000 GWh) of heat production was from biomass. Private houses not connected to the district heating grid mainly use firewood for heating (50 %) and considerably more in rural areas (>90 %), although the use of heat pumps is also rising.

Since 2012, development of micro-production systems (capacity per plant ≤15 kW) has been considerable, most significantly that of solar energy. The first connections to the grid were in 2012, with 186 micro-producers connected by 2014. In addition, there are a small number of autonomous producers not connected to the grid. The first commercial wind turbine connected to the main grid was established in 1997. The oldest wind farm still producing electricity is the Virtsu I farm, operational since 2002, with a total capacity of 1.8 MW from three wind turbines.

Source	Capacity (MW)
Solar energy	3.2
Wind energy	302.7
Biomass	87.5
Hydroenergy	7.2
Biogas	10.2
Waste*	17.0
Total*	410.8

* Since energy produced only from organic waste is considered renewable, the installed capacity of waste-based energy production units is not included in total installed capacity.

Table 1.8.1
RE installed capacity in 2014

Figure 1.8.1
A positive example of wind-farm landscaping of reclaimed land at the Baltic Power Plant Ashfield No 2, Estonia (Photo: Eesti Energia Ltd)

Data on Landscape Quality

Valuable landscapes were defined between 1999 and 2003 in the framework of the thematic plan Environmental Conditions Directing Settlement and Land Use. As a result, 323 valuable landscapes were designated at local, regional, and national levels (Palang et al. 2011). Each delineated landscape was supposed to receive a management plan; this, however, has been rather sporadic. Still, this material has been widely publicised, first with each county preparing a book that introduced their valuable landscapes, and finally in a book that describes them all. Recently, a new initiative has emerged to designate national landscapes. With respect to landscape quality assessments for planning, considerable progress was made during this exercise, an outcome of which is that landscape quality assessments have become part of planning processes. A new guidance document for planning is currently being compiled which will include how and what to assess when dealing with landscapes in planning processes.

The nature conservation law defines protected landscapes as areas meant for the conservation, protection, research, introduction, and regulation of the use of the landscape. Although the law does not define landscape, and the ELC has only recently been signed, this clause has been used for the protection of values not directly related to biodiversity, notably aesthetics, culture, and recreation. The trend seems to be to use the protection of landscapes for managing biodiversity values and addressing landscape-related issues through planning.

Interaction between Renewable energy and Landscape Quality

Wind farms are the only type of RE plants which have encountered serious community resistance, largely due to the belief that their effect on the visual landscape will be greatest. In a expert survey carried out within the COST Action, the respondents from Estonia rated landscape issues as 'very important' or 'quite important' reasons for resistance to wind energy development. Planning of wind farms in valuable landscapes is usually opposed (Vaab et al. 2010). Current practice is that visualisation materials are used as planning tools, which usually enables developers to reduce some concerns of the local community. GORWIND (http://gorwind.msi.ttu.ee/home/info) is a recent Estonian-Latvian cooperation project that looked into the framework of spatial planning to identify the best options for visualising onshore and offshore wind farms. There are examples of positive opinions towards windfarms. For example, of the inhabitants living near the coastline in Ida-Viru, Harju and Lääne Counties nearly 80 % of 260 respondents had positive attitudes towards wind farms (Nerep 2017). Although the negative effect on the aesthetics of the landscape is reported, the reasons for a positive attitude include a perception of the environmental and other benefits, such as tourism attractions. Opportunities for public participation in the processes of siting wind farms are guaranteed by Estonian Planning Law, under the procedures of national and local governments, specifically plans designated as being for 'objects of significant spatial impact' (Planning Act, §§ 4 and 95).

The main reason for the slow development of the RE sector is a lack of a stable long-term policy framework, with balanced and fair support regulations, combined with technical limitations of limited local grid capacity and scattered settlements with low energy demand for co-generation. There is also public caution towards landscape change in their living environment, and the NIMBY effect. In 2017, the National Development Plan of the Energy Sector to 2030 was adopted by the government. The plan provides targets for RE and energy efficiency operational programmes and a vision for the renovation of buildings.

1.9

FRANCE

Bénédicte Gaillard & Alexandra Kruse

Situation of Renewable Energy

The French Landscape Law was passed in 1993, which aimed at protecting and developing landscapes across the breadth of contexts (natural, urban, or rural). Twenty years later the national debate about renewable energy commenced. Only now, after France hosted the COP21 in 2015 and in order to meet the enacted climate obligations, is RE increasing its share of energy production.

In 2015, RE represented 9.4 % of primary energy consumption. Except for hydropower, of which France is the third biggest producer in Europe after Norway and Sweden, the use of RE is at an early stage. Compared to other countries, there are relatively few wind turbines, solar thermal, or photovoltaic panels. Since December 2017, the national energy company EDF has been contacting private households to promote the installation of photovoltaic panels on their rooftops.

Data on Landscape Quality

A national policy to publish a landscape atlas has been supported at the regional level by the DIREN (Regional directorates for the environment) (Davodeau n.d.). Since 2009, the DIREN have been progressively replaced by the DREAL (Regional directorates for environment, planning, and housing). For more information and to view the landscape atlas, see www.statistiques.developpement-durable.gouv.fr/lessentiel/ar/279/1129/atlas- paysage.html. Data on the environment are edited annually by the SOeS (Service for Observation and Statistics) and published by the Ministry of Ecological and Inclusive Transition. Environmental impact assessments are required in planning processes. The content of the assessment is described in article R.122.5 of the Environmental Code which refers to landscape but not to landscape quality.

Different types of protected areas exist in France, such as national parks, regional nature parks, and nature reserves. Although the legislation concerning the environment distributes the relevant authority to different levels of administration (state, regional, departmental, and municipal), the legislation concerning landscape gives an essential role to the national level, in terms of defining the legal framework for different policies for the management of natural areas.

Interaction between Renewable Energy and Landscape Quality

Environmental impact assessments have to be carried out when RE installations are planned. For example, the planning of offshore wind power plants requires an assessment that considers the landscape (Ministère de l'Environnement, de l'Energie et de la Mer 2017), as does the installation of onshore wind power and, on a case-by-case basis so do solar panels (Ministère de l'Environnement, de l'Energie et de la Mer 2017).

According to the Environmental Code, public participation has to be included into the decision-making process relating to projects, plans, and programmes. Public involvement has already led to several achievements, e.g. the Charte de pays (charter of pays, a strategic orientation paper resulting from the collaboration between elected officials and public and private stakeholders which provides a vision of territorial evolution for the next ten years and determines favoured development axes), a wind power charter, and directives and objectives for local development. The 'communauté de communes' was recognised as pioneering with the Ardenne metropole obtaining EU LEADER project funding. For citizens with a strong attachment to the landscape as a part of their heritage, landscape quality is a very sensitive issue with respect to the acceptance of renewable energy projects, mainly onshore and offshore wind farms.

Solar panels are more acceptable to citizens due to lower impacts on landscape quality. An impressive example are the on-ground solar panels at Les Mées, Alpes de Haute Provence. A construction of which faced a challenge of avoiding negative visual impacts for the village of Les Mées and Puimichel. The project required an investment of approximately 70 M euros between May 2010 and January 2011. The ground preparation and construction phase of the project employed 350 people. It is located on a 36 ha field, comprising 79,000 modules, with a total capacity of 18.2 MW. Annual production of the site is 26 GWh, providing electricity for approximately 9,000 families, and displacing the emission of more than 9,200 t CO_2 annually.

RE type	Installed capacity (MW) 2015	Year of the first plant
Wind power onshore	10,013	2000
Wind power off-shore[1]		
Marine energy	241	2008
Small hydropower	2,000	1830
Large hydropower	25,400	1900
Solar PV	6,191	1990
Solar thermo-electric	1.01	
(only on pilot sites)	2010	
Geothermal	17.2	1985
Biomass	365	2003
Biogas	332	2000

Table 1.9.1
Installed capacity and year of first installation of RE in France

1 In 2011/2012, four projects were attributed off Fécamp, Courseulles-sur-Mer, Saint-Brieuc and Saint-Nazaire, cumulating a power of 1928 MW
In 2013/2014, two projects were attributed off Tréport and the Yeu islands and Noirmoutier, accumulating a power of 992 MW
In 2016, two projects were announced in the frame of a third call for tender off Dunkerque and the Oléron Isla

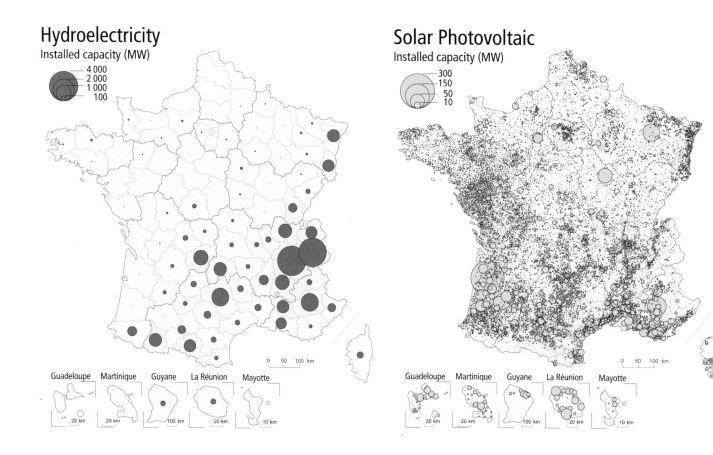

Hydroelectricity
Installed capacity (MW)

Solar Photovoltaic
Installed capacity (MW)

Figure 1.9.1
Installed capacity of RE in France
in 2015 (Source: Key figures of
renewable energy, edition 2016;
Ministry of Environment, Energy and
the Sea, February 2017 [in French];
Cartography: Daniel Herrero-Luque)

Biogas

Installed capacity (MW)

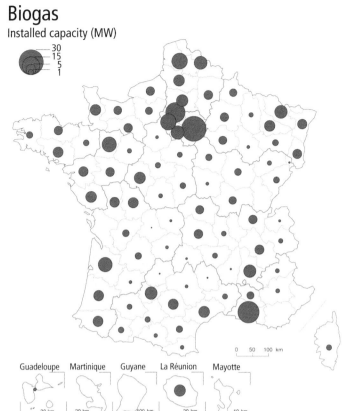

Wind energy

Installed capacity (MW)

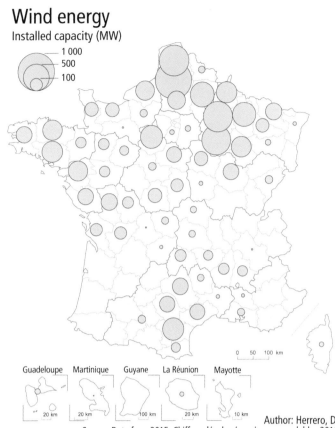

Author: Herrero, D.
Source: Data from 2015. Chiffres clés des énergies renouvelables 2016
- Service de l'observation et des statistiques.

GERMANY

Kim Philip Schumacher & Kathrin Ammermann

Situation of Renewable Energy

In Germany there are regional differences in the distribution of RE production. In northern Germany wind turbines are now an integral element of the landscape. In 1990 hydropower, mostly from river barrages in southern Germany, was the only noteworthy RE system. Driven by the ecological and anti-nuclear movements, Germany has decided to shut down all nuclear power plants by 2022 and to engage in a far-reaching transformation of energy production and consumption, the so-called 'Energiewende' (BMWI 2010). A key piece of legislation is the Renewable Energy Sources Act (EEG), passed in 2000. With the introduction of a feed-in tariff and feed-in priority it accelerated the development of RE production (Table 1). It provided the possibility for individual households to benefit directly from contributing to the production of renewable energy (e.g. photovoltaic equipment on roofs), which was seen as a major advantage. However, the EEG became costly and the latest amendment in 2017 resulted in the withdrawal of this practice. Earlier amendments in 2004, 2009, 2012, and 2014 also altered the development path of RE production several times (e.g. the biogas boom between 2004 and 2012).

Data on Landscape Quality

Germany has actively contributed to the development of the European Landscape Convention (ELC) but has not signed it, partly because of other legislation dealing with landscape issues. Currently, there is no national landscape assessment in Germany, but work is in progress on several approaches. Legislation for the protection and management of landscapes is a mosaic of many closely linked regulations with responsibility and legislation split between federal and state levels. The federal laws relating to the cultural landscapes in Germany are the Regional Planning Act and the Nature Conservation Act.

Federal states are responsible for most planning tasks (Bender & Schumacher 2008). This results in a mosaic of case studies and regional or local inventories of landscapes and landscape quality. The federal nature conservation law can protect landscapes because of their rarity, beauty, peculiarity (*Eigenart*), or their regional significance. This law, alongside landscape planning and impact regulation, has been at the heart of landscape policy in Germany since 1976. In planning processes, landscape, or landscape scenery and the impacts of planning on it, have to be considered in several assessment procedures, and a variety of appropriate

RE share in Germany	2014
of gross final energy consumption	13.5 %
of gross electricity consumption	27.4 %
of final energy consumption in heating/cooling	12.2 %
of final energy consumption in transport	5.6 %
of primary energy consumption	11.3 %

(BMWI 2015, 9)

Net installed electricity generation capacity in Germany in 2014	GW
Hydropower	5.58
Biomass	8.86
Wind onshore	37.56
Wind offshore	0.99
Photovoltaic	37.90

(Fraunhofer ISE 2016)

Renewables-based electricity generation in 2014

	Gross electricity generation (GWh)	Share of gross electricity consumption (%)
Hydropower	19,590	3.3
Wind energy	57,357	9.7
Photovoltaics	35,115	6.0
Biogenic solid fuels	11,800	2.0
Biogas	29,140	4.9
Others	8,377	1.4
Total	161,379	27.4

(BMWI 2015:11)

Table 1.10.1
RE in Germany 2014

assessment approaches and methods have been developed (Roth & Bruns 2016). Landscape quality is an issue in landscape planning at regional or other levels of administration, the details of which depend upon each federal state and their regulations and guidance on protection and development of landscapes.

'Landscape protection area' is one category of protected area in Germany. The establishment of these areas can be based on ecological or aesthetic values, importance for cultural history, or tourism. Such an area must have at least one protection goal, with a combination of goals favoured. This category of protection offers a number of options but there are problems with their effectiveness because many interests can conflict with the aims of the protection (agricultural and forestry use, settlement, traffic, energy production, etc.) (Bender & Schumacher 2008). In 2015 there were 8,598 landscape protection areas, equivalent to 27.7 % of the area of the country (BfN n.d.).

Interaction between Renewable Energy and Landscape Quality

Landscape quality (in the sense of *Landschaftsbild* / scenery) is especially relevant in relation to wind energy planning. In regional planning processes, areas can be designated for, or excluded from, wind energy development. Landscape has to be considered in project planning under the instrument of the Impact Regulation under Nature Conservation Act. This aims at dealing with the consequences of impacts on nature and landscape (quality). Negative impacts have to be avoided and minimised, compensation provided (e.g. financially), or offsetting arrangements made elsewhere.

Changes in landscapes and scenery are the principal arguments in public opposition against RE projects, with action groups claiming negative influences (Schmidt et al. in press). Landscape change caused by RE facilities is a significant factor in public perceptions of RE, especially wind power, but is not taken into account to the same extent in planning regulations, and planning processes. A current and growing issue with significant impacts on landscape quality is the associated construction of new high voltage power lines across the country to transmit electricity from wind energy production areas (e.g. on the coasts) into the grid system for use elsewhere.

National Overviews

Figure 1.10.1
Wind turbines in an agricultural landscape in Germany. (Photo: Hauke, © BfN–Kathrin Ammermann)

1.11

GREECE

Georgia Sismani, Nikos Papamanolis & Georgios Martinopoulos

Figure 1.11.1
Typical layout of photovoltaic and solar thermal system for hot water production on the rooftop of a dwelling in Greece (Photo: Georgios Martinopoulos)

Situation of Renewable Energy

Greece has a significant potential for the production of RE, mainly in the form of biomass, solar, wind, and hydropower. According to EU Directive 2009/28/EC, Greece has to achieve a target of 18 % RES in gross final energy consumption by 2020, which has been increased to 20 % by the Greek Government (Law 3851/2010). The overall target is broken down into subtargets of 40 % RES of gross electricity consumption, 20 % RES of final energy consumption for heating and cooling, and 10 % RES of final energy consumption for transport by 2020. Regarding electricity production, currently more than 30 % is produced from renewable sources, mainly through wind, hydro, and solar power conversion systems.

Hyrdopower was one of the first RES to be utilised in Greece, starting in the early 1920s and expanding quickly to a current total installed capacity of 3,017 MW, equivalent to more than 15 % of the total installed capacity of Greek electricity production. Over the last decade, small hydropower plants were constructed adding an installed capacity of 197 MW (Hellenic Operator of Electricity Market 2017).

Wind energy utilisation started in the early 1980s. In 2016 the total installed capacity was 2,370 MW, even if 23,000 MW had been requested already quite a few years ago (Papadopoulos et al. 2008).

Greece is one of the pioneers in the use of solar energy conversion systems. The two oil crises in the 1970s triggered interest in renewable sources, and subsequent increases in the price of electricity played a key role in the rapid development of the market. The first solar thermal systems were installed in the late 1970s. For more than fifteen years the country enjoyed high sales and had the largest area of solar thermal collection per capita in the EU, ranking tenth in the world in 2016. The first solar thermal collector for a domestic solar hot water system was produced in Greece in 1974, which was a simple open circuit collector, followed a year later by the first closed circuit system. At the end of 2015 the total surface area of installed capacity was 4.4 million m^2, equivalent to circa 3,500 MW. The first solar energy systems were installed in the 1980s with the greatest increases in capacity of EU countries.

The first photovoltaic systems were installed by the Public Power Corporation, mainly on the more remote islands, the main incentives of which were the distance from the main electricity grid and their silent operation. Other installations by the Hellenic Communications Organisation were to power stand-alone antennas and transmitter networks (up to 100 kW) and by the Hellenic Navy for district lighthouses (up to 70 kW). The dramatic increase in uptake of photovoltaics started in 2008 with the installation (<10 kW) on rooftops of individual households,

which were financed privately or through loans. Progressively, the status for permits became clearer for medium and large-scale installations, leading to a tenfold increase in installed capacity. These were considered to be one of the safest of investments, with a relatively rapid pay-back period and a high positive net present value. Currently, the total installed capacity of photovoltaics is almost 2,605 MW (Martinopoulos & Tsalikis 2018).

Historically, biomass in the form of wood and wood waste is an important source of fuel for heating in buildings, particularly of households in small communities that are located near sources of timber. The burning of wood in fireplaces or stoves also serves as a means of heating in many residential buildings in urban areas (Papamanolis 2015). However, only a few biomass energy projects for electricity generation are in development, mainly for the utilisation of municipal waste in biogas plants. The total installed capacity of biomass energy currently stands at 52 MW for the 13 projects in operation, producing a total of 222 GWh of electricity. Regarding geothermal energy, the majority of installed capacity is associated with heating applications, either in residential properties, glasshouses, or industrial warehouses, especially in the form of low enthalpy ground source heat pumps. Although the geothermal potential of the country exceeds 500 MW (from high enthalpy fields) and 1,000 MW (from low enthalpy fields) only 190 MW are currently operational (Andritsos et al. 2015).

Data on Landscape Quality

No official data on landscape quality are available. Although Greece is a signatory to the European Landscape Convention, there is a lack of awareness of landscape-related issues and few initiatives to place them on the public agenda (Terkenli 2011).

Interaction between Renewable Energy and Landscape Quality

The main issue with the utilisation of RE sources is the attitude of the local population. Although a location might exhibit high RES potential and adequate infrastructure, social opposition might set barriers. The greatest proportion of RES in Greece are systems in the built environment where opposition is minimal, or large hydropower plants which were built in the past decades. Most recent studies focus on the social acceptability of wind, small hydropower, and photovoltaic power plants. According to field surveys, there are high levels of acceptability of RE applications in Greece, recognising their environmentally friendly character and the benefits of these projects, although differences have been observed between attitudes on the mainland and those on the islands, where the tourism industry is significant (Kaldellis et al. 2015).

1.12

HUNGARY

Csaba Centeri, Robert Kabai, Béla Munkácsy, Márton Havas, Tamás Soha, & Csaba Csontos

Figure 1.12.1
Wind park in Western Hungary
(Photo: Csaba Centeri)

Situation of Renewable Energy

In Hungary there are 1,015 MW of installed RE production capacity (IRENA, 2017). Over the next few years circa 2,000 MW of new solar photovoltaic capacity will be installed. However, it appears that landscape protection is not considered as a limiting factor with most of the systems to be green field installation. At present, the number of operating RE plants is very low (Table 1) and there is no reliable prediction about their growth.

Data on Landscape Quality

Useful thematic databases are available containing information about various components of the natural and cultural landscape. Some, such as those relating to geological survey data, vegetation maps, scheduled monuments, and unique landscape features, are accessible on the web. The national classification of geographic units includes a complex description of landscapes (Dövényi

2010). However, a systematic national survey of landscape character has not been prepared. As of 1 January 2017, Hungary had 39 'landscape protection districts' on the list of national protected areas. The reason for the existence of these districts is the protection of good quality landscape and scenery (Deputy State Secretariat for Environment Protection 2018).

Interactions between Renewable Energy and Landscape Quality

The first visual impact assessment relating to a RES facility was prepared for the (unrealised) Nagymaros hydropower plant in 1981 (Csemez 1996). Now, impacts that include changes in landscape character are assessed as part of the Environmental Impact Assessment of the projects. The relevant Hungarian law, Government Decree No. 314/2005 (XII. 25.), lists the following

RE type	Year/ installed capacity	Year of the first plant	Number of plants	Installed capacity every five years from 1995	Comments, sources
Wind power onshore	2017: 329.3 MW	2000	172 turbines, 37 plants by 2011	1995: 0 MW 2000: 0.25 MW 2005: 17.27 MW 2010: 296 MW 2015: 329.3 MW	Hungarian Wind Energy Association 2011
Small hydropower (<10 MW)	2017: 15.4 MW	1895	37	57.1 MW	Hungarian Energy and Public Utility Regulatory Authority 2016
Large hydro-power (>10 MW)	2017: 41.7 MW	1956	4		
Solar PV on ground	2016: 13.4 MW	2013	many building-mounted	1995-2005: >1 MW 2010: 2 MW 2015: 168 MW	HITA (2012) IRENA RE Capacity Statistics 2017
Solar PV on roofs/facades					
Solar thermo-electric					
Geothermal	2015: 905,58 MWt 2017: 3 MWp	2017: first CHP geothermal powerplant inTura	In 2015, 1622 wells produced thermal water warmer than 30 °C.	2004: 342 MWt 2015: 905,58 MWt	Árpás (2005) HITA (2012) Data from 2013 and 2014: Tóth 2015 Data from 2016: MEKH*
Biomass	58.2 PJ 2015	2009	7 (5 operating)	2010: 40.74 PJ	HEPURA (2016) HITA (2012)
Biomass-pellet				2006: 500 t/y 2010: 190 000 t/y	Hungarian Pellet Association
Biogas	2009: 23 Mwe (Tóth 2010) 2014: 74 plants with 69.5MW capacity (Source: mvmpartner)	1954		2010: 0.32 PJ	HITA
				2010: 14MW 2012: 21 MW 2015: 43 MW	Hungary's Renewable Energy Utilisation Action Plan
				2000: 0 MW 2005: 2.1 MW	Ferenczi et al. 2009

*MEKH= Geothermal Assessment of Hungary

Table 1.12.1
Basic data of renewable energy production in Hungary

RES projects, thresholds, and other criteria that are used to determine whether the project will be subject to an assessment:

• Projects Which Will Be Subject to an EIA

Hydropower plant: without a threshold if located at a nature conservation area of national significance.

Wind power: if greater than 500 kW and located within a nature conservation area, Natura 2000 area or the surface buffer zone of a cave.

• Projects Which May Require an EIA Based upon the Screening of the Environmental Authority

Hydropower plant: a) greater than 5 MW, or b) without a threshold if located within the buffer zone of a water resource area, a nature conservation area, a Natura 2000 area, or the surface buffer zone of a cave.

Geothermal power plants and facilities: a) greater than 20 MW, or b) without a threshold if located within the buffer zone of a water resource area, a nature conservation area, or Natura 2000 area (except for household level facilities).

Wind power: a) tendering process (without tendering in the last 10 years); b) technical regulations (max. 2 MW capacity/turbine); c) 12 km buffer zone around the settlements (which means that there is not a single m2 for new wind turbines in Hungary). Hungary suffered from a bad decision (not undertaking public participation in planning) regarding the building of a large hydropower plant along the border with the former Czechoslovakia (agreed in 1977) and decided not to build on the Hungarian side leading to various conflicts (Fürst 2003). A very good example is the creation of a solar park as a means of restoring of an area of open cast mining (Mátrai Erőmű ZRt. n.d.).

1.13

ICELAND

Karl Benediktsson & Edda R.H. Waage

Situation of Renewable Energy

The RE situation of Iceland is very different from that of other European countries. Location, topography, and geology are the basic explanatory factors. The country sits in the path of cyclones that bring significant precipitation and strong winds. The centre of the country is an uninhabited highland plateau where many large glacial rivers originate. Iceland also sits on a mantle plume, giving rise to much volcanic and thermal activity. These facts of nature are the source of an abundance of geothermal, hydro, and wind energy. Together, the share of renewable sources in total primary energy production is 85–90 %, albeit including high thermal losses (Figure 1.13.1). Therefore, to a large extent the energy transition has already been achieved, although fishing and transportation still remain dependent upon fossil fuels.

Iceland's RE is used both in thermal form, as hot water, and converted to electricity, together making up more than half of the RE produced (Figure 1.13.1). The extensive use of geothermal water is unique. Public district heating systems (Figure 1.13.2) date back to the 1930s. With the rise of oil prices in the 1970s, the state sponsored an effort to extend such systems and establish new ones in many towns and villages. Some 90 % of all housing is now heated by geothermal water. Apart from space heating, geothermal water is used in a myriad of other ways, for greenhouse production, aquaculture, industry, outdoor swimming pools, and for the thawing of ice and snow from city pavements in winter.

Electricity production per capita is much higher than elsewhere in Europe (IEA 2015), over 75 % of this is used by heavy industry (Orkustofnun 2017a). Since the late 1960s, most decisions to construct large power plants (Table 1.13.1, Figure 1.13.2) have been tied to energy-intensive industries. Much of the loss of thermal energy reflected in the primary energy statistics (see Figure 1.13.1) occurs at geothermal power plants. However, three of these are dual-purpose, producing both hot water and electricity, with associated improvements in efficiency. Foreign energy companies and investors are showing increasing interest in large-scale wind energy development.

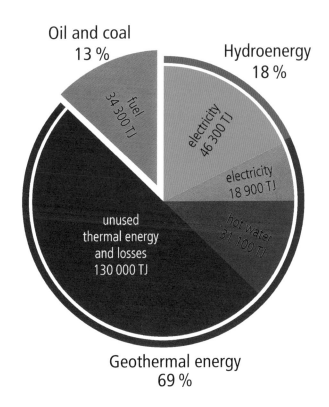

Oil and coal
13 %

Hydroenergy
18 %

fuel
34 300 TJ

electricity
46 300 TJ

electricity
18 900 TJ

unused
thermal energy
and losses
130 000 TJ

hot water
31 100 TJ

Geothermal energy
69 %

Figure 1.13.1
Primary energy sources
in Iceland and main
forms of energy use
(Sources: Hagstofa
Íslands 2017, Har-
aldsson & Ketilsson
2010a; Notes: Data for
2014). Minor quanti-
ties of wind energy
and bioenergy are not
shown. Estimates of
geothermal losses are
for 2009 (Haraldsson
& Ketilsson 2010a).

	Hydro	Geo-thermal	Wind
Year of first installation	1913	1969	2011
Number of plants	56	6	3
Installed power of largest plant	690 MW	303 MW	1,8 MW
Total installed power	1988 MW	665 MW	2,4 MW
Share of electric-ity produced	73.6 %	27.3 %	0.1 %

Sources: Orkustofnun 2017a,b.

Notes: Data for 2016. Only grid-connected power plants
included. Excluded are micro-hydro and wind turbine units where electricity is used locally.

Table 1.13.1
Electricity gener-
ation in Iceland

Data on Landscape Quality

Despite being noted for varied and visually arresting landscapes,
Iceland does not have a well-developed system for landscape
quality assessment. Despite some methodological experimenta-
tion, no systematic national mapping of landscapes has taken
place yet. However, landscape awareness is growing. Iceland
signed the European Landscape Convention in 2012. Legislative acts for land use planning, environmental impact assessment
(EIA), and strategic environmental assessment (SEA) all refer to
landscape as a topic to be given due consideration in the respective processes in planning. However, the coverage of landscape
issues varies, e.g. between municipal plans and EIA reports.

Historically, nature conservation paid significant attention to
landscapes, which was reflected in the establishment of numerous protected areas, the aim of which was to conserve landscapes perceived as particularly scenic and diverse. This period
was followed by one during which the focus changed towards
ecological and biodiversity concerns (Waage & Benediktsson
2010). From 2013, following a new Nature Conservation Act,
landscape issues have come back to the fore. In Icelandic culture, the very concept (*landslag*) centres on the 'natural' landscape (Waage 2013), whereas European conceptions of cultural
landscapes are less established.

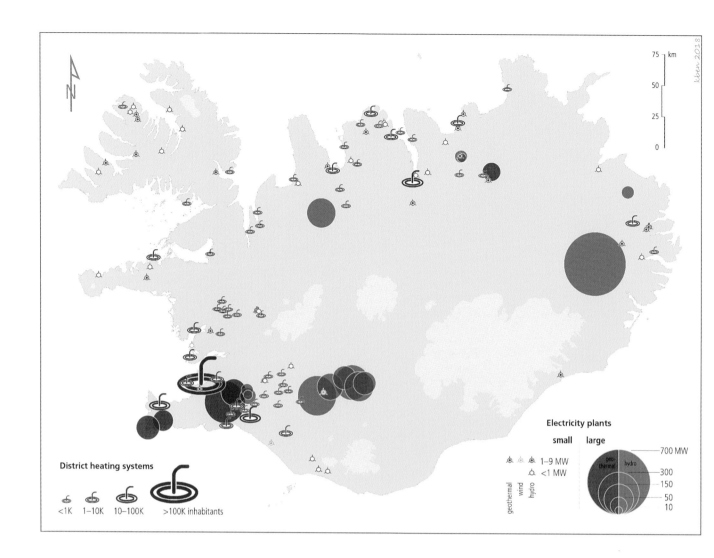

District heating systems

<1K   1–10K   10–100K   >100K inhabitants

Electricity plants

small   large

geothermal   wind   hydro   1–9 MW   <1 MW

geothermal   hydro

700 MW
300
150
50
10

Interaction between Renewable Energy and Landscape Quality

Landscape impacts of large-scale RE developments have been of great concern to many Icelanders, not least due to the encroachment of heavy industry. Landscape values have been central to some of the country's most divisive environmental conflicts. In an attempt at reconciliation, a special planning process, the Master Plan for Nature Protection and Energy Utilization, has been set up. Entering its fourth phase in 2017, the Master Plan proceeds by taking a number of potential energy project sites, analysing their characteristics, and evaluating the various impacts (Áætlun um vernd og orkunýtingu landsvæða 2017). The sites are proposed by the energy companies. Each of four expert committees deal with a distinct set of impacts, ranking the proposals in order of preference. One committee is tasked with assessing landscape values for each project site, and how these values would be affected. The results of the committees are combined to arrive at a threefold classification of project sites: for utilisation, conservation, and awaiting a final decision.

The landscape assessment of the Master Plan is based on a highly quantitative and quasi-objective methodology. Four attributes are assessed, all of which highlight visual characteristics: diversity,

Notes: Data are for 2016. Symbols for district heating systems represent where the population is concentrated, which may be some distance from the geothermal sources. There are numerous single-use geothermal sites (e.g. for swimming pools) and small local hydroelectricity plants which are not shown.

Figure 1.13.2
Public district heating systems and grid-connected electricity plants in Iceland (Source: Orkustofnun 2017b, Oddsdóttir & Ketilsson 2012, Haraldsson & Ketilsson 2010b)

rarity, size/pristineness, and visual value. The first three lend themselves to numerical analysis, whereas it seems unclear what 'values' form the basis for the assessment of visual value. Methodological experiments have been done to account for the subjective valuing by local residents, but such qualitative methods have not been integrated into the planning process. There has been no concerted effort to obtain the opinions of non-local publics about these landscapes, many of which are uninhabited.

The Master Plan process is well anchored in law, its results increasingly forming the basis for major decisions about energy projects. The question is whether such a site-based approach can prevent new conflicts from arising. To date this has not been fully realised: development at some sites classified for utilisation continues to be resisted. In part this may relate to the fact that a comprehensive national energy policy does not exist. The Master Plan also appears to be based on the belief that a depoliticised and expert-centred planning process can deliver consensus, but that in itself is a highly contestable claim.

1.14

IRELAND

Ken Boyle & Pat Brereton

Situation of Renewable Energy

In Ireland, as of 2015, 27.3 % of gross electricity generation comes from renewable energy sources. This equates to 7,857 GWh of electricity from renewables. In terms of total indigenous energy production from 49,381 GWh 20 % or 9,876.2 GWh is from renewable sources, the balance of indigenous energy production is from natural gas production and peat extraction. In the transport sector renewables account for 5 % of fuels used while in heating they account for 6.8 % of energy use.

Wind accounts for 12.6 % of the fuels used in electricity generation and electricity generated from wind and hydro (normalised) in 2015 accounted for 21.1 % and 2.5 %, respectively, of Ireland's gross electrical consumption. Biomass and renewable waste accounted for 1.0 %, landfill gas for 0.6 %, biogas for 1.0 % and 0.01 % from solar. Over 80 % of renewable electricity generated came from wind power, with installed generating capacity reaching 2,440 MW (SEAI 2017).

The first commercial wind turbine installed in Ireland was at Bellacorrick, Co. Mayo and started operation in 1992. There are currently 276 wind farms on the island of Ireland with 226 in the Republic of Ireland with an installed capacity of 2,878 MW. The Ardnacrusha hydropower station on the River Shannon became operational in 1929. Since Ardnacrusha was built, the Electricity Supply Board (ESB) has harnessed the majority of the country's hydroelectric power potential. The ESB now has eight contracted and 59 connected hydro generators, which provide 212 MW, 2.8 % of the total generation capacity. While much of Ireland's hydropower resource potential has already been tapped, some opportunities remain for small-scale decentralised generation (www.energyinstitute.org).

Ireland has one pumped hydro storage facility in Turlough Hill in Wicklow, which was built between 1968 and 1974 with a total capacity of 292 MW.

Solar farms currently account for about 1 MW of electricity generation but the sector is set to expand with 1.5GW in the planning stage throughout the country.

Data on Landscape Quality

There is currently no national landscape map. Individual local authorities undertake landscape assessment mapping on a county basis. This has led to a lack in continuity in assessment of landscape quality in some county boundary areas. The guidelines are currently under review.

The Heritage Council (https://heritagemaps.ie/) has a GIS database of landscape quality assessments based on local authority mapping. This mapping includes landscape character areas for counties surveyed and designated scenic areas, scenic routes, and scenic views.

Figure 1.14.1
Turlough Hill, Co. Wicklow. Ireland's only pumped storage power station, located in the heart of the Wicklow mountains national park (Photo: © ESB Archives 2016)

Interaction between Renewable Energy and Landscape Quality

Local authorities have, or are developing, renewable energy strategies that are integral to their development plans. All the current strategies are available at http://airo.maynoothuniversity.ie/mapping-resourccs/airo-research-maps/environmental-research-projects/nirsa-wind-strategy-mapping. The Sustainable Energy Authority of Ireland (SEAI) combines data on wind energy, landscape and habitat designations, and current windfarm locations at http://maps.seai.ie/wind/. This mapping has proven an effective preplanning tool for wind energy developers, assisting in site selection.

Landscape factors assume different significance for different types of renewables, as follows:
- Wind—highest, landscape is one of the primary 'drivers' of the wind energy strategy process
- Solar—medium, guidelines on solar farm development are currently lacking
- Tidal—medium to low [can be high in some locations and at some scales]
- Hydro—medium to low [can be high in some locations and at some scales]

- Waste to energy—low [community acceptance has been a major factor].
- Biomass—low [can be medium in some locations and at some scales].

Clare County Council (2016) noted the capacity to generate renewable energy has to be balanced with other considerations, including landscape characteristics, issues surrounding established landscape character and potential impacts, landscape impact, visual impact, mitigation, and cumulative issues.

Successful renewable energy projects are generally characterised by
- Compliance with development plan policy and strategies
- Compliance with regional and national guidelines
- Engagement in pre-planning
- Early consultation with prescribed bodies
- Continuous and meaningful community consultation

Unsuccessful renewable energy projects may be characterised by:
- Non-compliance with development plan policy and strategies
- Non-compliance with regional and national guidelines
- Inadequate or no engagement in pre-planning
- Inadequate or no consultation with prescribed bodies
- Lack of community consultation

1.15

ISRAEL

Na'ama Teschner & Rachelle Alterman

Situation of Renewable Energy

Israel was considered to have no significant fossil fuel potential until 2010, when several large beds of natural gas were discovered in Israel's economic waters. Also the country's geopolitical situation, which isolates its grid system from all neighbouring countries, thus creating an 'energetic island', is central to its energy policy (Teschner & Paavola 2013). Despite the historic reliance on imported fuels and geopolitical challenges, Israel's RE share in its total electricity production and consumption is still low in comparison to many OECD countries. Given its arid conditions and the lack of any major rivers, Israel has almost no hydroelectric capacity. In 2014, the total installed capacity of RE in Israel was approximately 800 MW (96 % solar energy), which constituted only 2 % of the total electricity production (Figure 1). This figure is gradually on the rise, as Israeli policy-makers make efforts to meet the country's commitment of 10 % RE share by 2020, and 17 % by 2035.

However, in one type of RE, Israel is a global pioneer: solar thermal panels for water heating. As early as in the 1960s, Israel revised its Planning and Building Law to make these compulsory for most new residential buildings (Teschner & Alterman 2018). The sight of these installations on rooftops is a trademark of Israeli cities. In recent years, solar farms have been promoted in the country's southern desert area, where land is somewhat more ample than in high-density regions. The largest solar farm, currently under construction after a decade of approval procedures, is based on thermal energy technology. A few more large-scale solar projects are in progress, with some regulatory improvements. The first working medium-size wind turbines were approved in the 1980s. The largest installation (1992) encompasses ten turbines in two sites in the Golan Heights. Currently, several projects with a total capacity of approximately 200 MW are under consideration, but in practice, as of 2017, only two new plants have been constructed, together constituting only 22 MW.

Data on Landscape Quality

About 18 % of the country's area is designated as nature reserves and national parks (heritage sites). According to the law, RE installations (or any type of constructions) are prohibited within declared nature reserves and national parks unless the Israel Nature and Parks Authority approves the plan, which means that it must not hinder the original designation of the area.

A county-wide assessment of landscape quality is embedded in the Comprehensive National Plan (no. 35), a statutory plan approved in 2005 and updated recently. This plan is legally binding, and designates land use zones. The plan sets up criteria based on landscape quality that distinguish between development-oriented and preservation-oriented zones. This assessment is based on evaluation of scenery and of ecological parameters, such as biodiversity of fauna and flora. Whenever development is proposed

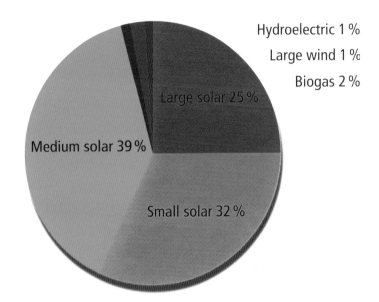

Figure 1.15.1
Installed RE
Capacity in 2014
(Ministry of Environ-
ment Website,
Last accessed: 12.3.2018)

within preservation-oriented areas according to the National Plan 35, the initiator must prepare a landscape-environmental assessment statement. The relevant planning authority may impose specific requirements, including impact assessment of proposed alternatives, description of environmental values, and scoping of agricultural land, heritage, and landscape values. Such assessment must be conducted also for the surrounding area. The analysis should gauge the anticipated impacts of the proposed development on the ecosystem, proposals for minimizing or mitigating negative impacts, and plans for reconstruction and rehabilitation. In addition to the provisions of National Plan 35, the planning law contains general rules about the types and sizes of proposed development that must conduct an environmental impact assessment. In cases where a full Environmental Impact Assessment is obligatory, landscape quality assessment will be conducted as part of the procedure.

Interaction between Renewable Energy and Landscape Quality

In Israel, the land-use planning regulations for RE siting are elevated to the national level. There are two special national plans, one for solar facilities and one for wind turbines. Both plans devote significant attention to setting the requirements of landscape quality assessments. With respect to solar installations, for example, any proposed RE project in open spaces, including agricultural zones, must be accompanied by such a document. In order to minimise visual impact, the assessment should also propose several alternatives for the location and spatial distribution of the facilities. The landscape assessment is to be undertaken by the developer, and will be reviewed by the planning authorities in considering the proposed project. Installations of rooftop photovoltaics are exempt from landscape quality assessments.

Landscape quality is a major factor in public acceptance of infrastructure in general and large-scale RE projects in particular. Representatives of environmental NGOs are members of every planning agency, at all levels of government. In several cases, the public representatives have voted against approval of RE projects located in open space due to anticipated excessive impact on landscape quality.

Environmental organisations lobby against large solar and wind farm installations and promote photovoltaic panels on rooftops. The Planning and Building Law sets minimum limitations related to visibility of rooftop installations up to 50 kW. However, in the high-density Israeli context, visual impacts of photovoltaic installation in urban settings are considered 'the lesser of two evils' compared to siting on open space (personal communication with environmental NGOs, 2017). So far there has not been a visual assessment or the potential accumulative impact of rooftop photovoltaics on the urban landscape.

1.16

ITALY

Michele Bottarelli, Raffaella Laviscio, Paolo Picchi, Alessandra Scognamiglio & Bruno Zanon

Situation of Renewable Energy

In Italy, the demand for energy over the last decades increased steadily until 2005, when it started to decrease due to an economic crisis (Italian National Energy Balance 2013). With regards to fossil sources, natural gas surpassed oil in electricity production in 2000; in 2016, coal covered 20 % and biofuels 10 % of the total production (199 TWh). In 2015, combined heat and power plants passed traditional power plants and covered 60 % of the production. Concerning electric energy, until the early 1960s hydroelectricity covered a good share of generation (82 %), but in the following decades there was a rapid increase of thermal generation.

In 2016 electricity demand reached 314 TWh, while the internal gross generation capacity reached 290 TWh. Production by renewables covers 108 TWh. Some technologies are rapidly evolving (SISTAN & Terna 2017), in particular photovoltaics and wind. Efficiency of hydroelectric plants have improved and especially small plants exploit the remaining bodies of water.

Italy implemented the EU Directive 2009/28 with a decree (DL 28/11) on the development of production and use of RE. It aims to reach 17 % of RE production in 2020 and integrates the diverse authorisation procedures by declaring that the assessment for the installation of renewable energy technologies (RET) must safeguard biodiversity, cultural heritage, and the rural landscape.

Moreover, the derective delegates the responsibility for authorising the installation of RET to regions, which are obligated to draw up specific guidelines.

The Italian government issued a first action plan in 2012, and in 2017 a new strategy was approved (National Energy Strategy 2017). It affirms that by 2030 Italy must

- Reduce energy consumption from 1372 TWh (in 2015) to 1256 TWh
- Increase energy consumption from renewable sources from 17.5 % to 28 %, in particular 55 % in electricity (from 33.5 %), 30 % in thermal energy (from 19.2 %), and 21 % in transport (from 6.4 %)
- Decrease energy costs and reduce dependence from other countries
- Stop energy production from coal
- Improve the quality of the oil refinery chain
- Reduce CO_2 emissions of 39 % in 2030 and of 63 % in 2050
- Invest in research, sustainable mobility, and resilient energy provision and delivery networks and processes.

The planned investments to improve networks, RE production, and efficiency are 175 billion euros by 2030.

Data on Landscape Quality

In Italy landscape is protected by the Code of Cultural Heritage and Landscape of 2004 (modified in 2008), which updates

Italy—Electric Energy: Number of plants and gross generation capacity, 2016

	Number of plants	Gross generation capacity in GW	%	Production in TWh	%
Hydroelectric	3927	22.7	19.4	44	15.2
Thermo-electric	5285	64.9	55.4	199	68.6
Geothermal	34	0.8	0.7	6	2.1
Wind	3598	9.4	8.0	18	6.2
Photovoltaic	732053	19.3	16.5	23	14.1
Total	744897	117.1	100.0	290	100.0

Source: Sistan & Terna, 2017

Table 1.16.1
Renewable energy production in Italy

previous laws (1939, 1985), to reflect concepts and definitions of the ELC. A large part of the territory is protected, if considered appropriate, both for its outstanding values and intrinsic characters of places. Regions must take care of the protected landscapes by elaborating landscape or territorial plans. Such plans must 'analyse landscape characteristics, *created by nature and history*'. They must define detailed frameworks and identify 'the measures for the correct insertion … of territorial transformation projects'. Few regions approved an updated landscape plan, but all regions manage landscape assessment procedures.

Besides the protected landscapes, there are 871 natural parks and protected natural areas, which cover more than 10 % of the national area, and 51 UNESCO sites. Other sectoral provisions cover forests, historical heritage sites, hydro-geological fragile areas, etc.

Interaction between Renewable Energy and Landscape Quality

Diverse authorities are involved in the assessment and authorisation of RE plants, but a key role is played by regions. They are responsible for the authorisation of RE plants. At the national level a Ministerial Decree of 2010 defined 'Guidelines for the authorisation of RE plants'. This document sets out the criteria by which regions should identify measures for an appropriate landscape integration, identify unsuitable areas, and define compensatory measures. The Ministry of Culture proposed guidelines for wind energy plants to guarantee landscape design principles for RET by considering the characters of places (morphological, formal, historical, and perceptive factors). Currently, the regions are the main promoters of directives and guidelines; in general, these are heterogeneous documents that consider landscape aspects only partially, aiming at streamlining the process, developing an implementation tool according to the energy policies and, when present, the regional energy plan.

Guidelines elaborated within regional landscape plans are different. They provide rules and formulate criteria to support design activity with particular suggestions oriented toward the treatment of landscapes. Guidelines specifically address renewable energy plants. This is the case for Lombardy and Apulia, whose documents identify the most suitable areas for installation and suggest specific studies to evaluate landscape compatibility, while providing examples of good and bad practices. Veneto and Sardinia regions, as well as some provinces, have elaborated guidelines and documents for the assessment of PV plants.

The regional directives and guidelines for renewable energies are both tools for design support and decision making. In general, they are oriented to facilitate the construction of plants providing criteria and parameters for the assessment of environmental compatibility.

1.17

LATVIA

Anita Zariņa & Margarita Vološina

Renewable energy type	Number of plants in 2014 (LAEF n.d.)	Installed capacity, MW in 2014 (CSP 2016)	Year of the first plant installed
Large-scale hydropower (> 10 MW)	3	1560	1936–1940
Small-scale hydropower (< 10 MW)	146	30	1924
Biogas	57	58	1983
Biomass	32	63	n.a.
Wind energy	58	69	1995

Table 1.17.1
Data on the most significant RE types in Latvia

Figure 1.17.1
The second largest onshore wind farm in Grobiņa (33 turbines, 2002) has become publically recognised landscape scenery, part of a new landscape identity in the Kurzeme region and a local tourism attraction (Photo: K. Reinis, Grobiņa Municipality).

Situation of Renewable Energy

More than one third of Latvia's total energy consumption is generated from renewable resources (Tabale 1), of which more than about 70 % is produced by hydropower plants (HPP). In terms of production, 55 % (2803 GWh) of the total amount of electricity comes from renewables, while the total installed capacity for electricity generation from RE reached 1780 MW in 2014 (LIAA n.d.).

The most important source of electricity is the cascade of three large HPPs on the Daugava River. Another significant RE resource is biomass, mainly fuel-wood, which is used in both housing and industrial sectors (LIAA n.d.). The first biogas station was built in Soviet times, while biomass plants are of more recent origin—the first large-capacity (23 MW) cogeneration station was constructed in Jelgava in 2013 (Prodanuks et al. 2016).

Latvia has remarkable wind conditions along the Baltic Sea (Bezrukovs et al. 2015) with a newly established power transmission line being a significant advantage for potential onshore and offshore wind energy projects. Most existing onshore wind farms are already concentrated there. Estimates for solar energy utilisation also show adequate potential for development, e.g. 20 kW

capacity solar panels installed in SW Latvia (Nīca) reached 1470 kWh from 1 kW in 2015 (LAEF n.d.). Yet solar energy projects are comparatively marginal, and there are only a few small-scale marine heat or geothermal energy initiatives.

A shift towards RE in Latvia is inevitable and is generally part of the national political agenda. At the same time, in most cases actual deployment of renewable resources is based on private initiatives, which demands more specific support through clear decisions and reliable procedures (including spatial planning) among various stakeholders.

Data on Landscape Quality

Landscape maps and a general description of larger landscape units are the most tangible results of the landscape assessment at the national level, developed by geographers in the 1990s (Ramans 1994, Nikodemus 2018). For various reasons, this assessment has remained as a scientific estimate and it has been used for landscape planning purposes only indirectly. The legal framework for the protection of valuable cultural and natural landscapes at the national level has been attributed to four national parks, nine protected landscape areas and 42 nature parks,

and it has a strong ecological emphasis on the management and development of these areas.

The National Landscape Policy (2013–2019, adapted after the ratification of the ELC in 2007) still lacks national guidelines that would consider landscape quality principles in planning processes. In landscape planning practice, including a few existing approaches, the focus lies mainly on identifying of landscape areas, as well as on assessment of their natural, scenic, and historical values. At local level, the Spatial Development Planning Law (2011) empowers municipalities to develop specific landscape plans, which so far have barely touched upon the issues of RE projects or their potential.

Interaction between Renewable Energy and Landscape Quality

RE projects implemented in recent decades are usually located outside valuable cultural landscapes, but the public concerns for landscape quality are gradually becoming as important as those of political, health, or environmental issues.

The legal link between landscape quality, RE initiatives, and planning could be strengthened through the environmental impact assessment that is compulsory for all RE projects that meet a particular production capacity or location criteria defined by legislation. But the guidelines for landscape impact analysis regarding RE are only elaborated for wind farm projects. Rigorous landscape analysis is required, but in practice the planning process is quite unclear; it differs from case to case, with arbitrary involvement of stakeholders often resulting in local protests. This highlights the need for concerted planning, performed through close cooperation between developers, local authorities, and lay stakeholders.

Public debates about visual landscape issues mainly concern wind farms, while various bioenergy crop fields are overlooked. Bioenergy projects directly impact the visual qualities of Latvian landscapes through diverse agriculture and forestry and by targeting biomass from overgrown landscape elements—roadsides, drainage ditches, and waterfronts. Small HPPs have attracted public attention mainly due to their negative ecological implications. All in all, a survey of societal attitudes towards RE shows that wind turbines (Figure 1), as well as woodchip cogeneration plants and large HPPs are the most supported RE constructions in Latvia (SKDS 2008).

1.18

LITHUANIA

Darijus Veteikis & Giedrė Godienė

Situation of Renewable Energy

The total installed RE capacity in Lithuania in June 2017 was 1813 MW produced by two large hydropower plants (1000 MW), and 2,653 other power plants with a total installed capacity of 813 MW. The latter include 2,340 solar plants (74 MW), 163 wind farms (511 MW), 12 solid biomass plants (65 MW), 39 biogas plants (35 MW), and 99 small-scale hydropower plants (128 MW) (MoE 2016).

Water and wind mills were introduced in the 13th century. At the start of the 20th century there were about 1000 wind and water mills, most of which disappeared with the development of electrification. In 2017, 21 onshore wind energy parks comprise 207 plants (MoE 2017b), mostly on the western part of Lithuania and within 30 km from the Baltic Sea. There is a tendency to increase the installed capacity and gauge of wind plants: up to 2.4–3.05 MW, 100–134 m height and 100–120 m diameter in the most recent five wind parks (Marčiukaitis 2017). In 2012, two 940 kW and 995 kW solar power plants started operating (Lietuvos rytas 2012). The largest solar park is in Sitkūnai, Kaunas district (2.56 MW, 10,666 solar modules).

According to the National Strategy for Energy Independence, RE is planned to reach 45 % of the energy produced by 2030 and 80 % by 2050. Wind energy will become the main source producing at least 55 % of all electricity from RE in 2030, 65 % in 2050 (MoE 2017a). In terms of green energy production, the strategy intends to promote active participation of energy users and exponentially increase the number of individual cases with RE production for personal needs. The predicted number of electricity-generating consumers for 2020 is 34,000 (almost 2 % of all electricity consumers) and will reach 500,000 by 2030 (MoE 2017a).

By 2020, wind will become the main source of RE, reaching 770 MW of installed capacity. The installed capacity of solar power plants will increase to 190 MW, biomass 175 MW, and biogas 35 MW, while the production capacity of hydroelectricity will remain essentially unchanged (128 MW) (MoE 2017a).

The ongoing intensive development of the wind parks faces growing difficulties related to the lack of suitable territories in terms of wind speed and grid connections, administrative barriers, lack of spatial planning regulations, or sites' environmental and aesthetic limitations. The need for prioritisation and careful planning of wind power plants is of major importance to reduce conflicts with local communities and preserve the aesthetics of traditional landscape and biodiversity.

Data on Landscape Quality

The key concepts on landscape ecological and cultural peculiarities at the national scale are set out in a National Master Plan (2002). A number of landscape-oriented projects including a National Landscape Study (2006) and a National Landscape Management Plan (NLMP) were implemented. The NLMP provides summarised analytic data of landscape character and quality. Its main solutions include the network of landscape management zones and 17 types of regimes to form and maintain optimal landscape structure; a set of landscape structure optimality indexes and a methodology to identify their values within individual management zones; regulations on protection of the visual aesthetical potential of landscapes; perspectives on development of landscape and cultural identity protection. One third of Lithuanian territory is indicated as aesthetically valuable and 14 % of it as extremely valuable scenery in 27 unique areas (MoEnv 2015). National landscape monitoring is being are carried out at the local level (100 randomly distributed 1.5x1.5 km sample areas, in addition to the seashore, Karst Region, and protected areas) and the national level (European Land cover CORINE 1995–2012). New remote sensing projects have been initiated. Landscape assessment is a mandatory part of spatial planning, construction, environmental impact assessment of planned activities, and in the designation and planning of protected areas (PA).

The system of PA was created on the basis of landscape character by the Law on PA (1992) following the principles: landscape conservation and protection by different types and levels of legislation; representation of typical and unique landscape complexes as well as all landscape regions; integrated protection of natural and cultural values; even distribution of PA throughout the country; creation of the national 'nature frame', ensuring connectivity among PA and containing the most natural and ecologically sensitive, Natura 2000 areas, etc.; and compensating for the anthropogenic pressure within the country. In 2017 PA covered 17.64 % of the territory (SSPA 2017). Although protection of the aesthetic values of landscape has no representation in Lithuanian legislation, it is nevertheless implemented in most PA.

Interaction between Renewable Energy and Landscape Quality

Installation of new hydropower plants, wind turbines, or solar panels is prohibited in PA areas. Strategic environmental impact assessment (SEIA) and environmental impact assessment (EIA) for the planning of RE objects is obligatory when the project falls within 27 areas with exceptional scenery (Mo Env 2015) or within their visual reach, as well as within 10 km from the PA. Also, the special territorial plans for wind power plants need to be designed in accordance with the NLMP and other planning documents. Nevertheless, the quality of projects and assessments varies a lot due to non-standardised methodologies being applied.

Landscape quality, especially in the areas of traditional landscape are often the objects of public discussion and the cause of unacceptable RE projects being blamed for disturbing the traditional scenery, decreasing the attractiveness of landscape for recreational activities, and also affecting rural tourism. On the other hand, a national poll in 2016 found that 54 % of Lithuanians support the development of RE, thereby supporting an optimistic future scenario for RE in Lithuania (Delfi Grynas 2016).

1.19

MALTA

Daniel Micallef, Brian Azzopardi & Renata Mikalauskiene

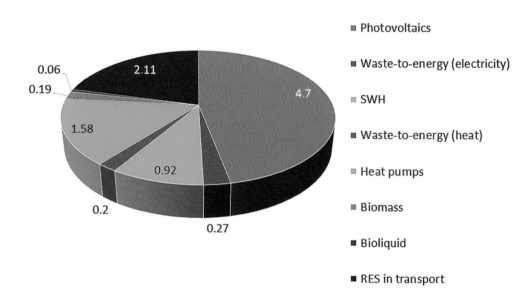

% of renewable energy by 2020

- Photovoltaics
- Waste-to-energy (electricity)
- SWH
- Waste-to-energy (heat)
- Heat pumps
- Biomass
- Bioliquid
- RES in transport

Figure 1.19.1
Renewable energy share projection by 2020 (Energy and water agency 2017).

Situation of Renewable Energy

The National Renewable Energy Action Plan II (NREAP) (Energy and water agency 2017) sets out the RE mix that is expected to deliver the 10 % target by 2020, maintaining the overall trajectory towards 2020 and the measures necessary to deliver these results. The technologies and mechanisms considered in this action plan include RE for electricity, including photovoltaic (PV) systems, micro-wind, and waste-to-energy; heating and cooling RE, including solar water heaters, waste-to-energy (heat), and heat pumps; RE in transport, including biodiesel produced locally from used cooking oil (UCO) and animal fat; and imported RE, including biomass and biofuel.

In 2014 the overall RE production in Malta was 4.68 %. Most of the systems are small-scale domestic systems starting with solar water heaters in the 1990s and followed by several research and training-based PVs.

Given its geographical location, solar energy is one of the most abundant resources to tap into, resulting in a strong emphasis on solar electrical and heat generation. Most of the RE share comes from roof-sited photovoltaics and solar water heaters, with the major applications being based in the urban landscape. Various small scale photovoltaic farms have also been proposed.

Data on Landscape Quality

Malta's landscape is made up of urbanised zones, historical sites, agricultural and vegetated areas, and coastal zones. Due to the island's size, the seascape is generally always only a few kilometres from urban or inhabited regions. There is currently no specific

Figure 1.19.2
Photovoltaic panels installed on an industrial site with a surrounding heritage. (Photo: CFL Group Engineering Malta Ltd)

legislation dealing with landscape quality. The only exception is the Development Control (DC) Policy of 2015, which mainly focuses on the quality of the urban landscape (Planning Authority 2015).

Interaction between Renewable Energy and Landscape Quality

Issues with landscape visual impact arise mostly in the context of roof-installed photovoltaics. This problem has been raised and addressed by the local Planning Authority by introducing the DC Policy of 2015 (Planning Authority 2015). According to this policy, when photovoltaic modules are installed on surface car parks or public spaces for instance, the height of the entire frame and modules should not exceed 3.4 m. When installed on roof tops, a height limitation of 1 m must be used, even at the expense of reduced efficiency, by having the panels mounted relatively flat. Figure 2 shows the installation of photovoltaics in an industrial zone surrounded by a historical site. This is an example of good integration practice within such highly diverse landscapes.

In 2009 a project description statement for the development of an offshore wind farm was submitted to the local planning authority which was subsequently rejected. The basis for the rejection was the negative effects on avifauna and marine biodiversity rather than visual impact. Several of onshore developments totaling 14.2 MW were also proposed but rejected on the basis of visual impact.

1.20
MONTENEGRO

Svetlana Stevovic, Jovana Jovanovic & Ivan Stevovic

Situation of Renewable Energy

In Montenegro there are nine RE plants: two large-scale hydropower and seven small-scale hydropower (Table 1). The total installed capacity is 658 MW and the annual production is 1911 GWh approximately. Renewable resources represent a share of approximately 75 % of the total capacity installed. The demand for electricity per household is high compared to other European countries, about 4,800 kWh/year (CETMA 2007a).

The estimate of the countrywide wind energy potential was first performed and presented in the form of maps showing average wind speed and average wind power. Subsequently, a more detailed technical potential evaluation was conducted to take into account all main restrictions. Wind measurements on the ground were used to refine the preliminary results and estimate the potential annual wind energy production. From the analysis, the wind speed turns out to be lower than 5 m per second in most of the country, However, the estimated values were increasing to 5–7 m/s toward the sea, reaching 7–8 m/s at the coast. Typical values of the actual wind potential are 100–300 W/m², increasing in the windiest areas at the ridges and tops of mountain ranges to more than 400 W/m².

Most high wind speed areas located in inner Montenegro lose their appeal due to the high altitude of the mountains. Most of these areas are not crossed by the road network either. Therefore, investments for improving road infrastructure and power lines would be necessary to allow transportation of wind farm parts to the construction sites. This means that, apart from few suitable locations, small turbines (750–1000 kW), implying small equipment, should be chosen in most sites. Medium-size wind farms, consisting of ten or more mills, would be more suitable than single machines.

On the basis of the Wind Tender for Wind Power Plants, two projects for the construction of wind farms are currently underway (CETMA 2007b):

1) Mozura, Municipality of Ulcinj and Bar, 46 MW, planned annual production about 100 GWh
2) Krnovo, municipality of Nikšić and Savnik, 72 MW, planned annual production about 160 GWh

Digital maps of global solar radiation over Montenegro were created to show the theoretical solar potential of the country. In

	Installed capacity (MW)	Production per year GWh (%)
Large hydropower HPP Perucica HPP Piva	5x38 + 2x58.5 = 307 3x114 = 342	930 (29) 960 (30)
Small hydropower (7 plants)	9	21 (<1)
Total	658	1911

Table 1.20.1
Installed capacity and
production of RE per year

Figure 1.20.1
Landscape of accumulation behind dam and HPP Piva constructed in 1975

particular, solar mapping is a means of showing the solar potential of buildings in order to identify which ones are suitable for retrofitting to solar energy, particularly solar domestic hot water. Montenegro has one of the greatest solar energy potentials in South-Eastern Europe: the annual amount of solar energy in Podgorica is about 1,600 kWh/(m²·d), The coastal regions enjoy more than 2,500 hours of sunshine a year and a very high level of solar radiation in summer, late spring, and early fall. The plains also exhibit a large number of hours of sunshine, though somewhat lower than on the coast. In the mountainous regions of the north, sunshine can be scarce.

After this assessment, it appears that the pure solar potential of the coastal and central regions of Montenegro is high, comparable to Greece and Southern Italy. The potential use of solar energy was assessed in two most promising sectors: solar thermal energy for households and for the tourism industry. If the development of solar thermal energy becomes an important governmental action in the future, the decision-makers need to have an overview of the market and possibilities to create the suitable set of regulations, laws, and support schemes.

Data on Landscape Quality

A national landscape assessment is in preparation. Specific assessments considering landscape quality for particular planning processes of certain strategic projects are done within EIA, ESIA, and SEIA. 13,812 km² (7.72 %) of the country's territory is under protection for landscape quality or scenery. These include 4 national parks (Skadarsko jezero, Lovćen, Durmitor II, and Biogradska gora), 40 natural monuments, 4 special natural areas, 2 nature reserves, and 1 other area, as per municipal decisions.

Interaction between Renewable Energy and Landscape Quality

Landscape quality is reflected in planning processes for RE through the preparation of specific documentation within EIA, ESIA, and SEIA. It is discussed in public hearings as well. Landscape quality is an obligatory issue in the public acceptance of renewable energy projects (Stevovic et al. 2015).

One of the highest arch concrete dams in the world (220 m high) is on Piva River. It was constructed in 1975, and today it represents the synergy between RE infrastructure and landscape quality (Figure 1).

1.21
NETHERLANDS

Berthe Jongejan, Henk Baas, Sven Stremke & Cheryl de Boer

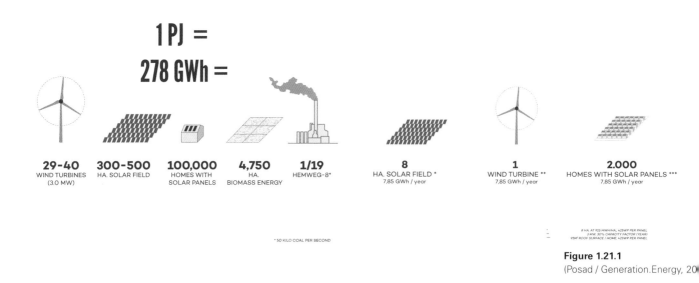

Figure 1.21.1
(Posad / Generation.Energy, 20

Situation of Renewable Energy

The Netherlands is facing a major challenge with regard to its energy supply. Fossil fuels will ultimately run out. Moreover, they increase greenhouse gasses. The Dutch National Energy Agreement therefore states that CO_2 emissions should be reduced by 80 to 95 % by 2050 and that RE should constitute 14 % of the total production in 2020 and 16 % in 2024. In 2016 the consumption of energy from renewable sources was 5.9 %. The total net electricity production in 2014 was 11039 GWh (Central Statistics Office, [CSO], 2018). The first wind turbines were installed in 1981 (0.25 MW) onshore and in 2007 (108 MW) offshore. In 2018, 2294 wind turbines had a total capacity of 4,2 GW (Bosch & van Rijn 2018).

Data on Landscape Quality

Dutch national law requires an environmental impact assessment. Strategic environmental assessments are also a tool which focuses on the consideration of environmental consequences in plans and programmes, with specific emphasis on the environment. The Netherlands' Commission for Environmental Assessment uses the Council of Europe's definition of landscape. As such, landscape can relate to both urban and rural settings, as well as to existing and new attributes. The Cultural Heritage Agency compiled descriptions of landscape character for 78 historic regions. These are intended to inspire municipalities and others to put current environmental changes in a broader time-depth perspective (Cultural Heritage Agency 2018).

- Different landscape mapping projects by national and provincial governments provide various types of data. The most relevant and comprehensive sources are listed as web services (Dutch National Spatial Data Service 2018).
- The Netherlands have assessments considering landscape quality for planning processes for 3 windturbines/15 MW or more. To determine the information required in the environmental assessment, three steps are necessary: 1. Describe or determine the ambition; 2. Describe the landscape qualities; and 3. Determine a tailor-made approach. Every province has developed maps and reports on landscape quality, primarily with a heritage aspect (Cultural Heritage Agency, landschap-innederland.nl, bronnen en kaarten)
- Certain landscapes are protected as World Heritage Sites (Beemsterpolder, Dutch Defense Line), or as cultural monuments (over 450 villages & townscapes, partly agricultural landscapes as well). Furthermore, landscapes can be protected by environmental/spatial planning instruments through the adaptation of provincial efforts. There are 20 national parks as well, but there is no strict legal framework regulating or protecting them.

Interaction between Renewable Energy and Landscape Quality

The transition to alternative forms of energy will have a major impact on the environment. This however has also occurred in the past. Peat extraction left behind large artificial lakes as well

	2010	2015	Forecast 2020
Wind power onshore	3739 GWh	5880 GWh	6000 MW
Wind power offshore	467 GWh	1036 GWh	4450 MW
Photovoltaic power	333 GWh	1436 GWh	-

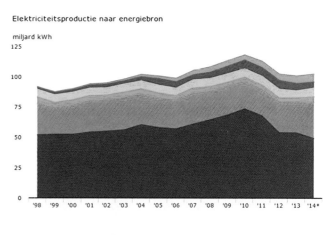

Elektriciteitsproductie naar energiebron

miljard kWh

◼ Wind, zon en water ◼ Biomassa ▨ Kernenergie en overig
▨ Andere fossiele brandstoffen▨ Steenkool ◼ Aardgas

Table 1.21.1
Wind and solar energy
consumption by source
and production forecast

Figure 1.21.2
"Electricity production by
energy source, CSO, 2018,
wind, sun and water, biomass,
nuclear energy, other fossil
fuels, hard coal, natural gas"

as new settlements along the larger and smaller canals. Following the invention of the wind mill, thousands of new structures soon dotted the open landscape. What is different now is the tremendous speed at which the landscape is changing. This acceleration increases the challenge to complete the transition to a climate-neutral lifestyle solely based on sustainable energy within the next 35 years. Some new energy sources, such as geothermal energy or heat-cold storage, will more or less blend in with the landscape and raise little protest. New wind farms and solar plants, however, will profoundly alter the environment. Finally, we should remember that energy production has also generated a wide range of landscapes and features which today are highly appreciated, such as the Kinderdijk windmills.

The national government is mostly involved in the development of large wind parks (> 100 MW). The national policy for onshore wind energy is focused on the nationally zoned areas for wind energy. Provinces are responsible for wind parks 10-100 MW. Each province has developed their own strategies and zoning categories. These actions are agreed upon through discussions with the national government.

Municipalities are responsible for wind parks < 10 MW. Many municipalities already have developed or are developing policies for RE. Some municipalities, particularly the larger ones, have produced a specific wind vision or policy. In Amsterdam, wind turbines must be at least 2 km away from the world heritage site. An extensive cartographic overview of all existing wind turbines in the Netherlands can be found at Bosch & van Rijn (2018).

Concerning solar energy, there is no specific national policy related to the placement of solar farms in rural areas. Policies are made at the provincial level. Generally, provinces try to prevent agricultural land from being used for solar farms. Different policies apply to farms up to 5 ha and those larger than 5 ha. Many municipalities stimulate individuals and companies to place solar panels on existing surfaces. In Amsterdam there is about 11 km^2 of suitable rooftop space. The placement of solar panels is primarily influenced through adherence to local regulations related to building and neighbourhood aesthetics and further through rules related to protection of the character of villages and cities (Huub van de Ven 2014).

Many communities attach importance to considering landscape quality when developing projects and plans. Nevertheless, schemes or projects to improve spatial quality or to make existing landscape the primary concern when implementing plans often fail to hold their own in environmental assessments. There is no national consensus on a definition of landscape quality. This intricate situation has complicated the development of processes and evaluation frameworks for the country as a whole. Experts have begun to argue for the movement away from landscape quality to environmental quality ('leefomgeving') which, by default, comprises other parameters such as smell and sound but also, in parts, functionality and environmental performance of the country.

1.22

NORWAY

Sebastian Eiter, Wendy Fjellstad, Pia Otte & Katrina Rønningen

Situation of Renewable Energy

The main source of RE is large-scale hydropower, accounting for 96 % of electricity production (Statistics Norway 2017). The first commercial station was established in 1885, and many were built by the early 20th century (Rusten 2013). A second wave of development followed in the second half of the century (e.g. Faugli 1994). By early 2017, the average annual production was 133.4 TWh, of which small-scale plants (<10 MW) accounted for 9.9 TWh (NVE 2017).

Most large-scale stations work with high pressure, i.e. a large altitudinal difference between mountain reservoirs and turbines. High annual precipitation, rough relief, and impermeable bedrock provide favourable conditions. About 90 % of the energy supply is owned by public bodies, half of it by municipalities or counties (LVK 2017). The largest, state owned producer accounts for 36 % of the capacity (NMPE 2015).

Interest in development of wind parks has increased recently. In 2016, 1.4 % of the electricity production was from wind power (Statistics Norway 2017). By the end of 2017, the capacity was expected to be 1400 MW, and concessions have already been given for a further 5450 MW (NVE 2017).

Bioenergy is starting to play a role. Forest harvesting could increase from 10 to 15 million m³ per year without reducing the stock (Granhus et al. 2014), although increased harvesting of slow-growing boreal forests has raised concern about negative climate impacts (Bright et al. 2014). In 2007, a target was set to achieve an annual production of 28 TWh by 2020 (Ministry of Environment 2007, Hansen 2013). Sustainability assessment of increased bioenergy harvesting is ongoing, taking into account carbon storage in soils, albedo effects, and emissions during harvesting, transport, and processing (NVE 2017).

Data on Landscape Quality

The Acts on Nature Conservation from 1970 and on Nature Diversity from 2009 have offered several types of protected areas, closely connected to IUCN categories. The most extensive protected areas are national parks, protected landscapes and nature reserves.

From the 1980s, a national plan for the management of watercourses graded hydropower development projects based on the degree of potential land use conflicts and on the plants' economy. Projects with high conflict potential or high costs could not be applied for. The plan was terminated by the government in 2016 (NVE 2016) to reduce bureaucracy.

A national reference system for landscape character (Puschmann 1998) is commonly used in spatial planning processes. It comprises 45 landscape regions, divided into 444 sub-regions and also grouped into 10 agricultural regions (Nersten et al. 1999). Planning consultants have developed it into more detail at regional or local level.

Year	All types of power			Hydropower			Windpower		
	No.	MW	TWh	No.	MW	TWh	No.	MW	TWh
2016	1125	33808	149.0	1066	31817	143.4	27	883	2.1
2014	1038	33697	142.0	977	31240	136.2	25	859	2.2
2010	810	31688	123.6	766	29693	117.2	14	425	0.9
2007	720	30076	137.2	676	28957	134.7	16	348	0.9

Interaction between Renewable Energy and Landscape Quality

Landscape impacts changed between the first and second wave of large-scale hydropower development. Early reservoir dams were constructed of bricks, water pipes were built above ground, and stations were monumental buildings, symbolizing the great opportunities perceived in the technology. More recent dams appear as huge, artificial heaps of stones, water is led through tunnels, and power stations are built inside mountains. Another landscape consequence is dried out watercourses, including iconic waterfalls. However, agreements can ensure a minimum amount of water during the tourism season. The Norwegian Water Resources and Energy Directorate has had an active approach to landscape adaptation and the visual impact of developments, although both the principles and the results have been disputed (Hillestad 1992).

Hydropower development triggered the transition of farm settlements into industrial 'one-company' towns. Legislation and tax systems have ensured that parts of the income generated remain locally. The concession acts ensure that water rights go back to the state after a period, to avoid permanently selling out national resources. Hydropower development was crucial for industrialisation, and the principle of sharing the benefits locally and nationally was important for the development of the welfare state. Large-scale hydropower development formed the background for environmentalism and nature conservation. However, most development happened before public participation was formally integrated into planning processes, although the Planning and Building Act from 1965 was a start. The 2008 revision of the Act implemented the European Landscape Convention, strengthening public participation (Eiter & Vik 2015). Former civil disobedience may have reflected the lack of formal citizen participation and power, e.g. the Mardøla action in 1970, and the mobilisations of environmentalists and indigenous Sámi reindeer herders against the Alta River development in the 1980s (Aasetre 2013). Landscape impact has also become a contested issue in terms of more recent, economically viable and politically supported small-scale hydropower developments (Hagen & Erikstad 2013).

Wind energy projects have been mainly large scale and top down. Although local groups are involved in plan hearings, the processes have been criticised a being predetermined by economic interests (Otte et al. in press; Bergens Tidende 2018). Environmental impacts affect historical dimensions of landscape (Jerpåsen & Larsen 2011), bird life (Christensen-Dalsgaard et al. 2010, Solli 2010), and wild reindeer and reindeer husbandry (Colman et al. 2008, Strand et al. 2017). Europe's largest onshore wind power project (1000 MW) was placed in a pristine natural area containing crucial grazing land for Sámi reindeer herding (Statkraft 2016). Increased production of bioenergy from biomass may impact landscapes in terms of siting new power plants, transport issues, and increased forest harvesting (Gundersen et al. 2016).

Figure 1.22.1
Large-scale hydropower stations in mountainous areas have always been the main source of electricity production in Norway. (Photo: Sebastian Eiter)

1.23
PORTUGAL

Ana Delicado and Luís Silva

Table 1.23.1
"Renewable energy
production in Portugal
as of 2017. Source:
INEGI/APREN 2018"

Figure 1.23.1
The wind farm of
Troviscal in Sortelha,
photo by Luís Silva

Figure 1.23.2
Solar power plant
at Amareleja, photo
by Ana Delicado

Situation of Renewable Energy

Portugal had one of the most ambitious targets in terms of the Europe 2020 strategy: 31% share of RE in gross final energy consumption. The latest figure, 28.5% in 2016 (Eurostat), indicates that it is well on its way to achieve the target. In 2010, that value was 24.2%. And if one considers just electricity consumption, the figure is 72%.

Large-scale hydropower is by far the main contributor, with 49.6% of the energy generated. In a country with just 92,212 km² (excluding Azores and Madeira), there are 46 dams. These dams are situated in the North and Centre regions. Onshore wind energy comes second. Wind farms are mostly located also in the mountain regions of the North and Centre of the country, although they are expanding to the south and southwest coast as well. All the islands of Madeira and Azores also have wind farms. Offshore wind energy is residual, due to the characteristic of the Portuguese coast: steep decline close to the shoreline. Biomass is third; however, it stands for just 3.9% of the RE mix. The biomass plants are located mainly at the northwest coast and in the centre of the country. Less than half of these are co-generation plants. Next is small-scale hydropower. Despite the natural conditions of the country, solar energy remains quite low in the ranking, with just 2.3% of the capacity installed. The number of solar power plants has steadily increased in the past few years, due to lower prices and more flexible policies. There are no thermal solar power plants. Biogas represents 0.6% of the energy mix. Most of the power plants are situated at the western coast, where the vast majority of the population lives. Geothermal energy is only available on the islands of Azores, with three power plants generating 0.2% of the total RE. Marine energy is no longer available, the two experimental facilities in Azores and Peniche are inactive.

Data on Landscape Quality

Portugal signed the European Landscape Convention in 2005 and issued the Landscape National Prize in 2012. Portugal has several protected landscapes, defined as "areas with landscapes resulting from the harmonious interaction between man and nature, which show a great aesthetic, ecological or cultural value." Officially, "the classification of a Protected Landscape is aimed at protecting the existing natural and cultural values, highlighting local identity and the adoption of measures compatible with the objectives of its classification."

At present, the country has two protected landscapes of national level in mainland Portugal, and it has 11 protected landscapes of regional or local level, all created in the past two decades. In

Types of RE	Installed capacity in 2017	Year of the first plant installed	Number of plants	Name of owners and/or installers
Large hydropower (LH)	6.75 GW	1922	46	Large company: EDP
Wind onshore energy (WON)	5.3 GW	1991	250	Large companies: ENEOP2, Iberwind, EDPr, Generg, EEVM, GDF Suez, EDF EN Portugal
Wind offshore energy (WOF)	2 MW	2011	1	Large company: EDP
Biomass (BM)	534.7 MW	1987	22	Small companies
Small and micro hydropower (SH)	470.5 MW	1906	163	Small companies
Solar PV ground-mounted power (SPVG)				
Solar PV on-roof power (SPVR)	317.5			
(does not include domestic generation)	2006		98	Small companies
Biogas (BG)	83.2 MW	2004	52	Small companies, municipal authorities
Geothermal power (G)	33.3 MW	1994	3	Small company: SOGEO - Sociedade Geotérmica dos Açores, S.A.
Marine energy (M)	0.7 MW	2005	2	Company: Eneólica Association: Wavec

addition, two landscapes are protected as World Heritage Sites by the UNESCO. There is also a protected landscape in the island of Madeira and six others in the Azores archipelago.

The Constitution includes landscapes within the objectives of territorial planning and landscape protection as a means to ensure nature conservation and cultural heritage. The Law on Environmental Policies (Law n. 19/2014) states that safeguarding the landscape requires the preservation of an aesthetical and visual identity, the authenticity of natural and cultural heritage and the systems that support sociocultural systems, contributing to the preservation of the specificities of the regions that constitute national identity. The legal framing governing territorial management (Decree-Law n. 80/2015) states that regional programmes must define landscape units and that building plans must take into consideration the integration in the landscape. Unlike other countries, there are no NGO solely devoted to landscape issues.

Interaction between Renewable Energy and Landscape Quality

The only regulation in Portugal that combine RE and landscape quality concerns is about Environmental Impact Assessment (EIA). However, the only RE infrastructures for which EIA are mandatory are hydroelectric dams and wind farms, and these latter only if above a certain number of wind turbines or located in a protected landscape. All other fall under general planning regulations. The Portuguese Environmental Agency provides instructions for drafting EIA studies of wind farms that include specific recommendations on landscapes. EIA studies must identify and characterise landscape units in terms of natural features and built elements; visual absorption, visual quality and landscape sensitivity must be classified, based on the preferences of local communities; and the main viewing points (roads, villages and towns, tourist interest points) of the future wind farm must be identified. Impact analysis must take into consideration the landscape units that are more affected, the visibility of the wind turbines and a quantification of the areas of high, medium and low visual quality and landscape sensitivity that will be affected, devising mitigation measures.

As in other countries, much of the controversies surrounding the siting of RE infrastructures are based on landscape concerns (Delicado et al. 2014, 2016; Silva and Delicado 2017). However, public discussions are not mandatory, so the use of more sophisticated visualisation techniques other than digitally manipulated photographs to which future wind turbines have been added, has never been done. Planning decisions are seldom influenced by public contestation and resistance (Delicado et al. 2014).

1.24
ROMANIA

Mihaela Hărmănescu, Mădălina Sbarcea & Maria Boştenaru Dan

Romanian region	Renewable energy potential				
	Wind	Biomass	Solar	Geothermal	Micro hydropower
Danube Delta	-	-	X -	-	-
Dobruja	X		X	-	-
Moldova	X	X	-	-	X
Carpathian Mountain	-	X	-	-	X
Lower Carpathians and hills	-	X	-	-	X
Transylvania	-	-	-	-	X
Western Plain	-	-	-	X	-
Southern Plain	-	X	X	X	-

Source: Based on Romania's Energy Strategy for 2007–2020 and National Meteorological Administration map.

Table 1.24.1
Romanian territorial regions' distribution of the national RE potential

Situation of Renewable Energy

Romania needs to ensure 24 % of gross energy consumption from RE sources in order to fulfil all provisions of the European Commission to reduce CO_2 emissions by 2020. Under the legislation enforced in 2005 and implemented in 2011, Romania met its target much in advance (2013). After the subsidised boom in RE production that led to a significant increase in energy costs for consumers, Romania, as other EU Member States, adopted a change of paradigm and, since 2013, gave up some of the incentives granted to RE investors (Câmpeanu & Pencea 2014). The investment attractiveness is favoured by the geographical location and local climate zoning, correlated with the energy potential distribution (Table 1.24.1).

According to the latest available data (in 2017 the data for 2016 were available on Electrica Furnizare SA), 42.38 % is the RE power capacity in Romania: 2.60 % of the net installed power capacity comes from solar power, 10.13 % from wind energy, 28.86 % from hydropower, 0.75 % from biomass and 0.05 % from other RE sources. Romania owns RE production units with a total installed capacity of more than 4,500 MW, which equals almost a quarter of the total capacity. Most of the RE production capacity is represented by wind units, especially in Dobruja, the second highest wind potential area in Europe (Dragomir et al. 2016). Solar energy amounts for a small part, but given the climate of Romania, with important solar regions in the south, there is significant potential for growth. There is a small ratio of energy from biomass as well. Slightly more than 50 % of the Romanian energy is produced from low carbon sources, with hydroelectric power almost equaling coal. The geography of Romania with one-third mountain areas helped to facilitate the installation of large-scale hydropower plants on rivers such as the Bistriţa, Argeş and Danube.

Dobruja has become an important wind farm area in Southeastern Europe. In the peak period of wind-energy investments, Tulcea County authorities were faced with an unprecedented number of proposed wind farm plans. As the evolutionary trend of RE development proposals shows, after the change in national RE subsidy policies, in Dobruja the number of proposed wind-farms and solar farms dropped dramatically.

Figure 1.24.1
Dobruja energy landscape
(Photo: Mădălina Sbarcea, 2017)

Data on Landscape Quality

In terms of landscape impact assessment, a framework is not clearly defined, but references can be found in Romanian legislation on spatial and urban planning (Law no. 350/2001, amended by several normative acts, including Law no. 190/2013 that introduced the necessity of protection of natural and anthropic landscapes). However, specific regulations for identification, evaluation, and management of landscapes have not yet been adopted, despite the fact that the ELC was ratified by Law no. 451/2002.

Interaction between Renewable Energy and Landscape Quality

Landscape quality is still not a central element in current RE planning, although environmental impact studies are mandatory in order to obtain environmental approvals. The issue of landscape value is difficult to tackle and integrate in strategic environmental assessments if qualitative standards or indicators related to landscape are lacking.

The ratification of the ELC was an important step in planning and preserving the Romanian landscape, but the introduction of landscape quality objectives and values into impact studies is very important for ensuring consistency of RE projects. Current practice illustrates the frequent use of some methods of analyzing and assessing landscape as a collection of environmental elements, yet a well-rounded framework for qualitative assessment of the impact of proposed projects on the landscape is imperative.

In Tulcea County, the Environmental Protection Agency requires the impact assessment procedure for each park that exceeds ten turbines and for any development that is overlapping with Natura 2000 sites. Having interviewed several architects that planned and authorised windfarms in Tulcea between 2007 and 2014, the authors found only one situation where authorities required a landscape-impact study. This particular requirement was motivated by a plurality of factors that converged towards the obvious need to take into consideration the impact that the development would have on landscape quality. Such factors included the close vicinity to cultural heritage landmarks or other important developments being planned in the area at the same time.

SERBIA

Marija Lalosevic, Dragi Antonijevic & Mirko Komatina

1.25.1
Preserved Landscape of the 'Natural Core of Belgrade', the confluence of the Sava River into the Danube, view from Kalemegdan fortress towards New Belgrade (Photo: Marija Lalosevic)

Situation of Renewable Energy

With the exception of hydropower, deployment of RE sources focusing on biomass plays a small role in Serbia's energy balance. However, due to the potential of RE, and as Serbia's strategic vision is to increase the share of RE sources and implement legal support for their adoption, it is expected that RE production will increase.

The natural potential in Serbia provides conditions for the use of RE, such as biomass, sun and wind energy, small-scale hydroelectricity, and geothermal energy. The potential of these energy sources is approximately 25 % of the country's present primary energy consumption.

The technical annual potential of RE production has been estimated to be more than 50 TWh, comprising biomass energy (31.4 TWh), hydropower (7.0 TWh), solar energy (7.0 TWh), geothermal energy (2.3 TWh), and wind energy (2.3 TWh) (www.obnovljiviizvorienergije.rs). In 2017, RE sources should provide 18 % of the total domestic production of primary energy. Of that 18 % renewable, 59 % is sourced from biomass, 40 % from hydropower, and less than 1 % each from biogas, wind, sun, and geothermal energy (www.mre.gov.rs).

According to the latest official data (2015), total annual primary energy production in Serbia was 125 TWh. According to sectors (TWh), this was comprised of coal 83.7, oil 12.9, gas 5.3, hydroelectricity 10.2, geothermal energy 0.07, solid biomass 12.8, biogas 0.07, solar energy 0.01, and wind power 0.0 (www.mre.gov.rs). The gross annual production of electricity (in TWh) produced by Serbian power plants amounts to fossil fuel power plants 27.1, hydroelectric power plants 10.7, electricity generated by combined fossil fuel-central district heating plants 0.05, solar electricity generation 0.01, wind electricity generation 0.000, other sources 0.3—giving 38.3 TWh in total (www.mre.gov.rs). In 2009, in order to stimulate construction of RE plants, the Serbian government adopted a decree supporting measures for producing electricity from renewable sources and from highly efficient, combined electricity-district central heating production. Subsequently, in 2013 and again in 2016, the government additionally increased the capacity of these plants. The current target quotas prescribed (until the end of 2020) for power plants using renewable sources are: 500 MW for wind turbines; 10 MW for solar energy plants; 1 MW for geothermal plants; 100 MW for solid biomass plants; 30 MW for biogas plants; 10 MW for landfill gas plants; and 3 MW for waste material plants.

Regarding the history of RE development in Serbia, it is noteworthy that the first hydroelectric power plant commenced operation in May 1900 in Valjevo, followed in August 1900 by the hydroelectric power plant in Uzice. This latter plant was based on Nikola Tesla's multiple phase AC system, and was built just a few years after the first AC hydroelectric power plant in Niagara Falls.

Data on Landscape Quality

The Spatial Plan of the Republic of Serbia 2011–2020 (2010) is the main planning document in the field of environmental and landscape protection management and is of the greatest strategic

importance to the country. The Spatial Plan foresees and stresses the need for detailed elaboration of the most significant, important factors for planning, protection, development, and land use in the sense of creating prerequisites for sustainable development adjusted to natural and anthropogenic values and conditions.

Serbia does not have, for now, an integrated national landscape assessment. There are protected areas for landscape quality and scenery, primarily consisting of the six national parks and the natural areas of outstanding significance. Serbia has set aside 516,350 hectares (i.e. 6.59 % of the total territory) as protected natural environmental areas.

Public involvement in the processes of developing and adopting planning documents is prescribed by the Law on Planning and Construction. Written submissions are permitted during the public review phase of draft planning documents. However, the process itself does not offer citizens the possibility of involvement in the final decision-making.

Interaction between Renewable Enegy and Landscape Quality

Serbian laws relevant to the RE sector and landscape protection are Law on Nature Protection (2009), Law on Environmental Protection (2004), Law on Planning and Construction (2009), Law on Energy (2014), Law on Efficient Energy Use (2013), and Law on Forestry (2010).

Relevant strategic documents are: the Spatial Plan of the Republic of Serbia 2011–2020 (2010); National Action Plan for the Utilisation of RE Sources (2013), Energy Sector Development Strategy of the Republic of Serbia for the period until 2025 with projections to 2030 (2015), and the Programme for the Implementation of Energy Strategy for the period from 2017 until 2023 (2017).

In the scope of utilisation of RE sources and in the context of the relationship between their use and landscape quality, Serbia needs to prepare strategic projects such as the Project for Intensive Use of RE Resources, and the Project for Increasing Energy Efficiency. Additionally, programmes must be defined which will enable investment in this branch of energy engineering (through concessions and similar plans), creating conditions for greater use of the renewables sector.

Finally, there is a lack of projects that are concerned with education of experts and the general public on RE sources, protection of land/environment assets, energy efficiency, and sustainable planning. In accordance with citizens' low level of knowledge and their lack of inclusion in decision-making in the sphere of RE resources, besides education, it will be necessary to develop support mechanisms for citizens within the decision-making system.

1.26

SLOVAKIA

Martina Slámová & Attila Tóth

Figure 1.26.1
A case study on the visual impact assessment of a wind farm in Tvrdošín (Jančura and Bohálová 2009)

Situation of Renewable Energy

A framework for national legislation constitutes the directive 2009/28/EC on the Promotion of the Use of Energy from Renewable Sources (RES). The key national legal standard on RES is the Act no. 309/2009 Coll. on Promotion of Renewable Energy Sources and Highly Efficient Cogeneration. In 2014, 23.0 % of total electricity was produced from RES and Slovakia is on the track to reach its target in 2020 (ŠÚ SR, MH SR 2014).

Currently, hydroelectricity is the most utilised RES in Slovakia (68.9 % of total electricity production from RES). Fast-growing constructions of small-scale hydroelectric power plants were limited by the Ministry of Environment. Solar energy covers 9.2 % of RES After a "solar boom" in early 2011 the financial support of large-scale photovoltaic (PV) installations was reduced. However, several PV installations have been established on agricultural land. Since 2014, biomass RES have been promoted (7.4 %) (ŠÚ SR, MH SR 2014). Slovakia was the last EU Member State that implemented wind energy into their national energy mix policies. Wind energy covers only 0.1 % of total electricity production from RES because the majority of localities with suitable wind conditions are situated within protected landscape territories (Tauš et al. 2005). The geothermal energy utilisation is of local importance, but there is a synergic correlation with regional development and local economy through recreation (e.g. Podhájska).

Data on Landscape Quality

Slovakia has several legislative tools influencing the implementation of RES from the perspective of landscape quality. Following the principles defined in the European Landscape Convention (ELC) coupled with social demand, the 'Method of Identification and Assessment of Landscape Characteristic Appearance' (Jančura et al. 2010) was developed. The method was designed for monitoring changes in visual quality in landscape types through control points. Correspondingly to this method, a directive of 'Standards and Limits for the Location of Wind Power Plants and Wind Farms in the Territory of Slovakia' (Enviroportal 2010; Chomjak and Tomič 2008) was created. It contains the Catalogue of Limits for wind farm locations available for the general public online.

Interaction between Renewable Energy and Landscape Quality

Large parts of non-protected landscapes constitute a matrix for protected cultural or natural heritage (Slamova et al. 2012). Landscape management is to be addressed to outstanding landscapes, usually legally protected, as well as to everyday landscapes which are of equal interest according to the ELC (CoE 2000). However, it is currently not a common practice to carry out visual landscape quality assessment within urban development processes, including RES in Slovakia. Landscape values should be considered in local master plans much more than they currently are (Tóth et al. 2016). Considering the progressive development of RES, the existing methodical tools on landscape quality assessment should be updated.

The proposal of the Tvrdošín Medvedie Wind Park:
visual zones
and monitoring control points
(Jančura and Bohálová 2009)

North

20 km

5 km

10 km

West

East

Visual zones – distances from monitoring points

South

Legend:

Limit A (strongest regulation)

Limit B

Limit C

Areas without limits at the national level

GIS spracovanie © Slovenská agentúra životného prostredia.
Centrum krajinného plánovania, prírodných a energetických zdrojov; 2009
Kartografický podklad SVM50 © ÚGKK SR 2000
Tematický obsah:
Slovenská agentúra životného prostredia, Ministerstvo životného prostredia SR, Ministerstvo hospodárstva SR, Ministerstvo pôdohospodárstva SR, Ministerstvo dopravy, pôšt a telekomunikácii SR, Ministerstvo kultúry SR, Ministerstvo obrany SR - Úrad vojenského letectva, Letecký úrad SR, Pamiatkový úrad SR, Štátna ochrana prírody SR, Letové prevádzkové služby SR, KURS 2001

Avaialble at:<https://www.enviroportal.sk/standardy-a-limity-pre-umiestnovanie-veternych-elektrarni-a-veternych-parkov-v-sr>

0   35   70   140 Kilometers

A view of Medvedie from Krivá in 2009

A proposal for a wind farm (7 km)

Nr. of monitoring locality	Name of the locality	Distance [km]	View direction
1.	Dlhá nad Oravou	8,8	NE
2.	Krivá	7,0	NE
3.	Nižná	4,1	NW
4.	Zemianska Dedina	1,7-2,8	N
5.	Tvrdošín, mesto	3,6	NNW
6.	Tvrdošín, cesta	3,08	NWW
7.	Bobrov	9,8	SSW
8.	Trstená prechod	13,2	SW
9.	Suchá Hora	21,9	SWW

A view of Medvedie from the town of Tvrdošín in 2009

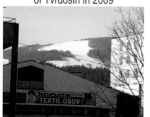

A proposal for a wind farm (3,6 km)

A proposal for a wind farm (8 km) from the village of Dlhá

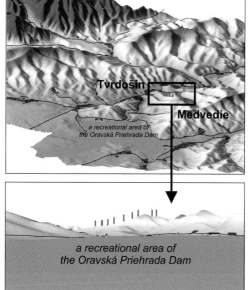

Tvrdošín

Medvedie

a recreational area of the Oravská Priehrada Dam

a recreational area of the Oravská Priehrada Dam

Authors of the case study on the visual impact assessment of the wind farm in Tvrdosin: Jančura, P., Bohálová, I. (2009): Assessment of the Suitability of the Tvrdošín Medvedie Wind Park in Terms of the Visual Characteristics of the Landscape. Pro Tempore, Kolégium, DKD; Slovak Environmental Agency Banská Bystrica. Unpublished.

1.27

SLOVENIA

Naja Marot, Tadej Bevk & Mojca Golobič

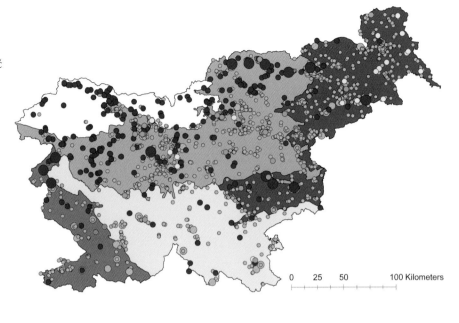

Figure 1.27.1
Distribution of renewable energy production units in landscapes regions of Slovenia

Production units by RE source

- Biomass
- Biogas
- Wind

Hydro
- ● <10 MW
- ● >10 MW

Solar
- ● <500 kW
- ● >500 kW

Landscape regions

- Alpine
- Interior karst
- Sub-Alpine
- Littoral
- Sub-Pannonian

Situation of Renewable Energy

The share of RE in domestic production of energy (154.114TJ) accounted for 31.4 % in 2014 (Vlada RS 2016). Altogether, RE contribute 21.5 % (European Commission 2017) to the gross final energy consumption; the 2020 goal is 25 %. The share of RE for heating and cooling has increased from 18.9 % in 2005 to 32.4 % in 2014, and the share of renewable sources for electricity generation increased by 5.2 % only, from 28.7 % to 33.9 %. The share of RE in transport is low, only 2.9 % in 2014. In electricity production (17.437GWh in 2014), the biggest renewables' share comes from hydropower (36.5 % in 2014). The oldest production unit Hrušica, located in the Alpine gorge Vintgar, was already installed in 1904. Nowadays, there are 19 hydropower plants of a larger size. For harnessing hydropower Slovenia exploits most of its largest rivers, namely Sava (8 power plants), Drava (8) and Soča (3). They run through contrasting landscapes, from the alpine valleys through hilly interior sub-alpine valleys to the Sub-Panonian.

Solar power plants have existed in Slovenia since 2001. Use of solar power has increased since 2009, with the biggest leap from 2011 to 2012, when installed capacity more than doubled, from 99MW to 221MW. In 2014 the electricity production from solar energy accounted for 1.5 % of the electricity production. Currently, there are 3381 solar power plants with a total installed capacity of 260MW (PV portal 2017). They are dispersed throughout the country and mostly installed on rooftops (Figure 1). However, regions having the most solar installation produce the least solar-rooted energy.

Although adequate wind energy potential can be found in Alpine, sub-Alpine and karst regions, the majority of interest as well as the two existing wind turbines (3.2 MW of installed capacity, since 2014) are located in the latter. Biomass mostly serves as a source of heat production, 57 % of the heat produced for the households derives from it (SORS 2017). Altogether RE contributes 34 % to heating and cooling in terms of gross final energy consumption (European Commission 2017). There is no geothermal or tide/wave power plant in Slovenia

What is needed to advance deployment of RE sources in Slovenia at this stage are clear strategic plans on energy infrastructure, which will soon be formulated in a renewed Action plan for RE, as well as a renewed national Strategy for Spatial Development. Although energy transition is a national effort, local communities should be incentivised through various instruments, including shared ownership and management, to proactively participate in it and in this way minimise local opposition.

Figure 1.27.2
Roof-installed PV in a
typical Slovenian landscape
(Photo: Csaba Centeri)

Data on Landscape Quality

Based on a comprehensive national landscape assessment, Regional Distribution of Landscape Types in Slovenia (Marusic et al. 1998), with the objectives to establish an inventory of landscapes, evaluate them, and provide guidelines for conservation guidelines for management and development, areas of outstanding landscapes and management guidelines were prepared in the late 1990s. Unfortunately the results of these studies were not integrated into policy until 2004 when areas of nationally important landscape character were designated and included into the national strategic spatial development strategy. Outstanding landscapes, however, never received legal status as protected areas. Nevertheless, some of these landscapes were subsequently protected through cultural and natural heritage conservation mechanisms which also enforce certain limitations in terms of planning processes. On the national level, no other legal requirements pertaining to the landscape, apart from an environmental impact assessment, regulate planning processes, and these considerations are left to the discretion of planners and limitations are enforced by other sectors. As landscape is not covered by its own national policy, the subject is included across several others, such as the national spatial development strategy, nature conservation law, law on the protection of cultural heritage, and the rural development programme. However, municipalities have

an option to enforce landscape protection through municipal spatial plans and to enforce the use of RE on the basis of the local energy concept.

Interaction between Renewable Energy and Landscape Quality

Similar to other countries, landscape is among the main issues in discussion of the sites for RE technologies. Experiences with wind farms show that landscape is an important factor brought up by the public, right after nature conservation—particularly the protection of birds. While solar panels have not yet been deployed to a broader extent and mainly focused on roof instalments, there is little evidence on their interaction with the landscape. When it comes to hydropower plants, landscape is again seen as an important factor due to improvement of flood protection, and creation of new recreation areas and facilities, but it still shadowed by nature conservation efforts.

Slovenia needs a national landscape policy that could benefit not just RE deployment but address other spatial development conflicts as well. Similar to experiences in other countries, the opposition of local inhabitants, although expressed as concern for landscape and nature, is often driven by the lack of opportunity to participate in decisions about management of and profits from RE development.

1.28
SPAIN

Verónica Hernández-Jiménez, Daniel Herrero-Luque, Blanca del Espino, Richard Hewitt & María-José Prados

Figure 1.28.1
Electric pole facing wind turbines in Osuna, at the surroundings of the Protected Landscape of the Campiña de Sevilla. (Photo: Juan José González)

Situation of Renewable Energy

After many years at the forefront of RE implementation, Spain is experiencing a situation of considerable uncertainty with respect to the future of this sector. The country's ability to comply with international climate change commitments (2020 to 2050) is now seriously under question. However, there are major differences in capacity for RE across regions which strongly reflect the different territorial, economic, and administrative circumstances (Martinez et al. 2016). To comply with the international climate change commitments, regional characteristics and civil society support for the RE sector are likely to play a key role. Those regions have increased system resilience and reduce the vulnerability of the energy transition to system shocks.

However, if progress is to be made at the required speed, a change of direction at the national level is required. In addition to Law 1/2012 removing direct subsidies to RE, two further laws on RE have been deployed. The first (Law 24/2013) makes prosuming[1] not viable by abolishing the feed-in-tariff (FIT) and replacing it with a per kWh charge. FIT was one of the key pillars of the Spanish renewables boom (del Río 2008). The second (RD 900/2015) substantially increases charges to grid-connected consumers with accumulation (battery storage). So it seems that the government doesn't want to encourage self-consumption either. These laws seem largely aimed at shoring up consumer demand and raising revenue to reduce the government debt to the electricity companies, known as the tariff deficit, before actively continuing to promote RE.

Data on Landscape Quality

The landscape quality legislation, management, and assessment has a certain tradition and this has been transferred to the regional administrations. At the same time, landscape-related competencies are sometimes afforded by environmental and territory, sometimes by cultural heritage laws and also other protection instruments. At the national level, both the environmental and heritage administrations have authority for the main general regulations about landscape assessment, maps, and data (Mata Olmo & Sanz Herraiz 2004, Jimenez [dir.] 2009). Additionally regional governments have created specific agencies for landscape assessment, designed in close cooperation with academic specialists. The availability of regional datasets, maps, and atlases is their main output, together with academic proposals concerning the definition of landscape quality analysis and preservation. Data and maps are based on regional landscape units and delimitation divided into environmental and cultural landscape criteria. The detailed scale of maps allows a deeper knowledge and analysis of data but sometimes makes systematic data compilation difficult. Spatial and urban planning include natural and cultural landscape categories to be preserved together with other concepts regarding landscape quality or landscape scenery, such as visual

field, harmony or perspective. Until then, only the Historical Heritage Law (16/1985) included a pioneering concern about the possibility of protecting areas because of their landscape values. It had allowed some regions to include cultural landscapes as protected areas. However, measures designed for protecting environmental preservation rules are often stronger than cultural protection ones. Environmental protection had been established by national laws (4/1989, 42/2007) and now 'protected natural spaces' are well defined and typified. Between them, the designation of protected landscapes becomes a useful tool facing the conservation of landscapes with cultural values, and it is developing. Related to these legal frameworks, environmental impact assessment and environmental impact statements report landscape quality preservation together with physical, cultural, and even social analysis.

There are also regional regulations where landscape vulnerability and landscape impact reports are mandatory, sometimes linked to environmental approaches. In terms of cultural landscape, heritage regulations usually require landscape impact reports for interventions into protected areas. In conclusion, spatial planning includes environmental impact assessment and environmental impact statement. These reports mainly refer to landscape quality by physical attributes more than aesthetical and psychological ones.

Interaction between Renewable Energy and Landscape Quality

Development applications are typically handled at local level (municipality), but are subject to requirements mandated from the regional level. Landscape vulnerability and landscape impact reports are one such regional level requirement. These are usually carried out as part of the ordinary environmental assessment process in any development application. In terms of cultural landscape, heritage regulations usually require landscape impact reports where interventions fall within protected areas. At the same time, landscape impact reports are also required, for example, to obtain authorisation for new power plant developments. Following European Union directives, spatial planning includes environmental impact statement and environmental impact assessment. However, while these reports mainly refer to landscape quality in terms of physical attributes, aesthetical and psychological attributes also need to be taken into account for landscape quality protection. In practical terms, this means that many of the most important impacts of renewable energy developments on landscape in Spain are not properly integrated into current regional spatial and landscape planning procedures.

1 *Prosumers* are citizens or organisations who both produce and consume energy, typically householders with RE installations who "feed-in" to the electricity grid.

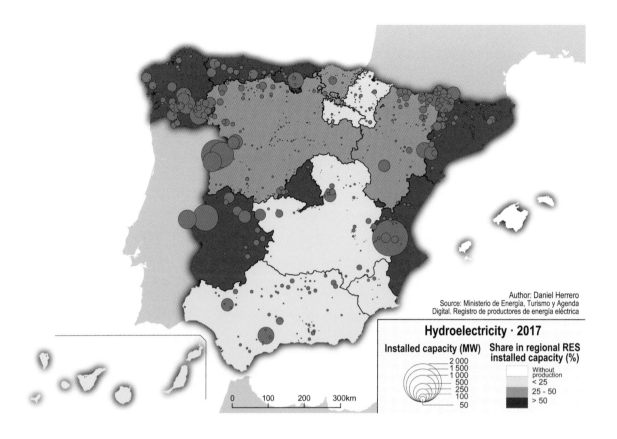

Author: Daniel Herrero
Source: Ministerio de Energía, Turismo y Agenda
Digital. Registro de productores de energía eléctrica

Hydroelectricity · 2017

Installed capacity (MW)

2 000
1 500
1 000
500
250
100
50

Share in regional RES
installed capacity (%)

Without production
< 25
25 - 50
> 50

0   100   200   300km

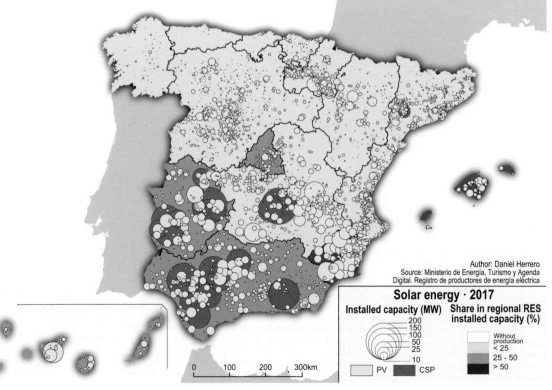

Author: Daniel Herrero
Source: Ministerio de Energía, Turismo y Agenda
Digital. Registro de productores de energía eléctrica

Solar energy · 2017

Installed capacity (MW)

200
150
100
50
25
10

Share in regional RES
installed capacity (%)

Without production
< 25
25 - 50
> 50

PV   CSP

0   100   200   300km

1.28.2
Renewable Energy
Installed Capacity,
2017. Source: Regis-
tro de Productores
de Energía Eléctrica
(2017). Ministerio
de Energía, Turismo
y Agenda Digital.
By Daniel
Herrero-Luque

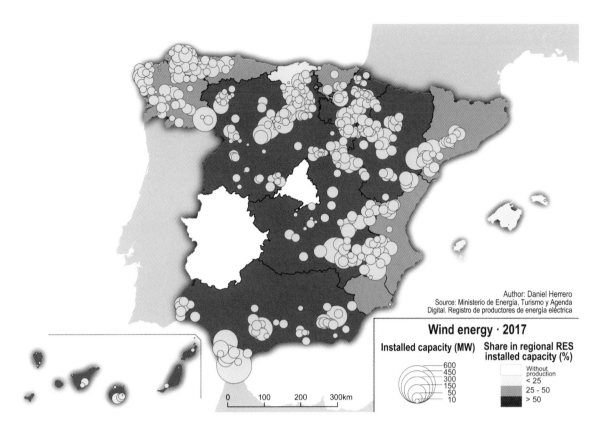

Author: Daniel Herrero
Source: Ministerio de Energía, Turismo y Agenda
Digital. Registro de productores de energía eléctrica

Wind energy · 2017

Installed capacity (MW)

Share in regional RES
installed capacity (%)

- 600
- 450
- 300
- 150
- 50
- 10

Without production
< 25
25 - 50
> 50

0   100   200   300km

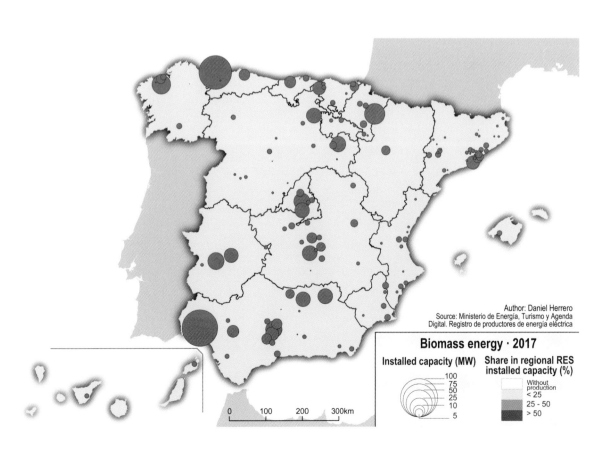

Author: Daniel Herrero
Source: Ministerio de Energía, Turismo y Agenda
Digital. Registro de productores de energía eléctrica

Biomass energy · 2017

Installed capacity (MW)

Share in regional RES
installed capacity (%)

- 100
- 75
- 50
- 25
- 10
- 5

Without production
< 25
25 - 50
> 50

0   100   200   300km

1.29
SWEDEN

Alexandra Kruse

Situation of Renewable Energy

Sweden is energy independent (Swedish Government 2010). The total production is 159 TWh, which is 125 % of the country's own usage. In 1980 a phase-out of the use of nuclear energy was decided by the country (Swedish Government 2010). According to the national action plan of 2010, 'the vision is that in 2050 Sweden will have a sustainable and resource efficient energy supply and without any net emission of greenhouse gases into the atmosphere' (Swedish Government 2010, 3). Reduction of the dependence on oil and its direct impact on climate change, while enhancing competitiveness towards developing new technology, industry, and commerce, are also national aims (Swedish Government 2010). The COP21 target to produce 50 % of the national energy consumption from renewable sources by 2020 was already achieved by 2015 (53.9 %; Eurostat 2017). Sweden realised at the same time the biggest share of renewable energy of the gross energy consumption within EU-28 (Eurostat 2017).

Measures to reach this target address different sectors and include

- Energy tax on electricity and fuels
- CO_2 taxation
- Green certificate system for RE
- Emissions trading of CO_2 in EU targeted instruments
- Information and education, innovation and R&D
- Various specific programmes and support schemes

The results can be summarised as follows:

- Increased ambition for the electricity certificate system to 2020 (finance 30 TWh to 2020 compared to 2002)
- Tax reduction for micro-generation of electricity
- Reduced percentage of investment support for solar power, but increased overall budget

- Support for private households' storage of own electricity production
- Close cooperation with Norway
- Development of a strategy on solar power

Worldenergy (2016) gives the following data on installed RE production capacity: Hydro: 16.2 GW with 2057 plants in total, of which 1050 MW is small hydro (5th position after IT, FR, DE, ES); Biomass: 282 kW through municipal solid waste (electricity generating capacity of 4,990 TJ), wood: 33,720 TJ, with a generating capacity of 2,652 kW; Solar: 85 MW, Geothermal: 5.6 GW; Wind: 6.03 GW. Marine energy: There is a wave energy test facility called Islandsberg on the west coast of Sweden.

Sweden focuses not only on changing energy production, but also on new attitudes among all inhabitants. Several measures are envisaged and promoted by the state:

a) Energy-efficient households: on 1 January 2008, a new law on energy declarations entered into force. The government is investing in information and advice on how to save energy. Each of the 290 municipalities has an energy adviser (SI 2015).

b) Deregulation of the power market: since 1996, customers can choose their supplier, today, about 200 companies sell electricity (SI 2015).

c) Industry: a growing number of businesses invest in renewable energy or work on important energy economics or the use of green energy.

d) Transportation: electric cars are supported. In 2015 there were 12,000 rechargeable vehicles: 42 % electric and 58 % plug-in hybrids (SI 2015).

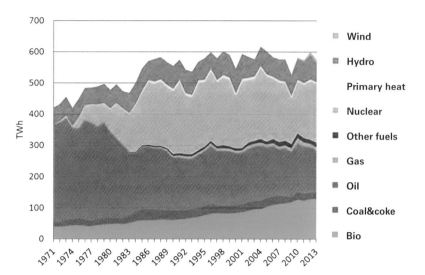

1.29.1
Energy production by source. Although already decreasing, 2013, Sweden's energy production was still based on nuclear power (around 40 %). (Source: López, Eva Centeno, Swedish Environmental and Energy Ministry 2016)

Several research and technology projects are being conducted, mostly joint efforts between industry, government, academia, and nature conservation authorities, e.g. ElectriCity Gothenburg: Electric buses, powered from wind and hydropower. are 80 % more energy-efficient than diesel-powered buses (SI 2015, 4). The Biosphere Reserve Lake Vänern tests different, environmentally friendly mobility solutions for individual use as well as for public transport, including electrified streets and a hydrogen service station (project ElectriVillage). A solar park and huge rechargeable batteries will be connected to the local grid. Further research and development is concentrated on fuel cells. Swedish researchers are seeking ways to produce hydrogen and trying to imitate photosynthesis in plants (Swedish Institute 2015, 2).

Data on Landscape Quality

The overarching law of the land is the Natural Resource Act (NRA), which is an umbrella law. The Planning and Building Act (PBL) is the most important executive law. Both acts came into force in 1987. The overall aim of the NRA is defined thus: 'Land, water and other aspects of the physical environment are to be used in a way that promotes their good long-term utilisation from ecological, social and economic points of view' (NRA 1987). These laws are followed up by general plans, district regulations, and detailed plans, where more specific laws come into force in particular circumstances. Property plans and permissions are at the detailed end for each construction, which also include suitability assessments of the land for the planned construction (Hall et al. 1994, 144f). The NRA includes landscapes protected by law includes so-called landscapes of national interest, which can be

- Particular value for tourism and recreation
- Costal and archipelago protected from industry that interferes with the natural environment
- Mountain areas
- Rivers protected from exploitation for hydro-electric power

Twenty-one historical cultural landscapes have been defined (Sporrong 1994), while several recent approaches to grouping them exist (Helmfrid et al. 1994). Since 2003, the National Inventory of Landscapes in Sweden, NILS monitors the landscape; this includes field inventory and aerial photo interpretation (SLU 2017).

Interaction between Renewable Energy and Landscape Quality

Sweden has signed the ELC. From 1990 to 1997 the *National Atlas* was published with special focus on cultural landscapes and descriptions of all geographical areas (Helmfrid 1994). The atlas describes landscape genesis, development, patterns, settlements, and transformation processes. Resource exploitation and management with respect to landscape and natural values are also included (Baudou 1994). In 2013, an Integrated Landscape Character Assessment (ILCA) was introduced and designed to be used in spatial planning (Trafikverket 2015).

Further links:
http://www.energimyndigheten.se/en/facts-and-figures/statistics/
(Accessed 9 January 2018)
https://sweden.se/society/energy-use-in-sweden/
(Last update: 12 January 2018, viewed 12 January 2018)

SWITZERLAND

Stefanie Müller, Christiane Plum & Marcel Hunziker

Stefanie Müller, Christiane Plum & Marcel Hunziker

Figure 1.30.1
Hydropower plants are perceived as a part of the Swiss landscape: Reservoir of the hydropower plant Wägital in Schwyz, Switzerland. (Photo: Matthias Buchecker)

Situation of Renewable Energy

The Swiss composition of RE production types reflects the dominant landscape characteristics of the country. There are three main landscape types in Switzerland: the Swiss Plateau—a dispersed settlement belt, characterised by a flat to a smooth hilly surface area—and on both sides, mountains, the Jura and the Alps.

The alpine landscape region constitutes 60 % (FDFA 2017), the largest area, and due to its large river system it is one of the main RE providers. Hydropower covers more than half of electricity production (56.4 %) (SFOE 2015a). Other types of RE production contribute 2.2 % to overall electricity production (SFOE 2015[a]). Hydroelectricity has a long history. The first plant was installed in 1816 (SFOE 2015b). At the beginning of the 20th century around 7000 small-scale hydropower plants were in operation (SFOE 2015b). But with the advent of low-cost electricity from large-scale power plants, many of the small-scale plants ceased production. In 2014, around 1000 small-scale (overall capacity: 2360 GWh) and around 590 large-scale hydropower plants (overall capacity: 36960 GWh) were in operation (SFOE 2015b, SFOE 2017a).

In 2014 solar panels with an overall capacity of 1460 GWh were installed on building roofs and facades (SFOE 2015b). The production of energy by solar power is driven by household investments and there is no large-scale solar power plant, either in operation or currently planned.

Efforts to build large-scale onshore wind energy farms have been limited for a long time mainly due to topological reasons.

Nevertheless, the first small-scale wind energy plant was installed already in 1986 (SFOE 2017b). Ten years later the first large-scale wind energy plants were installed on Mt. Crosin (located in Swiss Jura). Nowadays, large-scale wind mills are also installed in high-alpine regions. In 2014, 59 large and small-scale wind mills on 37 sites with an overall capacity of 100 GWh are operating (SFOE 2015b). However, the aim of the Swiss energy strategy 2050 to install up to 800 large-scale windmills (SFOE 2012) is still highly contested.

Energy production based on biomass and geothermal sources is contributes to a relatively small amount of the total RE production. In terms of biomass, typically heat of wood combusting is used for electricity and the heating of houses (overall capacity: 10700 GWh) (SFOE 2015b). Only 97 biogas plants are in use (overall capacity: 1210 GWh) (SFOE 2015b). The situation is similar for geothermal energy production. Despite having a high potential there is no large-scale geothermal energy production yet. However, the estimated overall capacity by geothermal large-scale plants in 2030 is 800 GWh (SFOE 2016). On the household level, electric heating pumps based on geothermal technology are increasingly being used. By 2014, 240,887 geothermal electric heating pumps had been installed, producing with an overall capacity of 3500 GWh (SFOE 2015b).

Data on Landscape Quality

Until the 1960s the preservation of landscape quality was seen as a civic duty. In 1966 the implementation of the Law of Nature and Cultural Heritage made landscape protection a duty of the

state government. Followed by these nation-wide efforts, in 1969 the regulations and, in 1979, the laws for spatial planning were enforced (VLP-ASPAN 2014). These laws and regulations express the challenge of providing the functionality of a landscape in terms of economic developments, but also of maintaining the immaterial values of the landscape (Bundi 2016).

One of the first instruments were the 'regulations for the inventory of landscape and historical monuments' (called BLN regions) (Rey et al. 2017). Since 1977—regularly revised—this inventory has designated landscape areas and buildings of high quality at national scale (Rey et al. 2017). In 1984, a national landscape monitoring programme called 'landscape under pressure' was started, and replaced in 2007 by the 'landscape observation programme' (called LABES) (Rey et al. 2017). LABES assesses, documents, and evaluates the state and development of Swiss landscape based on physical information and public perception of landscape (Rey et al. 2017).

Interaction between Renewable Energy and Landscape Quality

The planning of RE infrastructure is polarised between an utilitarian and a conservative understanding of landscape. Landscape quality is thus a dominant crux within all three planning stages: on the national, the cantonal, and the local level. An RE project cannot be realised before an iterative landscape assessment across various spatial scales has been conducted (ARE 2017).

Nevertheless, public opposition against new RE infrastructure remains high. This is especially the case for highly landscape-intrusive RE infrastructures, such as wind energy plants. Traditional highly landscape-intrusive RE infrastructures, such as large-scale hydropower plants, are more accepted by the public in cases where there are no concerns about ecological issues (Figure 1.29.1).

This might be due to the fact that the public has become used to large hydropower plants in the Swiss mountain landscapes since many decades (early 20st century). A recent case study actually reveals a high appreciation of such a landscape despite of the high visibility of the hydropower infrastructures. Often large hydropower plants are, however, not perceived by the public as renewable energy (Hunziker et al. 2014).

With respect to new highly landscape-intrusive RE infrastructures landscape-fit, indeed is an essential condition for acceptance. Opposition however, is not only due to a perceived misfit between landscape and technology. Moreover, a perceived lack of integration of the projects in the local context fuels opposition against new RE infrastructure initiatives (Müller et al. 2017).

This shows, how crucial a coherent and shared RE vision is for a successful energy transition. It furthermore shows the importance of the promotion of grassroots participation in the planing of RE projects, in order to make site decisions more comprehensible and future RE infrastructures in Switzerland not only landscape-suitable, but also place-integrated.

1.31

TURKEY

Emel Baylan

Situation of Renewable Energy

The energy demand of Turkey, a developing country, is increasing. In 2014 16.1 % of the total electricity production was achieved from hydropower, 3.3 % from wind, and 1.42 % from waste and geothermal sources. Hydroelectric production started in 1902 at a small-scale plant in Tarsus. The country's first large-scale hydropower plant was built in Istanbul in 1913. Construction of Turkey's first Republican-era hydroelectric plant began in 1926, and this micro-plant began producing in 1929. In 2014, Turkey's hydroelectric power capacity was 23,643.2 MW. In this year, more than 450 hydropower plants were in operation and another 222 were under construction. Moreover, licenses were given to 300 hydroelectric plants with operating capacity of above and below 10 MW (Anonymous 2015).

The first wind energy plant was established in 1998 with an operating capacity of 17.4 MW. In 2014, 93 wind power plants (WPPs) in operation represented a total capacity of 3,636 MW, and another 85 plants of 1,162.8 MW in total were under construction. Further WPPs under license had a total capacity of 5,435 MW. Investment in wind power production is increasing significantly. The regions of Turkey with the highest total operating capacity of wind power plants are Marmara (1,240.88 MW), Mediterranean (543 MW), Central Anatolia (222 MW), Aegean (131.1 MW), Black Sea (80 MW), and South-Eastern Anatolia (27.5 MW). The country's onshore wind power potential is 48,000 MW, but only 3,424.48 MW of this potential has been realised so far, equaling only 7.5 % of the potential capacity. Turkey's total operating capacity for wind power plants has been targeted as 20,000 MW by the year 2023 (Anonymous 2014, Şenel & Koç 2015).

The first geothermal power plant was established in 2009 and has an operating capacity of 8.5 MW (Anonymous 2009). In 2014 the country had 15 geothermal plants with a total capacity of 405 MW. In the same year, there were eleven geothermal power plants under construction. Turkey's first biomass power plant was founded in 2016. There are currently six biomass power plants under construction. The country's first solar power plant was established in 2011 and had an operating capacity of 500 kW. As of May 2014, 38 solar power plants were in operation with a total operating capacity of 8.5 MW (Anonymous 2014). It is clear that RE production in Turkey has mostly concentrated on hydroelectricity and wind energy.

National Overviews

Data on Landscape Quality

Although many different national and regional maps exist which delineate landscape features such as soil, forest, and water resources, this data has not been examined or analysed in the framework of national landscape assessment. Researchers, ministries, and other governmental bodies, however, conduct landscape assessment studies related to their project areas.

Precautions are taken at provincial and regional levels to ensure that landscape quality in protected areas and fragile natural areas in particular does not suffer due to planning operations (Anonymous 2017). The idea of landscape quality in Turkey has only recently been acknowledged, following the implementation of the European Landscape Convention, of which Turkey is a signatory. Consequently, such analysis of landscape quality has not yet been adapted at the planning and design stages.

The Interaction between RE and Landscape Quality

Landscape and landscape quality are generally treated in terms of 'scenery', included in environmental impact studies in the planning and project stages of RE projects. The other qualities that make up landscape quality, such as fauna, flora, geomorphological, hydrological, soil, climate, and cultural values, are not evaluated as landscape qualities, but as environmental qualities. The possible negative impacts are not evaluated explicitly in terms of landscape quality, but of the environment: the interaction between pollutants and their receiving environments or between the project area and nearby protected areas and fragile natural areas.

During the preparation stage of RE projects, public participation is permitted in the evaluation of environmental impacts. However, it is limited to information meetings organised by the project owner's firm for the purposes of 'manipulation' or 'therapy' (Arnstein 1969) of the local residents. Indeed, the local residents do not have the opportunity to reject the project. As such, landscape quality is not an issue in public acceptance of RE projects. Even so, as a result of the damage to the environment and living spaces of the local residents by the many hydropower plants built at the beginning of the 2010s, ignoring natural processes and the resources needed by the local residents for their livelihood, important popular protests took place at local and national level. These protests forced the related ministries and other governmental bodies to consider the environmental damage caused by hydroelectric power projects.

1.32
UNITED KINGDOM

Gisele Alves, Sennan Mattar & David Miller

Situation of Renewable Energy

The UK Committee on Climate Change (CCC) is an independent statutory body. In 2014, it reported that the UK would meet its immediate emission reduction targets due to increased renewable electricity generation and a change from coal to gas power (CCC 2014). The Energy Act 2013 has a binding decarbonisation target which, developed from The Climate Change Act 2008, sets out to achieve an 80 % reduction in greenhouse gas emissions from 1990 levels by 2050. The focus on decarbonisation rather than RE generation was centred on the desire to promote gas power as a 'low carbon' alternative to coal (HM Government 2011).

By 2016, the gross final energy consumption from renewable sources in the UK was 8.2 %, up from 1.1 % in 2004 (Eurostat 2018). The contributions of different types of RE have changed significantly, with hydropower providing the greatest proportion in 2000, dropping to fifth largest by 2016 (Figure 1.32.1). Considerable emphasis is now being placed on marine renewables, particularly on tidal power, and expansion of offshore wind energy. The devolved administrations of Scotland, Wales, and Northern Ireland have established their own policies and targets.

Figure 2 illustrates the generation of RE. It shows an increase in the contribution of bioenergy, and seasonal differences with wind contributing more in winter (Q4 and Q1) and solar in summer (Q2 and Q3).

In Scotland and Wales, in particular, there is encouragement for community-led RE development. This forms part of an increased emphasis of policy-makers on sustainable development

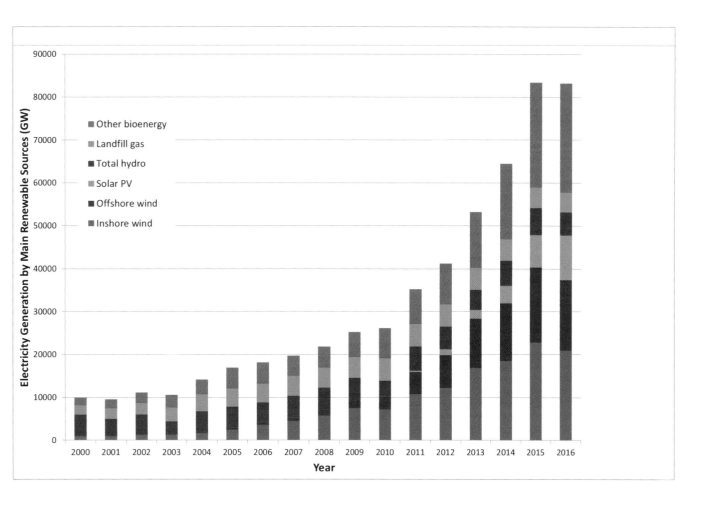

Figure 1.32.1
Electricity generation by main renewable sources (Source: Department for Business, Energy and Industrial Strategy 2017a, 158).

and diversification in rural areas, and social and environmental justice.

The types of RE generation varies across the UK reflecting the distribution of resources (Figure 1.32.3). In 2016, 66 % of renewable generation was from England, and 23 % from Scotland. The greatest amounts of RE were from onshore wind, principally from Scotland and England. Solar PV is becoming increasingly significant, increasing in England by 29 % and in Wales by 18 % between 2015 and 2016.

Data on Landscape Quality

At a UK level, there is neither single body responsible for landscape, nor a single dataset which represents its characteristics. Responsibility is divided across the devolved administrations for Scotland (Scottish Natural Heritage), Wales (Natural Resources Wales), and Northern Ireland (Northern Ireland Environment Agency), with the relevant UK government department having responsibility for England (Defra).

Several public policies make explicit reference to landscapes and their enhancement, protection, or management. Examples are the Northern Ireland Landscape Charter (Northern Ireland Environment Agency 2014) and the Scottish Land Use Strategy (Scottish Government 2016).

Across the UK, data on landscapes have been generated through the mapping of landscape character. This has followed the approach set out by The Countryside Agency and Scottish Natural Heritage (2002). The outputs are spatial datasets at national or local authority levels. This mapping has been undertaken by

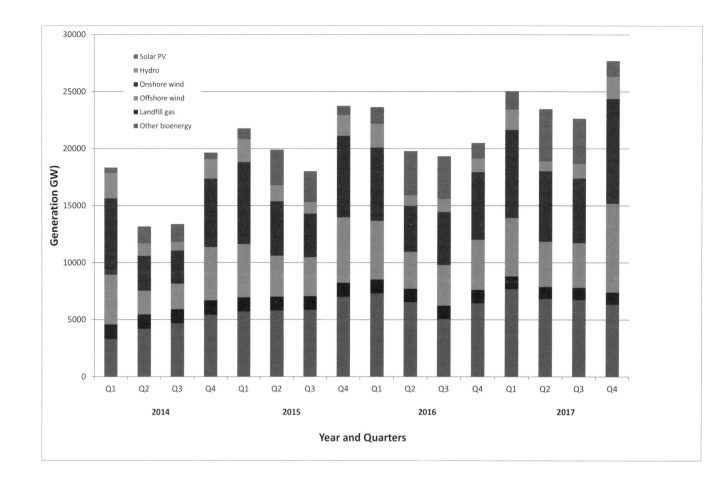

Figure 1.32.2
Renewable energy generation, by type, quarterly between 2014 and 2017 (Source: Department for Business, Energy and Industrial Strategy 2017b, 10).

each country following different initiatives, with the first national coverage produced for Scotland in 1999 (Hughes and Buchan 1999).

Other approaches have been used, notably that of LandMAP information system (Bullen 2017), which provides a baseline of landscape characteristics, qualities, and influences for Wales. Related assessments of landscape character have been produced for selected coastal areas of Scotland, Wales, and England. Data on landscape quality are also embedded in mapping for

i) Geographically defined purposes such as mapping of special landscape qualities (Cairngorms National Park Authority 2015); landscape sensitivity and capacity (Cornwall Council 2014)

ii) Thematic purposes, such as RE, forestry, and housing, and historic landscapes (e.g. Historic Land Use Assessment for Scotland

All of these datasets are publically available, complying with the requirements of enabling public access to environmental information.

Interaction between Renewable Energy and Landscape Quality

The planning and regulation of RE in the UK is devolved to the Scottish and Welsh Governments, Northern Ireland Assembly, and Defra for England. Each has its own regulations and guidelines on the development of RE, taking account of landscape issues in different ways.

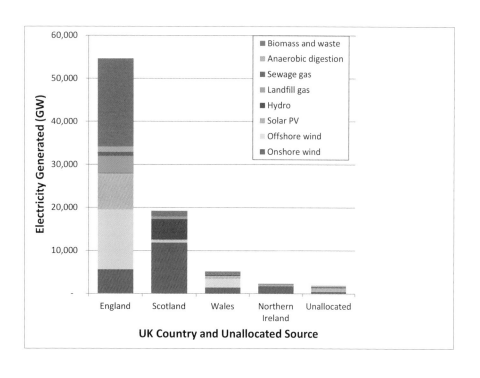

Figure 1.32.3
Renewable energy generation
for the United Kingdom in 2016,
by country (Source: Department for Business, Energy and
Industrial Strategy 2017b, 11).

All developments in the UK are required to provide an Environmental Impact Assessment (EIA) (Ministry of Housing, Communities and Local Government 2014). These planning processes implement the EU Directive on Environmental Impact Assessments (European Union 2011). Specific proposals are required to consider 'cumulative impact' in terms of its own and the overall impact of similar developments in the area. This may be in combination with visual and landscape considerations, the planning process requires a cumulative landscape and visual impact assessment. This requires developers to provide evidence that their proposal does not exceed acceptable change beyond which further development would be inappropriate in landscape and visual terms (e.g. PCNPA 2013).

Planning and statutory authorities across the UK request adherence to the *Guidelines for Landscape and Visual Impact Assessment 3* (The Landscape Institute and the Institute of Environmental Management and Assessment 2013). In many areas of the UK, landscape capacity studies have been prepared to guide the development of RE, in particular wind energy. These provide an indication of the extent to which a landscape type can accept a specific change without adverse impacts on its character. Scottish Natural heritage worked with many of the local authorities to produce digital datasets or reports on the capacity or sensitivity of landscapes to wind turbine development (e.g. Miller et al. 2010 Ironside Farrar 2014).

FORMER YUGOSLAV REPUBLIC OF MACEDONIA

Biljana Risteska Stojkoska & Aleksandra Dedinec

Situation of Renewable Energy

Since its independence in 1991, the Former Yugoslav Republic of Macedonia has faced many challenges in the energy sector. Due to the predominant role of lignite-fueled power plants, which have become the main emitter of greenhouses gases, air pollution has dramatically increased in many cities with Skopje becoming one of the most polluted cities in the world. Therefore, it was a priority for the country to consider major investments in the use of RE resources. However, due to the turbulent political situation over recent years, liberalisation of the energy market for households was postponed until 2020, which also caused a delay in most of the processes in the energy sector. As a consequence, RE represents still only a small portion of the total installed capacity.

Data on Landscape Quality

Although national institutions are committed to the process of harmonisation with EU legislation in all segments of society, there is still considerable work to be done regarding analyses of landscape quality and its interaction with RE resources. The only relevant data about landscape quality were published recently by Filipovski et al. (2015), much of which is not incorporated in the legislation. The only available official document is a rulebook on cadastral grading, along with determining and changing cadastral class culture (Katastar 2017).

Interaction between Renewable Energy and Landscape Quality

The following procedures apply for the development and construction of power plants utilizing RE sources (Ministry of

	Installed capacity (MW)	Number of plants	First plant installed
Small hydro-power	70.0	50	1927
Large hydro-power	561.7	10	
Photovoltaic power	14.7	90-100	2010
Wind power	36.8	1	2014
Thermal power	824 (+210 cold reserve TE Negotino)	3	
Combined heat and power	280	3	2015

Source: The Energy Regulatory Commission of the Republic of Macedonia, 2017, State Statistical Office of the Republic of Macedonia, 2017, and Energy Agency of the Republic of Macedonia, 2017

Table 1.33.1
RE Sources in the Former Yugoslav Republic of Macedonia (2014)

Figure 1.33.1
Wind Park Bogdanci, the first of its kind in the Former Yugoslav Republic of Macedonia, completed in February 2014. (Photo: ELEM, Macedonian Power Plants, 2017)

Economy 2017). If the land parcel where the investor intends to construct a photovoltaic power plant is classified as agricultural land, i.e. arable land, pasture, or forest, the investor may submit a request to the Ministry of Transport and Communications, or the municipality, to have the relevant agricultural land to be declared as construction land by the Ministry of Agriculture, Forestry, and Water Economy to be permanently converted into construction land (Law on Agricultural Land, Article 49). If the land parcel is privately owned, the investor needs to purchase the land, sign a long-term contract leasing the land from the government, or obtain a concession for the land use (Law on Construction Land, Articles 13–40).

Photovoltaic plants are subject to the requirement of preparing and submitting an environmental impact elaborate study. They cannot be constructed or start their operation before a positive decision on the environmental impact elaborate study is obtained in accordance with the Law on Environment. In the case of hydroelectric power plants, the state determines the place of construction. Therefore, there were reactions in the public (in various forums and workshops on energy and renewble sources) against building plants on highly fertile agricultural land.

Although Macedonian institutions are seriously committed to the process of harmonization with EU legislation in all segments of the society, there is still much work to be done regarding adequate analyses of the landscape quality and its interaction with the renewable energy resources.

2

INTRODUCTION: ENERGY LANDSCAPE AND LANDSCAPE QUALITY

Although different countries approach landscape in different ways, the policies for protecting it have developed since the end of the 19th century along three main lines of thinking: the picturesque paradigm, the environmental paradigm, and the cultural paradigm (Frolova et al. 2015b). The third way of conceiving landscape is reflected in the European Landscape Convention (ELC), which defines landscape as 'an area, as perceived by people, whose character is the result of the action and interaction of natural and/or human factors' (Council of Europe 2000, 3). This new way of conceiving landscape has affected the approach to energy landscapes in Europe.

The first use of the expression 'energy landscape' appeared in 2002 (Pasqualetti et al. 2002). Subsequently landscapes which were viewed as denaturalised and instrumental space (Pitte 1983) are increasingly perceived as 'holders' of sensibilities, thoughts and utopias deeply rooted in a territory (Frolova et al. 2015), although there remains the challenge of establishing it as a well-delimited and unified topic of study (Pasqualetti and Stremke 2018). The RELY Glossary (Kruse & Marot 2018) defines an energy landscape as one characterised by one or more elements of the energy chain (e.g. energy extraction, assimilation, conversion, storage, transport, or transmission of energy). 'The outcome can be a multi-layer energy landscape comprising combinations of technical and natural sources of energy within a landscape. In the framework of RELY, the energy landscape definition is focused on renewable energy and the impact on landscape quality' (Idem. 16). In RE landscapes, energy infrastructure is perceived and treated as part of the landscape, even if these landscapes may face social opposition (Bender 1998). This perception It reaches beyond the expert view of landscape as a purely material entity that has also been in use for several decades (Frolova et al. 2015b). It takes into account the importance of the perceptions of landscape by the people who share, value, and use it (Olwig 2007). Based on the conceptual framework established by Pasqualetti and Stremke for energy landscape analysis (2017, 5), we posit the following characteristics of RE landscapes:

RE landscapes tend to be supplemental, unlike conventional landscapes which tend to be extractive (Pasqualetti and Stremke 2018). RE tends to have lower energy density and therefore usually requires more land per final unit of power provided (Pasqualetti and Stremke 2018), and its relative visual impact (per MW) can be greater (Wolsink 2007, Wüstenhagen et al. 2007). The spatial appearance of an energy landscape is determined by the amount of space required for energy development, the compatibility of energy with other land uses (Pasqualetti and Stremke 2018), and size of elements of energy infrastructure. Public attitudes toward RE landscape can change with time. RE landscapes have less significant permanence than conventional energy landscapes (Pasqualetti 1997). RE landscapes hold greater potential for post-energy use because of lesser environmental impact, while many interventions are reversible in nature (Stremke 2015).

The COST RELY Glossary on RE Terms defines landscape quality as 'the perception of the holistic environmental, cultural, sensory and psychological characteristics of a landscape with respect to their benefits or significance to people. It is relative, not absolute, requiring interpretation in the context of geographic scale (i.e. local, regional, national) and, or human experience' (Kruse & Marot 2018, 33). Each form of RE transforms the landscape in its own specific ways and can therefore effect landscape quality in different forms (directly and indirectly), although their impacts could be both positive and negative. In addition to the type of RE, the impact also varies depending on the context and scale of development and the methods used. Lessons explored in different chapters of this book show the complex, interwoven nature of the processes through which the joint assembly of a RE capacity and a culturally shared landscape can be achieved. This chapter presents a systematic review and meta-analysis of empirical research findings on landscape quality change through RE development. It focuses on specific RE production systems (hydro, wind, solar, geothermal energy, bioenergies) and their impacts on landscape quality (Kruse & Marot 2018). Our research is based on empirical data and findings from the COST Action TU1401 members' research projects/practical planning projects. We also used data from Eurostat 2015 (http://ec.europe.eu/eurostat/web/main/home) and data from Bundesamt für Energie BFE Switzerland (http://www.bfe.admin.ch). It might be noted that no attempt was made here to assess the quality of the data on which this broad-brush overview is based. The Eurostat database was considered

the most appropriate set of data to use, enabling a reasonably well standardised overview. Eurostat provides clear explanations of the exact meaning of the various categories of data that are entered into its database. Even so, national data may vary in coverage and quality. It is likely, for instance, that some traditional forms of RE usage—notably the use of firewood—are underreported. Also not all direct geothermal usage may be recorded accurately.
The general scope of landscape change, quality and dynamics was taken into account, allowing the research approach to provide 'future-proof' findings, by presenting a systematic review of the nexus between RE production systems and Europe's landscapes and by providing a pan-European documentation and synopsis of landscape quality, and character assessment methods.
First of all we carried out a statistical study of different kinds of RE development in different COST Action countries. Observed development in installed capacity and number of RE acquisition facilities did not translate directly into extent of influence on landscape and was analysed in parallel with its landscape impacts and its assessment in different EU countries. This permitted us to develop an overview of methods of landscape assessment for RE development in Europe.

Marina Frolova & Csaba Centeri

INTRODUCTION OF RE TYPES AND THEIR IMPACTS ON LANDSCAPE

Karl Benediktsson, Marina Frolova, Csaba Centeri & Benjamin Hennig

In this chapter, a broad overview is provided for describing the state of RE production and the share of RE in total energy consumption for all the countries that have participated in the COST Action RELY. It should be noted that, apart from the EU countries, these include several non-member states. The basis for the overview is data for 2015 from the Eurostat database. A series of cartograms for the main sources of RE as well as total RE production by country is provided in Figure 2.1.1

The cartograms were created using a density-equalising algorithm based on Gastner and Newman's (2004) approach. In the transformed maps, geometric accuracy is sacrificed, but the area of each country corresponds to the quantity being mapped while at the same time aiming to preserve each country's shape. This enables a quick grasp of the geographical distribution of the variable in question (Hennig 2013). The complementing pie chart provides guidance to each energy type's share in the overall RE production in Europe. Marine energy production was excluded from this map series due to its negligible overall share of significantly below 1 %.

The development of RE capacity has been influenced by a range of complex cultural, contextual, socioeconomic, political, and physical factors (Ellis et al. 2007), which have led to an uneven pace and extent of development. For the majority of the countries, RE production still accounts for less than 15 % of domestic energy consumption. The share of RE is lowest in Luxembourg, Malta, Bel-

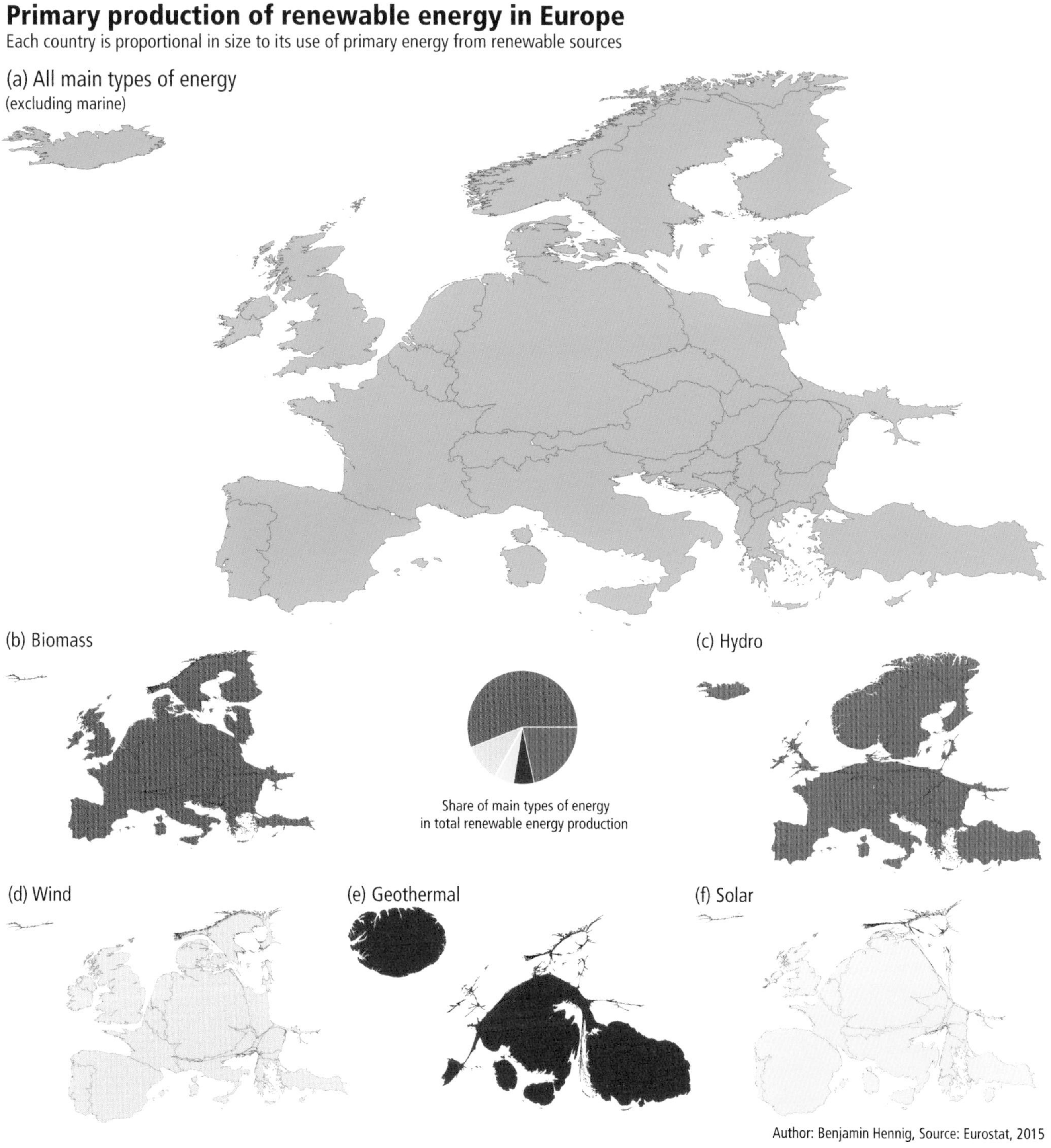

Primary production of renewable energy in Europe

Each country is proportional in size to its use of primary energy from renewable sources

(a) All main types of energy
(excluding marine)

(b) Biomass

Share of main types of energy
in total renewable energy production

(c) Hydro

(d) Wind

(e) Geothermal

(f) Solar

Author: Benjamin Hennig, Source: Eurostat, 2015

Figure 2.1.1
Cartograms showing total
RE production (a) and of
the five most important RE
types (b-f) by country, 2015

gium, and the United Kingdom. The share is considerably higher in several countries in the south, east, and north of Europe. About half of the energy consumed in Latvia comes from RE, and in Iceland the share is about 75 %.

The total quantity of RE primary production—in all its types—in those countries amounted to 10,5 M TJ. The amount of production of each RE source is very uneven geographically, as environmental and other conditions vary considerably. The five most important of these sources will now be identified and their geographical distribution outlined very briefly. The reader should consult the map and cartograms for a more nuanced picture. The order follows the quantity of energy produced from each source.

Not covered in this geographical overview are two energy sources. *Marine energy,* which includes e.g. tidal and wave installations, is still a very small part of the RE mix, albeit a source for which considerable technological development is taking place, and which could become significant in several parts of coastal Europe in the future. *Environmental thermal energy*—where heat pumps are used to concentrate thermal energy from sources at ambient temperatures—is not covered here ei-

ther, but this is already a substantial energy source in certain countries. Although such systems need considerable electricity for their operation they are classified as an RE source as the thermal output is greater than the electrical input. Both these energy sources will receive some attention at the end of Chapter 3, but as data for environmental thermal energy installations are very limited and they have in fact very limited landscape impacts, they were not subject to further analysis in the COST Action RELY.

As seen from this broad-brush overview, simple locational/geographical factors—among them latitude, topography, and geological characteristics—explain some of the differences in the RE mix between the various countries of Europe. This also has impacts for policy and investment strategies. Relative abundance of some energy sources may lead to other sources that are less technologically settled or economically viable being overlooked. With the urgency of the transition to RE becoming ever more evident, however, it is likely that those countries to which this applies will put increasing efforts into developing other available RE sources also in addition to the most obvious ones.

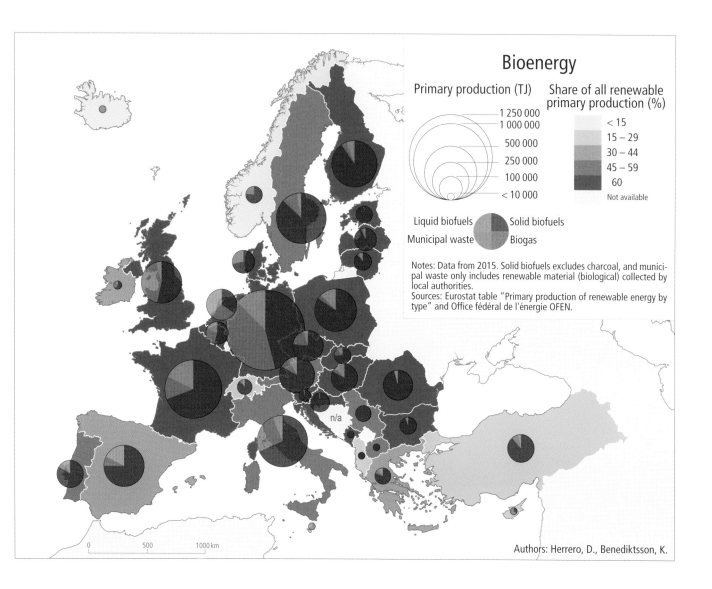

Bioenergy

Primary production (TJ)

- 1 250 000
- 1 000 000
- 500 000
- 250 000
- 100 000
- < 10 000

Share of all renewable primary production (%)

- < 15
- 15 – 29
- 30 – 44
- 45 – 59
- 60
- Not available

Liquid biofuels   Solid biofuels
Municipal waste   Biogas

Notes: Data from 2015. Solid biofuels excludes charcoal, and municipal waste only includes renewable material (biological) collected by local authorities.
Sources: Eurostat table "Primary production of renewable energy by type" and Office fédéral de l'énergie OFEN.

Authors: Herrero, D., Benediktsson, K.

2.1.1
Bioenergies

Paulo Brito, Mateusz Slupinski,
Marina Frolova & Michael Roth

Bioenergy is one of the most important RE sources; its production is growing fast and it can be stored in liquid, gaseous, and solid states (McKendry 2002). Germany, Italy, France, the Benelux countries, Spain etc. have become important producers in Europe (Solomon and Barnett 2017) (Figure 2.1.1 and 2.1.1.1).

Biofuel technologies are classified into first generation technologies produced from food crops (grains, sugar beet) derivatives, and advanced or second generation technologies produced from non-fossil, non-food materials (e.g. switchgrass, wood waste, algae). Bioethanol production is limited in Europe due to the lack of available land, while the biodiesel market and feedstock production (mainly rapeseed oil) are larger (Solomon and Barnett 2017). The leaders of the bioethanol production within the COST Action countries are: France (975 million litres), Germany (920), and Belgium (557), while the top three producers of biodiesel/hydrogenated vegetable oils (HVO) are Germany (3,808 million l), France (2,681) and the Netherlands (1,954) (Flash et al. 2016). Advanced/second generation biofuel production has taken off only since the past six years in the EU, due to favourable

Figure 2.1.1.1
Bioenergy production across European countries. Authors: Karl Benediktsson & Daniel Herrero-Luque.

Figure 2.1.1.2
Oilseed rape landscape
(Photo: Bohumil Frantál)

policies related to their lower greenhouse gas emissions than transport fuels. Since 2007 several HVO thermo- and biochemical plants have been built in Finland, the Netherland, Spain, and Italy.

As for *biomass for heat and power*, this is generated through direct combustion or through the production of biogas. Forestry products are mainly for direct combustion, while a wide range of inputs are used for the production of biogas. Main pellet producers among the COST Action countries in 2014 were Germany (2,100 gigatons, or GT), Sweden (1,490 GT), and Latvia (1,300 GT) (Flash et al. 2016). As for biogas, Germany with its 8,928 biogas plants of total capacity of 4,177 MW is the leader in biogas production, accounting for 65 % of total EU production. It is followed by Italy (2,100 plants/900 MW) and the Czech Republic (507 plants/358 MW).

Direct Landscape Impacts

The advantages of using bioenergy come with inherently problematic properties (immobility, low-energy density, scattering, etc.) that characterise these power sources (García-Frapolli 2010).

The most important landscape impacts of biofuel production and production of electricity from bio resources are visual impacts on landscape and land-use change. Bioenergies induce direct and indirect land-use changes when pre-existing agricultural activity is converted into new, often more intensive forms of agriculture (Figure 2.1.1.2).

Bioenergy transforms pre-existing agricultural landscapes and their related social practices (Gundersen 2016). The biomass processing facilities could vary in size from local generating stations to larger biorefineries (Calvert et al. 2017), so their landscape impact is scale-dependent. However, even the small biogas structures covered with plastic foil and the silos cause visual impacts and landscape change. Medium and large installations are comparable to regular buildings used in intensive farming and agriculture (Figure 2.1.1.3). The influence on the landscape through monocultures of energy plants depends on the individual composition of a the substrate used at a bioreactor. Biomass does not always have negative impacts on the landscape, but can also have positive effects if, for instance, different flower mixes were used.

The low-energy density implies the necessity to produce large amount of biomass (1.5 t bio is equivalent of 1 t of coal and 2–4 t to replace one unit of petroleum/natural gas [Calvert et al. 2017]). Therefore, there is an important resurfacing of infrastructure and activity associated with biomass distribution and conversion. Biogas tends to be produced at a large industrial scale, which effects landscape character (Bluemling et al. 2013, Carrosio 2013, Ferrario and Reho 2015).

Figure 2.1.1.3
Biogas plant installations: a. small, b. medium, c. large bio-gas plant (Photos: a. Wolsfeld, Germany 2011 by Slupinski M., b. Biogas plant, Altscheid, Germany 2011 by Slupinski M., c. Biogas plant in Sweden [Swedish Gas Centre 2007])

Indirect Landscape Impacts

Plantations normally require large areas of arable land and impose new systems of value on a landscape (Calvert et al. 2017). Indirect impacts of bioenergy on landscape quality are multiple: effect on soil, gaseous emissions, unfamiliar smell, and water contamination (Hastik et al. 2015, Sokka et al. 2016). The withdrawal of organic matter from the forests may also have negative landscape impact. Large-scale exploitation may have medium-term impact (Holland et al. 2015), since changes in soil lead to changes in ecosystems, altering their flora and fauna, and to loss of biodiversity. Rare species may need a long time for regeneration after the extraction of bioenergy (Johansson et al. 2016).

In addition, plantations can significantly alter the intervisibility in the landscape by replacing low-height crops to above-eye-height-crops). Thus, formerly open landscapes are converted to a different landscape character. Increased traffic of biomass transport and processing rise conflicts with local residents.

Mitigation Strategies

These negative impacts can be avoided, reduced, or mitigated by adhering to three principles of landscape design: conservation of ecosystem and social services, consideration of local context, and monitoring outcomes and adjusting plans to improve performance measures over time (Gundersen 2016). In order to minimise negative impacts of bioenergy production on landscape through limiting land-use impacts of close-loop biorefineries it is important to encourage the production of energy crops onto land considered marginal or abandoned for agricultural purposes, although this has been proven to be problematic or even socially regressive (Calvert et al. 2017). There have also been debates on whether to develop more centralised biofuel production units or smaller, decentralised ones (Kienast & Gregersen

2017, Ravena et al. 2007). Another more general strategy in bioenergy policy is to favour development of advanced or second generation biofuels (e.g. lignocellulosic ethanol, Bio-SNG, synthetic biofuels) that use a wider range of feedstock including lignocellulosic material, waste and residues or stimulate the production of algae biodiesel and do not compete with food production. For more effective landscape management and protection the bioenergy related policies should be integrated into agricultural, forest, and environmental protection policies.

Potential Positive Impacts

The production of second-generation biofuels from the valorisation of domestic and forest waste is a route with very positive impacts in terms of landscape and environmental value. The recovery of waste gives it value, reducing the negative impacts of dumps and landfills. On the other hand, the use of forest residues and forest cleaning biomass makes it possible to prevent forest fires, which are particularly significant in the countries of Southern Europe and which have drastic consequences in terms of the landscape and environment.

Public acceptance of bioenergy installations and their landscape changes is related to more global issues, such as competition between energy and food production, environmental degradation (GHG emissions, soil and water degradation, biodiversity loss, etc.) and its social consequences (land rights infringements, food security impacts, etc.), and perceived unfairness in the distribution of benefits and disadvantages along the RE production (Carrosio 2013, Palmer 2014, Ferrario and Reho 2015, Frolova et al. 2015b).

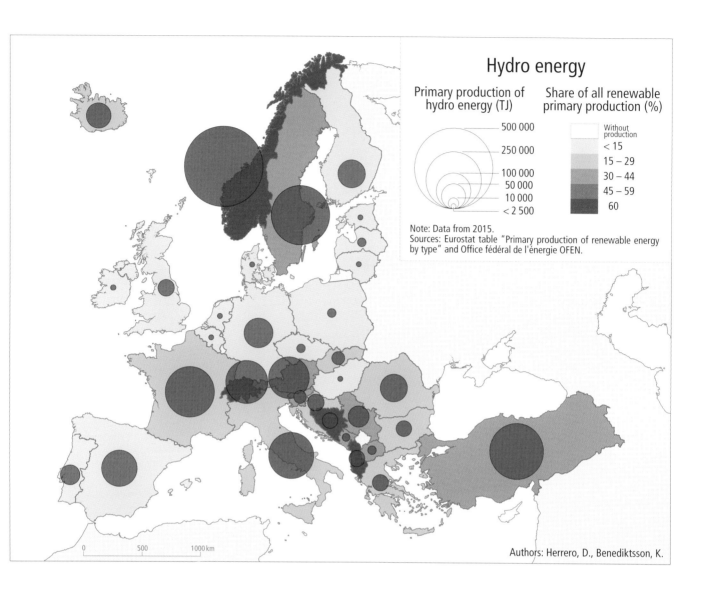

Hydro energy

Primary production of hydro energy (TJ)

500 000
250 000
100 000
50 000
10 000
< 2 500

Share of all renewable primary production (%)

Without production
< 15
15 – 29
30 – 44
45 – 59
60

Note: Data from 2015.
Sources: Eurostat table "Primary production of renewable energy by type" and Office fédéral de l'énergie OFEN.

Authors: Herrero, D., Benediktsson, K.

2.1.2
Hydro Energy

Marcel Hunziker & Dominik Braunschweiger

General Overview

Since ancient times, humankind has harnessed the kinetic energy of moving water as a means of transportation, for the purpose of irrigation, and for the operation of various mechanical devices such as mills, cranes, or lifts. Ever since the rapid progress of electrical engineering during the late 19th century that sparked the Second Industrial Revolution, hydropower has also been used to generate electricity.

As a renewable source of energy, today hydropower is more important than ever. In 2012, 43.5 % of the electricity production from renewable resources within the EU were produced by hydropower plants (Eurostat 2015) (Figures 2.1.1 and 2.1.2.1). With a total installed capacity of around 1,066 gigawatts (GW), hydropower plants are currently the biggest producer of RE in the world (Bundesministerium für Wirtschaft und Energie 2015). Nevertheless, a lot of potential still remains untapped: about 75 % of worldwide and 47 % of European hydropower potential are currently unused (IPCC 2011).

Figure 2.1.2.1
Hydro energy production across European countries. Authors: Karl Benediktsson & Daniel Herrero-Luque.

Figure 2.1.2.3
Direct landscape effects of
power lines, Austria, 2016
(Photo: © Csaba Centeri)

Figure 2.1.2.4
Direct landscape effects of the dam and the reservoir of the hydropower station at Sfikia, Aliacmon River, Greece, 2017 (Photo: © Csaba Centeri)

Direct Landscape Impacts

As for direct landscape effects of large facilities, their construction involves building power stations, damming rivers, and creating artificial reservoirs (Figure 2.1.2.2 and 2.1.2.4); during their operation they can cause a large fluctuation of the water level (Cohen et al. 2014, Hastik et al. 2015).

Plants that rely on large water reservoirs require the flooding of huge areas. In addition, big power plants and dams by themselves, as well as power stations and transmission lines (Figure 2.1.2.3.), are huge structures and their presence constitutes substantial change in landscape features (Hastik et al. 2015).

Meanwhile smaller hydropower schemes often utilise natural differences in altitude, small flows, or the decline in the pipes from water infrastructure to provide power for small communities. Small hydropower plants also often utilise run-of-the-river designs, which do not require large impoundment dams (although they may require a small, less obtrusive dam). Instead, a run-of-the-river plant diverts a portion of a river's water into a canal or pipe to spin turbines. Since run-of-the-river designs mostly do not need large reservoirs, their impact on landscape is comparatively small. Consequently, run-of-the-river designs have risen in popularity lately and even some larger projects use them (Ferrario and Castiglioni 2017). Nevertheless, individual small plants are spread out over large areas and most of their infrastructure, including penstock/pipelines, is usually located above surface or at least visible from the surface while some of the infrastructure of large hydropower plants is generally located underground. Therefore, the visual landscape impact of a large number of smaller hydropower plants could exceed that of one larger plant with equivalent output (Abbasi and Abbasi 2011, Koutsoyiannis 2011, Bakken et al. 2014). However, non-visible parts of hydropower landscapes are also very important. Ferario & Castiglioni (2017: 831) list numerous impacts of hydropower on landscapes that aren't immediately visible. For example underground pipes, turbines, pumps, but also surface water 'swallowed' by derivations and pipelines kilometres away and landscape elements eliminated by hydropower development, such as villages, mills and sawmills, meadows, pastures, huts, roads and trails, and entire valleys flooded by artificial lakes.

Indirect Landscape Impacts

As for indirect landscape effects, water diversion for electricity generation can lead to drying up of large watercourses, and the damming of lakes and rivers can lead to the erosion of the shoreline, thereby destroying soils and biota. Increased water discharge can cause riverbank erosion downstream of power plants (Rosenberg et al. 1995). Rapid flow variations due to hydropower plants can affect both physical and chemical qualities of water (Cushman 1985, Evans et al. 2009). These drastic changes in water-related ecosystems (Cushman 1985, Čada 2001, Evans et al. 2009) normally lead to unfavourable landscape quality changes.

Mitigation Strategies

As a study by Bottero (2013) illustrates, landscape impacts of hydropower projects largely determine public perception and evaluation of a project. In fact, depending on the landscape in question, damages to the landscape can account for a significant portion of a project's total costs. Thus, the key to increasing public acceptance of hydropower projects is the successful management of landscape impacts: dams, power stations, and transmission lines together with the accompanying infrastructure are usually considered disturbing (Hastik et al. 2015, Frolova 2017). Meanwhile, the perception of other visual elements may depend on the original state of the landscape and its cultural value. Frolova et al. (2015a) for example show that artificial lakes are often considered attractive,

many adverse impacts of hydropower plants can be mitigated: diverse solutions such as fish ladders help mitigate the impact on the species that would otherwise be threatened by the drastic changes to their ecosystems (Čada 2001). Power stations and power lines as well as the accompanying infrastructure are considered especially disturbing to the landscape (Hastik et al. 2015, Frolova et al. 2015a) and thus they should be placed underground to reduce their visibility whenever feasible. Utilising existing old infrastructure such as abandoned mills for the construction of small hydropower plants may help with both reducing the monetary cost of a project and reducing the impact on the landscape (Stevovic et al. 2016). Additionally, based on their comparative analysis of power plants in Norway, Bakken et al. (2014) show that landscape impacts of large hydropower plants may be reduced significantly, for example by establishing reservoirs from natural lakes, or by building large run-of-river plants which do not rely on reservoirs at all.

Potential Positive Impacts

Landscape impacts of hydropower plants need not necessarily be all negative. Large dams and artificial lakes can often become major regional attractions, boosting tourism and local income (Hastik et al. 2015, Frolova et al. 2015a). For example, hydropower infrastructure now plays a significant part in the local environment and became an important feature of the landscape in many European mountains (Frolova 2017).

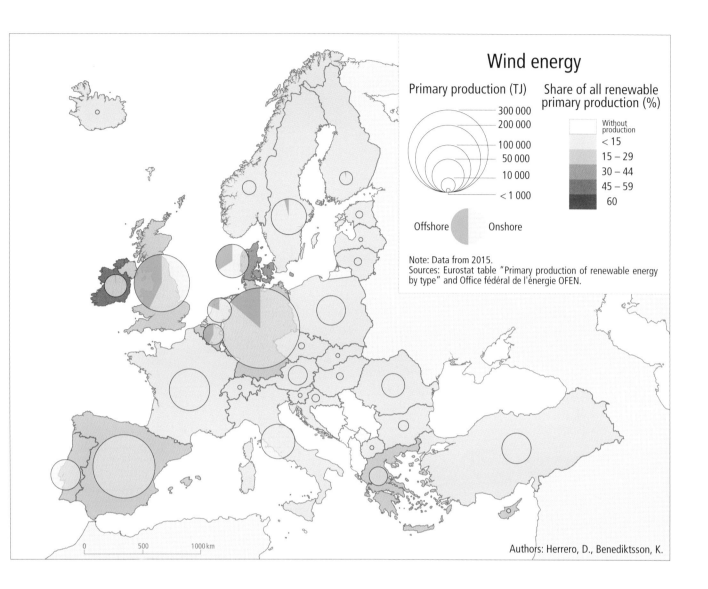

Wind energy

Primary production (TJ)

- 300 000
- 200 000
- 100 000
- 50 000
- 10 000
- < 1 000

Offshore   Onshore

Share of all renewable primary production (%)

- Without production
- < 15
- 15 – 29
- 30 – 44
- 45 – 59
- 60

Note: Data from 2015.
Sources: Eurostat table "Primary production of renewable energy by type" and Office fédéral de l'énergie OFEN.

0   500   1000 km

Authors: Herrero, D., Benediktsson, K.

2.1.3
Wind Energy

Marina Frolova & Georgia Sismani

General Overview

In 2015, the installed wind power capacity in the EU was 142 GW: 131 GW onshore and 11 GW offshore. Wind power was installed more than any other form of power generation (44.2 % of total capacity) (EWEA 2016). Germany has the largest installed capacity in the EU (45 GW), followed by Spain, the UK, and France. Sixteen EU countries have over 1 GW capacity installed, while nine of them have more than 5 GW (Figures 2.1.1 and 2.1.3.1).

As for offshore wind energy, this is a relatively young but continuously growing industry. The result has been in large-scale deployment of offshore wind farms (OWFs) in many EU coastal countries, in particular in the UK, Germany, Denmark, Belgium, the Netherlands, and Sweden.

Direct Landscape Impacts

Wind energy landscapes are characterised by considerable height (up to 160 m) of wind turbines (WTs), making their visual or perceived impact on landscape very pronounced (Figure 2.1.3.2) (Hurtado et al. 2004, Wolsink 2007, Möller 2010, Torres-Sibille et al. 2009).

The most common classification of wind farms (WFs) is based on number of WTs and capacity:

Figure 2.1.3.1
Wind energy production across European countries.
Authors: Karl Benediktsson & Daniel Herrero-Luque.

large size (1 or more WTs, > 1 MW), medium size (1WT, 0.5–1 MW), miniwind (1 or more WTs, < 0.5 MW), or microeolic (1WT, < 0.01 MW) (Ruggiero and Scaletta 2014).

However, landscape and visual impact of WFs are strongly influenced not only by the size (Ruggiero and Scaletta 2014), design, and layout of WTs, but also by the make and model of the WTs (SNH 2009). It also depends on the existence of multiple WFs cumulative impact. The development of associated infrastructure such as roads, transmission lines, ancillary buildings, etc. has significant impact on landscape, too (SNH 2009). Finally, night lights, shadow flicker, and stroboscopic effects from WTs (Kil 2011) could affect their perception.

The appearance of a wind energy landscape depends on the position of WTs, landscape type, WT's size, and proximity to WT (Danish Energy Agency 2009). For example, WTs in costal zones and mountain ridges are highly visible and change landscape character. Contrarily, in other visual contexts, such as large and uniform inland plains and evenly sloping terrain (due to its large dimension), in farmlands (due to their visual complexity), or areas with large technical installations (due to their industrial character), WFs can be perceived as just one more element in landscape and will not modify landscape's spatial characteristics. Other direct landscape impacts of WFs are land use conflicts, which are less significant for small-size installations, but are essential for large-scale WFs (SNH 2009).

Figure 2.1.3.2
Massive random distribution of wind turbines in Peñaflor de Hornija (village), Spain (Photo: Daniel Herrero-Luque)

Figure 2.1.3.3
Offshore wind farm in
Egmondaan Zee, Nether-
lands, September 2013.
https://pixabay.com

Landscape impacts of OWFs have been investigated mostly in recent years. Up to now most
OWFs have been located relatively close the shore,
making landscape impact assessment crucial for local communities, as OWFs may affect or compete
with other human activities (e.g. tourism, fishing,
shipping) (Sismani at al. 2016). The most considerable direct landscape impact of OWFs is visual
(Figure 2.1.3.3), although it is less significant compared to land-based WTs. offshore winds tend to
flow at higher speeds, thus allowing the production
of more electricity than onshore. Therefore offshore
wind turbines (OWTs) require shorter and less visible towers (Bilgili et al. 2011).

The visual impact of OWFs is reduced as the distance from the shore increases and is considered
negligible for OWFs located more than 8 km from
the coast (Henderson et al. 2003). Apart from the
number and size of OWTs, visual perception may
differ during the day (Sullivan et al. 2012) and according to the local environmental conditions and
the movement of the blades (Bishop and Miller
2006).

Indirect Landscape Impacts
Negative indirect landscape effects of onshore WFs,
which can affect landscape quality, are related to hazards to birds and bats, noise pollution, and destruc-
tion and degradation of habitats (Wolsink 2007,
Ruggiero and Scaletta 2014, MEEDDM 2010).
WFs may also affect underground and surface waters and produce electromagnetic disturbance.

OWFs' indirect negative effects are again related to impacts on local ecosystem (birds and marine life), noise (mostly during construction phase), and coastal erosion (due to change of local wave climate) (Bergström et al. 2014, Tougaard et al. 2008). Consequently, the impact of OWFs may differ according to the type of OWTs. As fixed bottom OWTs are usually chosen for shallow waters near the shore, their installation and operation may cause greater impact to the coast compared to floating OWTs, which are located at larger distance from the shore. In the case of floating OWTs, noise and visual impact are reduced even more and the local sediment transport patterns are less affected.

Mitigation Strategies

Landscapes that previously contained large technical installations (industrial activities, harbor areas, etc.) can more easily assimilate a WF, due to thematic association with industrial structures (Danish Energy Agency 2009). Similarly, WFs could be placed into other visually complex contexts, such as power lines and towers, agricultural buildings, houses, and roads (DEHLG 2006).

The colour of WTs is important for mitigation of their landscape impact. Numerous studies show that 'no single color of WT will consistently blend with its background and it is more important to choose a color that will relate positively to a range of backdrops seen within different views and in different weather conditions' (SNH 2009, 8).

Many landscape and visual impacts of WFs could be minimised by the appropriate selection of design, layout, and location (MEEDDM 2010), by avoiding their visibility from sensitive viewpoints (SNH 2009), and by technical monitoring and specific restoration actions (MEEDDM 2010).

In case of OWFs, some potential negative effects could be mitigated through strategic planning (Bergström et al. 2014) and appropriate site selection (Lindeboom et al. 2011). Landscape and seascape character types can provide a good basis for designing guidelines of WFs. All the associated elements, other than WTs, should also be located and designed to respect the character of surrounding landscape (WEDG 2006).

Potential Positive Impacts

Although landscape is often cited as an argument in the conflicts around WFs, instead of being considered as a problem for local inhabitants, WTs can even form a positive part of a local landscape and sense of place and affirm an identity in a given landscape (Frolova et al. 2015b). From the aesthetic point of view, WTs can be perceived as sculptural elements in the landscape, evoke positive association where related to modern structures, and be associated with technological efficiency, progress, environmental cleanliness, and utility (WEDG 2006).

As in case of onshore WFs, prior experience of the public with OWFs may significantly influence their perception towards them and thus, at some point they could be considered as part of the local landscape (Ladenburg 2009, Ladenburg and Dubgaard 2009). Many studies indicate that OWFs may also lead to ecosystem benefits, as consequence of reduced pressures from shipping, commercial trawling, and dredging in the area. This may enable the establishment of large areas of seabed, and consequently, creation of a new habitat (Gill 2005, Inger et al. 2009, Wilson and Elliott 2009). Thus, may also be a potential increase in local biodiversity (van der Molen et al. 2014).

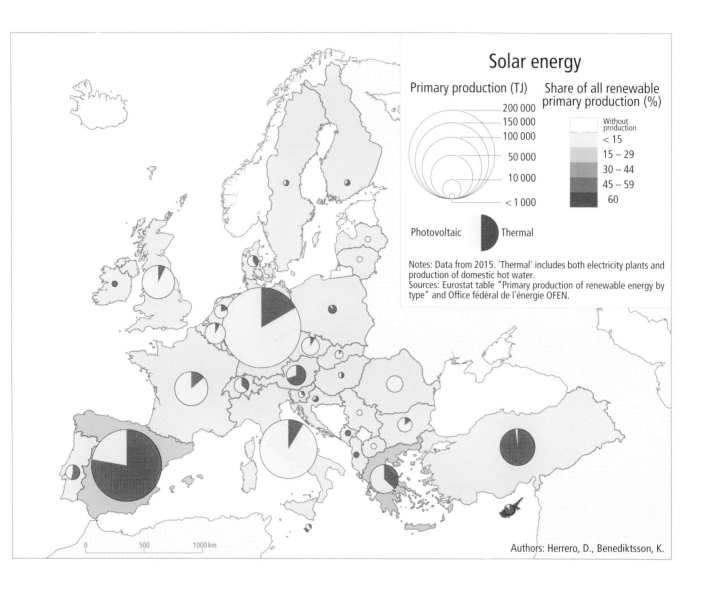

Alessandra Scognamiglio,
Georgios Martinopoulos,
Emilio Muñóz-Cerón & Marina Frolova

2.1.4
Solar Energy

General Overview

Solar energy systems compete with conventional fuels mainly in two applications: electricity and domestic heat generation (hot water and space heating). The most common solar system for electricity production (off and grid connected) is photovoltaics (PV) as solar thermal power (STP), also known as concentrated solar power (CSP) systems, are at an early deployment stage (Río et al. 2018). Solar thermal collectors are the systems most widely used for domestic heat generation.

At the end of 2016 the worldwide installed PV capacity was about 303 GW_p, with a market growth in 2016 of about 50 %, as 76 GW_p were added (IEA 2018). The leading country is China, followed by Japan and the USA (representing 26 %, 14 %, and 13 % of the cumulative worldwide capacity, respectively); within the EU (20.5 %), Germany and Italy are leading with 14 % and 6 % respectively (Figures 2.1.1. and 2.1.4.1). The targets set by most EU countries for 2020 were vastly underestimated due to the decrease of PV prices coupled with the incentives provided during the previous years. Moreover, an increase of solar PV electricity

Figure 2.1.4.1
Solar energy production across European countries.
Authors: Karl Benediktsson & Daniel Herrero-Luque.

is expected (RE Directive), as more countries are shifting towards net metering incentives as a mean to meet the targets of the Energy Performance of Building Directive (REN21 2017).

STP is still a minority solution. Nevertheless, Europe has a leading position, since Spain (the only country among the COST action participants with STP) is the world leader with a total capacity of 2.7 GW out of 4.8 GW installed globally (Río et al. 2018).

Solar thermal systems (ST) represent a mature market, for the residential sector (domestic heat generation). The worldwide installed capacity at the end of 2016 is 456 GW$_{th}$; in Europe it reached 34.5 GW$_{th}$ (REN 21; Weiss et al. 2017). Regarding newly installed capacity, Germany represents almost 40 % of the European market (38 %), followed by Italy and Greece (both about 9 % each).

Technical Characteristics

There are two principal types of PV topologies: on-ground PV (generally large systems) and building added/integrated PV (generally small systems). In the first case PV module arrays are installed optimally oriented and tilted on ground, through supporting structures and foundations, to maximise the solar radiation capture. As for the second topology, PV modules are added (BAPV) or integrated (BIPV) onto/into the building envelope and therefore no additional land use is required for their installation. The design of BIPV or BAPV is very multifaceted.

Regarding STP, it uses a large array of mirrors and/or lenses (parabolic troughs, solar power towers, linear Fresnel concentrators, and Stirling parabolic dishes) to concentrate solar energy onto a focal point (Roldán-Serrano 2017).

The most common ST typology for space heating and hot water production systems is the one using flat plate collectors (about 84 % of the market)

(Weiss et al. 2017); as for PV modules, these can be in/on building mounted and on ground.

Direct Landscape Impacts

Land use and visual impact are unanimously acknowledged as the main direct impacts of these solar technologies. These impacts are more critical when the systems cover large areas of land, being installed on ground instead of in/on buildings. This is the case of large scale PV (Hastik et al. 2015) and of STP. Even if an appropriate design can reduce the weight of impacts on the landscape, there are only a few studies on this topic.

Regarding PV, landscape impacts of large ground PV have been investigated in numerous publications (Beylot et al. 2014, Hernandez et al. 2014, Scognamiglio 2016, Turney and Fthenakis 2011). Publications focus also on specific aspects, such as integration of PV with the landscape (Kapetanakis et al. 2014), land use (Dijkman and Benders 2010, Fthenakis and Kim 2009, Horner and Clark 2013, Lakhani et al. 2014), visual-aesthetic impact (Mérida-Rodríguez et al. 2015a,b, Torres-Sibille et al. 2009), water related aspects, and glare.

Concerning STP, despite there being very few studies on their direct landscape, the glare effect from mirrors, the visual impact of the tall vertical cooling towers, and the columns of steam released into the atmosphere have been acknowledged as the main ones (Andrés-Ruiz [de] et al. 2015).

Mitigation Strategies

Appropriate siting (e.g. former mines, former landfills, or industrial areas and sites where the visibility of solar systems is low), integration into buildings (with consequent reduction of transmission infrastructure), use of appropriate technological components, and good design are the main mitigation strategies for solar technologies.

Figure 2.1.4.2
Large solar PV coupled with agriculture. 2.6 MWp ground mounted PV; geranium and passion fruit are cultivated in between the modules (covering 4.99 hectares of a former sugar cane field) in Ravine des Cabris, Reunion Island (FR). Design Akuo Energy. Image courtesy of Akuo Energy.

The dual use of land (e.g. PV + agriculture, PV + other energy systems) is being investigated as a possibility for implementing PV in agricultural areas (Tsoutsos et al. 2005, Scognamiglio 2016).

The repercussion in the landscape of ST is very limited (Comodi et al. 2014, Martinopoulos 2016, Piroozfar et al. 2016, Tsilingiridis et al. 2004), as most ST is installed on buildings, with impacts comparable to the ones caused by BAPV or BIPV installations.

Possible Positive Impacts

Compared with traditional fossil fuels, the use of solar technologies brings very positive impacts in terms of landscape and environmental value. Solar technologies can be used in combination with other traditional land uses, and can perform several functions at the same time.

Studies conducted in Italy, Germany, and the UK show that the public acceptance of solar technologies is improved when the design is considered pleasing (Palmas et al. 2012).

There are many options for designing solar technologies, e.g. PV, so as to generate positive landscape impacts. For instance: 1. PV can coexist with agriculture and grazing (with possible increase of crop production thanks to the shade provided by modules, and possible grass maintenance cost decrease) (Figure 2.1.4.1); 2. The supporting structures of PV can be used as land stabilisation elements; 3. The pattern of the PV fields can be designed so as to ensure the spatial definition of a certain area (e.g. public parks, bike lanes, walking paths); 4. PV modules can provide shade in spaces where this is needed; 5. PV can be designed so as to meet certain given ecological and landscape objectives (Scognamiglio 2016).

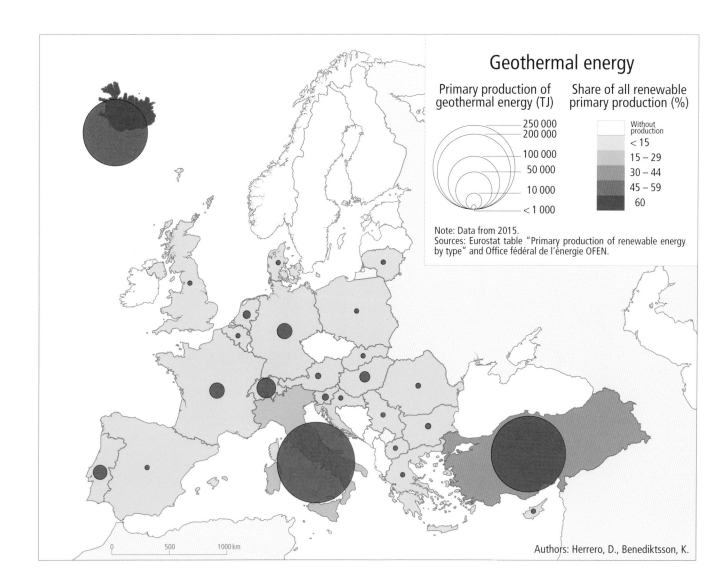

Geothermal energy

Primary production of geothermal energy (TJ)

250 000
200 000
100 000
50 000
10 000
< 1 000

Share of all renewable primary production (%)

Without production
< 15
15 – 29
30 – 44
45 – 59
60

Note: Data from 2015.
Sources: Eurostat table "Primary production of renewable energy by type" and Office fédéral de l'énergie OFEN.

Authors: Herrero, D., Benediktsson, K.

2.1.5
Geothermal Energy

Karl Benediktsson

General Characteristics

Geothermal resources are categorised as either low or high enthalpy, with temperature of 150 °C at surface pressure often used to separate the classes (Martín-Gamboa et al. 2015). They are either used directly, e.g. for space heating, or for the production of electricity. Geothermal energy contributed some 6% of all RE in Europe in 2015 (Eurostat 2015), but this is very geographically concentrated, with most of the production in only three coun-

tries: Italy, Turkey, and Iceland (Figures 2.1.1 and 2.1.5.1) albeit potentially available in many other parts of the continent. A high geothermal gradient (the rise in temperature with depth) is an indicator of geothermal potential. This characterises several regions in Europe (Hurter and Haenel 2002), especially in the three countries already mentioned, as well as parts of Greece and most of Hungary. Large areas of France, Spain, Serbia, Macedonia, and Romania also have rather high geothermal gradients.

Technical Characteristics

Geothermal fluids from high-enthalpy fields are suitable for electricity production. The first such power plants used dry steam (without a liquid component) taken straight out of the ground, but

Figure 2.1.5.1
Geothermal energy production across European countries. Authors: Karl Benediktsson & Daniel Herrero-Luque.

most newer installations use single flash or double flash technology, where the fluid is taken to the surface under pressure and then 'flashed' to steam (DiPippo 2015). The use of high-enthalpy fields for space heating and similar purposes requires the use of heat exchangers. Water from low-enthalpy geothermal fields is often suitable for direct use, e.g. for heating of buildings or bathing purposes. It can also be used for electricity generation, although mostly on a small scale using 'binary systems' with a secondary working fluid to drive turbines. This technology, which is still developing, could considerably enlarge the role of geothermal energy for electricity production in countries without high-enthalpy resources.

The true renewability of geothermal energy resources is open to question (Barbier 2002). Low-enthalpy systems based on naturally flowing hot water are indisputably renewable. However, if greater volumes of steam or fluids are extracted from a subsurface reservoir than are flowing into it, the situation is similar to mining (Arnórsson 2011). This is especially a concern in large projects making use of high-enthalpy fields to produce electricity. To ensure the long-term renewability of a geothermal project, the size of the reservoir thus needs to be very carefully assessed beforehand and closely monitored after use has commenced.

Figure 2.1.5.2
Geothermal pipelines at Nesjavellir, SE-Iceland. The power station is just to the left of the picture (Photo: Brynja Rán Egilsdóttir)

Direct Landscape Impacts

Substantial landscape impacts can result from geo-thermal infrastructure development (Kristmanns-dóttir and Ármannsson 2003). In case of direct use, the visibility of pipelines is the most common impact, but natural geothermal surface features (e.g. hot pools) may also be altered. At high-enthalpy fields developed for electricity production, the landscape impacts are more prominent and diverse. Already at the research stage, wells are drilled for assessing the characteristics and potential of the reservoir. If development goes ahead more extraction wells are drilled, and usually also wells for reinjecting spent fluids. Each well pad may cover up to half a hectare of land, and sometimes even more, and may be located at some kilometres distance from the power station itself. Access roads need to be built for heavy drill rigs and other machinery. The angular form of the pipelines, with their thermal expansion loops, can be visually jarring in the landscape, contrasting sharply with natural forms (Figure 2.1.5.2).

The generation station is a rather complex assemblage of industrial-looking steam separators, cooling towers, and pipes, in addition to the buildings housing the turbines and generators themselves (DiPippo 2015). Geothermal electricity generation thus usually results in a landscape that has a decidedly industrial character (Figure 2.1.5.3).

Figure 2.1.5.3
Industrial landscape at Hellisheiði geothermal power station (Photo: Csaba Centeri)

Indirect Landscape Impacts

Beyond direct physical alterations of the landscape, geothermal energy production has various other environmental impacts (Kristmannsdóttir and Ármannsson 2003, Shortall et al. 2015). The development of high-enthalpy fields in particular can have a range of undesirable consequences. Some of these may indirectly affect landscape quality. Hillside stability is sometimes weakened by thermal changes in the soil and landslides have occasionally been the result. Subsidence of land can also occur as fluids are extracted. Hot springs, fumaroles, and other evidence of geothermal activity on the surface may change in appearance or even disappear. Reinjection may also trigger earthquakes, although they are hardly large enough to affect landscape.

A major issue is the discharge of myriad chemicals in geothermal steam or spent fluids (Kristmannsdóttir and Ármannsson 2003). This involves a large range of elements and compounds, such as arsenic (As), boron (B), mercury (Hg), lead (Pb), cadmium (Cd), iron (Fe), zinc (Zn), manganese (Mn), lithium (Li), aluminium (Al), ammonia (NH_3), carbon dioxide (CO_2), and hydrogen sulphide (H_2S) which is a signature compound of geothermal steam. The problems vary from site to site. H_2S in particular can have adverse effects for both human health and local ecosystems (Shortall et al. 2015). Furthermore, chemical pollution may alter vegetation close to the power plant, unless fluids are reinjected and the steam subjected to chemical scrubbing. Other, more subjective, impacts may also be problematic

Figure 2.1.5.4
Hot springs and fumaroles
at Námaskarð, Iceland
(Photo: Csaba Centeri)

such as noise from blowing wells and foul smell of hydrogen sulphide that can detract from landscape quality as perceived by the public.

Mitigation Strategies

More often than not, the landscapes where geothermal fields are found are highly regarded, as their special surface formations, including hot springs and fumaroles, are comparatively rare, and unusual colours give a unique appearance to such landscapes (Figure 2.1.5.4).

Often such landscapes appear quite 'natural' before they are developed. This points to the need of careful and cautious planning when geothermal energy infrastructure is put in place. Mitigation strategies may involve the reclamation of vegetation disturbed at the research and construction phases, with local species if possible. Some of the unavoidable long-term negative impacts can be minimised by good engineering and design. Examples include using a single well pad to drill several separate wells with directional drilling technology, putting pipelines

underground, choosing colours that harmonise well with the landscape, and designing aesthetically pleasing and inconspicuous shelters for wellheads and mufflers.

Possible Positive Impacts

Given the characteristics outlined above, it is hard to envisage positive landscape changes with geothermal energy development. The technology is rather new and has not (yet at least) become valued as 'industrial heritage'. In some cases, however, un-foreseen landscape impacts have turned out to be positive. The best example perhaps is the 'Blue Lagoon' in Iceland, which has become a major tourist attraction and spa. Many tourists who bask in its peculiarly blue, silica-rich, and warm geothermal water do not realise that it doubles as a spillwater pond for the nearby geothermal power station.

PAN-EUROPEAN POLICY ASPECTS OF RE AND LANDSCAPE QUALITY

Csaba Centeri, Marina Frolova & Daniel Herrero-Luque

2.2.1
Overview of European Policies on RE

Based on the EU Directive 2009/28/EC, national plans and laws have determined ambitious national targets for RE sources (RES) by 2020. In addition, the European Commission has also set ambitious EU goals on climate and energy issues for 2030 as an increase of the share of RE in final energy consumption by at least 27 % at EU level (European Commission 2016, European Council 2014).

The energy transition depends on the opportunity to attract high amounts of capital (Masini and Menichetti 2012), and to make RE technology competitive and profitable without external supports. Due to that it was necessary to establish: (i) a stable and efficient policy and regulatory framework and (ii) a well-functioning RE market (IEA-RETD 2016).

In this way, regulatory policies play a key role, because they allow the creation of a new regulatory framework and include all measures that assign a 'reward' to companies that produce and/or enter the network of electricity from renewable sources (Carfora et al. 2018). In regulatory policies the most common features are the repurchase rate (FiT: Feed in Tariff), the net metering, the green certificates (Tradable REC), and the forced share (Renewable Portfolio Standard). The feed-in tariff (FiT) scheme has been the most commonly applied mechanism for supporting RE investments in most European countries (Jenner et al. 2013, Lüthi and Wüstenhagen 2012).

The main advantages for FiT are their high levels of effectiveness and the low risk premiums and secure returns to RE investors (Abdmouleh et al. 2015, NREL 2010).

In 2014, 17 of the 28 EU Member States had adopted a feed-in tariff, ten a feed-in premium, six a quota scheme, and three a tender scheme (Strunz 2016).

The main threat of development is the change of legislative framework, the alterations of the FITs and even retroactive actions that may lead to an uncertain economic environment, negative effects on RES project's profitability, and a less favourable investment environment (Frantál and Prousek 2016). Angelopoulos et al. (2017) indicate that several retroactive changes of policies have been already taken in several countries of the EU (e.g. Belgium, Bulgaria, Czech Republic, Greece, Italy and Spain), leading to negative impacts on the RE investment environment (Baraja-Rodríguez et al. 2015). The Croatian feed-in tariff scheme closed in December 2015, Bulgaria's closed in February 2015, and that of the Czech Republic closed in January 2014.

2.2.2
Overview of European Policies on Landscape

The common platform and the origin for most of the national policies on landscape is the European Landscape Convention. As of June 2013, 38 Council of Europe member states had ratified the European Landscape Convention: Andorra, Armenia, Azerbaijan, Belgium, Bosnia and Herzegovina, Bulgaria, Croatia, Cyprus, Czech Republic, Denmark, Finland, France, Georgia, Greece, Hungary, Ireland, Italy, Latvia, Lithuania, Luxembourg, Moldova, Montenegro, Netherlands, Norway, Poland, Portugal, Romania, San Marino, Serbia, the Slovak Republic, Slovenia, Spain, Sweden, Switzerland, the Republic of Macedonia, Turkey, Ukraine, and the United Kingdom. A further two states have signed but not ratified it: Iceland and Malta.

This high number means that Europe is very well covered by parties who expressed their interest in protecting European landscapes. The ELC itself is aimed at shaping national landscape policies.

2.2.3
Integration of RE and Landscape among the Energy, Territorial, Environmental, Agricultural and Other Policies

The ratification of the ELC implies the integration of landscape issues in urban and regional policies, as well as in policies which may directly or indirectly affect landscape.

Additionally, the Gothenburg Strategy for sustainable development enhances the relevance of integration and synergy among policies, taking into account economic, social, and environmental effects in every decisional process.

Finally, the strategic environment assesment (SEA) process is a fundamental reference for environmental policies and planning activities. The tool should be adopted in every planning process that may produce environmental externalities, and it implies a certain degree of participation as a way to reduce and control environmental impacts. ELC and SEA are strictly connected: even if the ELC does not directly mention the SEA process, the Convention asks governments to adopt suitable tools for environment and landscape protection; on the other side, the SEA process should imply at least the dimension of environmental landscape in its evaluation.

In Southern Europe the targets vary from 17 % in Italy and over 20 % in Spain and France to 31 % in Portugal. While these targets are subject to constant evolution, as has happened in the recent 2030 EU framework (UE 2014), they have already led Member States to define, adopt, and implement ambitious RE policy frameworks, which have had profound social, economic, and environmental consequences (Warren et al 2012). For instance, feed-in tariffs for renewable energies—most often wind power or solar PV—have been introduced in Spain (1994–1997), France (2001), Portugal (2001), and Italy (2005).

While successful, the development of RE capacity has been influenced by a range of complex cultural, contextual, socioeconomic, political, and physical factors (Ellis et al. 2007), which have made it rather uneven, with the pace and the extent of development varying greatly from one Member State to the next.

There are numerous examples of taking into account the impacts of RE, e.g. in Germany visual sensitivity is determined by the landscape's relief and is characterised as the percentage of the area within a circle with a 5 km radius that is visible from the circle's centre. This measure is assigned to five categories ranging from low to high visual sensitivity.

LANDSCAPES IN EXISTING IMPACT ASSESSMENTS FOR RENEWABLE ENERGIES

Robert Kabai

RES facilities have impacts on landscapes to varied extents. Environmental (ecological) impacts are generally considered moderate and were also beyond the scope of the research project. This chapter therefore focuses on the impacts of RES facilities on landscape character.

2.3.1
Impacts on the Visual Landscape

The most obvious effect on character is the appearance of RES facilities in the landscape. The visual impact has therefore been a main concern since the advance of large scale hydropower projects. Later wind farms, as the most widespread RE structures worldwide, have become an intensively researched topic in the scientific literature. Nowadays, the visual impact of photovoltaic power installations is also an emerging topic, while there are relatively few articles dealing with that of other RES facilities. In 1994 Lange (writing about Switzerland) mentioned 'Technical difficulties in the visualization of landscape change, methodological problems in the evaluation of landscape beauty, and the lack of clarity in integrating visual aspects in the planning process' as major reasons for the insufficient consideration of visual aspects in practice (Lange 1994, 99). Since then, however, a lot of progress has been made in this field. With the development of methodology and advance of computer technology there are sophisticated tools to assist the Visual Impact Assessment (VIA) procedure.

Visual impact assessments may be classified as based either on expert or public preference approach. Regarding the expert approach, professional practice still relies on qualitative methodologies to a great extent, while the scientific world seems to aim at quantifying the impact through the development of specific indicators. 'The expert approach seeks to devise ways of measuring physical attributes of the landscape to reflect visual quality. This process is carried out by skilled and trained experts in the field, in a procedure which generally involves verbal or numerical characterisation of the landscape parameters' (Torres-Sibille 2009, 240).

Nowadays the methodology for assessing visual impacts is a rather developed and standardised procedure. Having the basic data of the proposed development available, the main steps are very much the same. Based on the Guidelines for Landscape and Visual Impact Assessment (LI-IEMA 2002) and general professional practice, these could be summarised as

- Identifying potential zones of visual influence
- Identifying main receptors (and related viewpoints) in these zones
- Assessing the sensitivity of the visual receptors
- Assessing the existing visual resource from the viewpoints
- Defining the appearance of the proposed development
- Assessing the scale or magnitude of the effect: the changes of the views resulting from the development based on the design proposal
- Judgement on the significance of the effect

In their articles, both Jombach et al. (2010) and Jerpåsen and Larsen (2011) present a valuable set of criteria for assessing visual impacts. However, some of the criteria seem to appear repeatedly under various names. Using several other sources, as well (LI-IEMA 2002, Torres-Sibille 2009), Table 2.3.1.1 provides a revised list, omitting or clarifying some of the denominations used, and assigning them to the relevant phases of the procedure. Assessing the existing visual resource and its change in the view is related to general landscape assessments, which is why it it has not been included here.

Public preference approach relies on the subjective judgement of the participants affected by the project. In the case of an existing RE structure, this would be based on real experience. For judging the impact of development proposals which do not exist at the time of survey, researchers apply visual simulations. This means that reliability of such public preference surveys depends on the technology and quality of simulations.

Based on the public preference approach, there are also economic methods to evaluate the public judgement of development proposals. Using the contingent valuation method, Mirasgedis et al. (2014) estimate people's willingness to pay for avoiding the visual impact of a projected wind farm in the neighbourhood by alternative solutions. They compare the results with those of other similar studies. The price people are willing to pay may be considered as an indicator of their visual preference. Gibbons (2015) measures the impact through the comparison of sales prices of houses that are either affected or not affected by the view of wind turbines. Comparisons suggest that wind farm visibility reduces prices, and the implied visual environmental costs are substantial.

Regarding their practical applications, visual impact assessments usually appear as part of environmental impact assessments (EIA) and strategic environmental assessments related to RES development proposals. These are standard procedures known worldwide, and within the Euro–pean Union (based on 2011/92/EU and Directive 2001/42/EC respectively) are very much the same. Assessments of visual impacts make also an integral part of landscape capacity studies that lay emphasis on locational aspects of RES infrastructure development.

Phases	Criteria
Identifying zones of visual Influence	• Topography • Land cover • Height of development • Height of receptor • Distance
Identifying main receptors	• Distribution of population (permanent, visitors, passers-by) and related locations
Assessing the sensitivity of the receptors	• Expectations of receptors (depending on experience and context)
Defining the visual attributes of the proposed development	• Form • Scale and mass • Line • Height • Colour • Texture • Number and arrangement of multiple structures • Movement, light, and shade effects
Assessing the scale or magnitude of the effect	• Visibility 1. Visual dominance: extent in the view, contrast with the background • Visibility 2. Angle of view • Visibility 3. Full/partial • Visibility 4. Typical weather or atmospheric conditions • Number of receptors (people) affected • Duration of viewing • Potential mitigation
Judgement on the significance of the effect	• Cumulative impacts

Classification based on the publications of Jerpåsen and Larsen (2011), Jombach et al. (2010), LI-IEMA (2002), Torres-Sibille (2009)

Table 2.3.1.1
Summary of assessment criteria for evaluating visual impacts

Although the visual sense is the dominant human sensory component, our experience of landscapes and their change is not merely visual. Some of the research therefore aims to take other sensory information into consideration when surveying impacts of developments on people. An example is the project of Swiss researchers aiming at integrating audio effects into simulations of wind farm proposals. Although this was not yet achieved in real-time, the researchers have developed a sophisticated procedure to create visual-acoustic simulations (Manyoki et al. 2014).

Moreover, the perception of various stimuli received by the senses is influenced by additional factors such as personal experience and attitude. Jerpåsen and Larsen argue that beyond physical conditions the assessment procedure 'should be widened to also include local storylines, experiences and local values attached to the landscape'

(2017, 207). These are obviously important considerations when we want to fully understand public responses to development proposals. Inclusion of these aspects highlights the need to move from a purely visual to a more comprehensive approach considering the impact on landscape character (as experienced by people).

This approach considering the perceived impact on character in two research papers by Devine-Wright (Devine-Wrignt and Hows 2010, Devine-Wright 2011). Articles deal with the public acceptance of two RE projects: an onshore wind farm (Devine-Wrignt and Hows 2010) and a tidal energy converter (Devine-Wright 2011). Based on the theory of social representations, the articles investigate how symbolic meanings associated with the development and the places affected influence place attachment, and what public response is produced. The tidal energy case study also proved that NIMBY-type opposition is not the only potential response to development projects, but a RE development project could also be interpreted to maintain or even enhance a place's character and thus place attachment (Devine-Wright 2011).

3 LANDSCAPE POTENTIAL AND VULNER-ABILITY TO RENEWABLE ENERGY DEVELOPMENT

With the Industrial Revolution, the development of energy generation and transport infrastructure has hugely accelerated the pace at which humans alter the face of the earth, as well as the (less visible) chemical composition of its atmosphere. With increasing wealth and the off-shoring of our most polluting industries, many people in 'developed' countries have grown to love those landscapes that remain least affected by modern machinery, often referring to them as 'unspoilt'. Although human aspirations to conserve the landscape and conserve the climate both draw on the same underlying ethic, concerns about the development of onshore low-carbon energy technologies (a key element of the quest for a low carbon energy transition in the 21st century), are often framed as a conflict between development and conservation. This framing can be attributed to a number of factors, from local dissatisfaction about planning decisions and resistance to the dominance of for-profit, private sector actors in renewable energy development, to the (almost unavoidable) tensions between the established regime of landscape protection (consisting of existing laws, institutions, professionals) and the growing niche of climate change mitigation and adaptation.

Over the course of the 20th century, different national strategies for landscape protection have been developed. Protected areas were identified and delineated through politically sensitive processes that combined qualitative descriptions with scientific narratives. Over time, experts developed methods to map 'landscape character' and describe 'landscape quality'. These categorisations are based on scientific knowledge (e.g. of the underlying geology, ecological processes) and/or values elicited from some members of the public. Since knowledge shapes values and access to knowledge is privileged, it is important to recognise and acknowledge that such categorisations, despite the best intentions, are by definition neither fully objective nor fully democratic. But in (relatively) affluent social democracies, sustainability dilemmas need to be managed somehow and to some extent. Practitioners to whom this management task has been assigned are inevitably drawn towards the development of categorisations and the use of existing instruments they know to be blunt. The RELY project seeks to provide guidance in the careful use of such instruments, so that our energy systems can become much cleaner without sacrificing much of the value of our landscapes.

Starting from the position that it is rare to find objective and scientific reasons for banning renewable energy developments, it is nevertheless possible to examine best and worst case scenarios for developing renewables and conserving landscapes. The worst cases include the building of renewable energy facilities in physically inappropriate locations, like on hill-top peatlands which are easily damaged (releasing more carbon dioxide than the wind turbines might off-set) or along active geological faults where wind turbines are likely to get damaged. Wind turbines along migration hotspots can put the lives of birds and bats at risk, while access roads and the ditches for the cables can cause erosion of sensitive soils on steeper slopes. It is also possible to identify rare and iconic locations that are culturally less suitable for development of renewables. Internationally famous landmarks like the Meteora Rock formations, Mont Saint-Michel, the Lorelei Rocks on the Rhine River, or the islands of the Venice Lagoon would not look the same anymore as in the post-cards we know if they had many wind turbines as neighbours. Some wider geographical areas may not look entirely unique from an international perspective but are nationally rare and precious, and might for that same reason be protected. Examples could include Hungary's Lake Balaton, Poland's Zakopane Mountain and Denmark's cliff coast at Møns Klint. These are but small parts of the national territory so a local ban on renewable energy generation will not have much effect on national targets, thus we can concentrate renewable energy generation in other locations.

In most countries we can indeed find locations that are perfectly suited for renewables because of high potential energy yield in highly anthropogenic landscapes that are devoid of any measures of aesthetic protection. Examples may include wind turbines along tall sea dykes and around heavy industry complexes and large container ports, or geothermal energy plants in quarries or abandoned mining areas. But most of our landscapes are neither characterised by overwhelming iconic scenery nor by (post) industrial stigma. These diverse landscapes are valuable in a number of ways to both local people and visitors, but these lands will need to accommodate most of the renewable energy generation required to achieve a low carbon society, starting at the current (low) levels of deployment and increasing renewable energy generation over time. This raises questions about pragmatic approaches and (second) best practices as we learn to install renewables sensibly, and with sensitivity to local socio-economic and geographical context, both of which are subject to change. This chapter seeks to further a pan-European approach to the mapping of potential conflicts and synergies between the maintenance of landscape functions and values on the one hand and the realisation of renewable energy potentials on the other.

Dan van der Horst & Bohumil Frantál

WHAT FITS WHERE? LANDSCAPE APPROACH TO RENEWABLE ENERGY DEVELOPMENT

Tadej Bevk & Mojca Golobič

3.1.1
Introduction:
From Issue to Opportunity

The concept of landscape has been increasingly gaining momentum as a research topic in various scholarly disciplines. It is a familiar and intriguing concept, but also complex and often ambivalent. The European Landscape Convention positios landscape as both a subject and an object, at the intersection of natural and cultural forces (Council of Europe 2002). As an arena where multiple interests clash and need to be reconciled it spawned a specific mode of problem-oriented thinking. Landscape 'provides a systematic basis for understanding the spatial patterns and processes we see around us' (Swaffield 2005, 6). Approaching a problem through a landscape perspective takes into account multiple layers and seeks to align local, regional, national, and global objectives (Reed et al. 2017, 489). As an inherently complex entity it thus provides a framework for understanding a problem. With landscape being an increasingly vocal issue in renewable energy (RE) debates, this chapter turns the tables and looks at RE development through a landscape perspective as a way of comprehensively approaching the energy transition. To this end various dimensions of landscape will be briefly surveyed to offer some insight into how such knowledge can contribute to finding suitable landscape contexts for RE use.

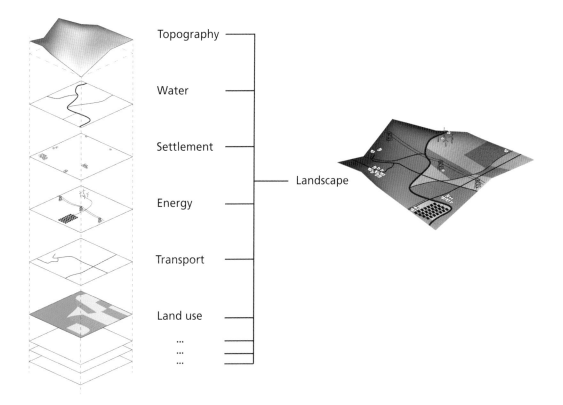

Topography

Water

Settlement

Energy

Transport

Land use

...
...
...

Landscape

Figure 3.1.2.1
Landscape is a system
with multiple sub-systems.
Analysis tends to look at
them separately, but it
is important to remem-
ber each change will
influence landscape as a
whole. (Source: author´s
conceptualisation)

3.1.2
Landscape System

The basic assumption in the landscape approach is
that landscape is a holistic entity, a whole which is
more than the sum of its parts (Antrop 2002, 17).
As such it is not defined only by inclusion of different landscape features, but also by relationships
between them. It is a system of subsystems: water-
sheds, ecosystems, societal organisation, energy production, distribution, etc. (Figure 3.1.2.1).
These are dependant on one another both vertically
(e.g. deforesting a plot will change water runoff),
as well as horizontally (e.g. changing river flow in
one part of the landscape will affect downstream areas). Employing landscape thinking requires us to

scale-up the problem and think beyond a project's
plot immediate border as any new intervention will
cause ramifications to the whole system. Evidence
shows RE development is often perceived as a disturbance in the system (Smardon and Pasqualetti
2017, 121). However, through understanding the
relationships in the landscape it is possible to an-
ticipate the disturbance and properly react to it.
Or better, by knowing the weaknesses or faults of
the existing system we can use the RE itself or its
by-product, to support and enhance it. In this light
RE development should avoid disturbing or breaking existing relationships and look to strengthen
weak ones and establish new relationships. To name
a few examples: photovoltaics can provide electricity in remote areas, hydropower dams can be used to

Figure 3.1.4.1
Photovoltaic panels in Vrhovo (Slovenia) are installed on a levee, following the curve of a river and thus emphasising a landscape structure. (Photo: Tadej Bevk)

mitigate severance of floods, and biogas plants can use produced heat to provide hot water.

3.1.3
Landscape Dynamics

There are multiple processes unfolding in the landscape system. They are caused by both natural and human factors. The processes add a temporal and dynamic aspect to a landscape, which is constantly changing with societal needs and wishes, natural evolution, succession, and adverse events. The changes are of different rate—some are quick and seem abrupt while others can take multiple gener-

ations. The landscapes we live in today are results of mainly anthropogenic aspirations to secure resources for living, but they are also a canvas for all the future changes that will occur. Landscapes will continue to change. Keeping this in mind, developments should not only be considered in light of status quo, but of future scenarios, going 'beyond the present-day challenges and envision future conditions' (Bürgi et al. 2017, 7). Even deciding to do nothing will have an impact on the future of the landscape. Not developing RE in one landscape might have ramifications in another landscape, possibly with greater consequences, as production of energy might be provided by other means. RE developments should be grounded in the broader historic view to show the landscape has changed in the past with the society and also provide a future perspective. In this regard, RE technology that is reversible (e.g. wind turbines) should take advantage of this fact and make specific future plans for decommissioning and use of the plot after the object's lifetime expires. Employing future-oriented problem-solving will also help put RE developments in a broader sustainability agenda, as scenarios can be efficiently utilised to show differences between future states of landscapes in a RE-based energy systems versus fossil fuel based.

3.1.4
Landscape Character

Through the interaction of natural conditions and societal management, landscapes across the world gained specific characteristics, which make them recognizable and distinctive from one another (Swanwick 2004, 111). Landscape character describes a rather homogenous unit of landscape, in which commonly occurring patterns of landscape features and structures can be observed. It is important to note that landscape character differentiates one landscape from another by explaining the specifics of each locality rather than ranking them as better or worse. While this character is grounded in the physical setting and evolution of an area, it also incorporates values and perceptions, but is not by itself an evaluation. Landscape character is always unique, but RE objects are more or less globally generic structures conditioned by efficiency of energy conversion (especially wind turbines and solar panels). However, when dealing with a cluster of such production units more design possibilities emerge, as they can be laid out in different manners. Existing patterns provide a ready-made guidelines for designing the development (Figure

3.1.4.1). Matching the scale and layout to the prevailing landscape structures and patterns can mitigate the visual misfit of RE facilities or even more clearly expose the landscape character. A linear layout of RE objects in a landscape with prevailing linear structures will blend more easily with the setting than a rigid raster structure. Conversely, in a gridded landscapes the grid should be used as main structural principle.

3.1.5
Landscape as Resource

While landscape is more than a resource for humans to exploit, it is also full of economic relationships between people and land. Even though today's society is detaching from its physical environment, land ownership and related benefits are still among key factors of societal organisation, considerably influencing the functioning and appearance of landscape. The physical benefits provided by land (food, water, minerals, etc.) are joined by perceptual benefits of landscape (scenic views, naturalness, identity, touristic attraction, etc.). Taking the economic standpoint, a new land use to be developed must prove to be beneficial to the relationships in the landscape. And vice versa, a landscape with poor economics can prosper from new land uses. Landscapes abundant in RE sources are intrinsically more suitable for RE development, but the landscape approach considers also other aspects, such as distribution of benefits and damages (Figure 3.1.5.1). As we all live in a landscape it is also a common good and as such is subject to the tragedy of the commons. Development's impact does not stop at the border of a site, but affects landscape as a whole, an example of this being visual annoyance. Positioning landscape as a common resource to which a broad range of stake-

Figure 3.1.5.1
Photovoltaic panels in
Desinec, Slovenia. The solar power plant is built on
rocky shallow soils unsuitable for farming and pro-
vides additional income to
the farmer. Mowed grass
from between the panels
is also used to feed livestock. (Photo: Tadej Bevk)

holders are entitled opens up the discussion and
engages local communities in a more actively participation in the planning processes (Wolsink 2007,
2702).

3.1.6
Concluding Thoughts: The Narrative of Energy Landscape

Reading academic literature on the interaction between RE and landscapes, one is led to believe they
do not fit well together, with landscape degradation
being one of the most reported environmental issues in the topic (Apostol et al. 2017, 9). As landscapes are a relatively stable backdrop to people's
daily lives, abrupt changes can be perceived as disturbance if they're not grounded in the landscape
context and future perspectives. In this perspective,
the opposition evoked by RE developments might
not be caused by selfish not-in-my-backyard motives, but rather by the uncertainty of what will
become of a familiar setting in the future (Devine-
Wright 2005). This is an important realisation as
it points towards future-oriented problem-solving.
The objective is thus not to conserve everything
as it is, but to change it in a way that will respect
existing values and create new ones. Landscapes
are already full of narratives as can be seen by this
overview of various landscape dimensions. Today,
energy landscape is also becoming a recognised

Figure 3.1.6.1
An energy landscape in
Burfell, Iceland. The vertical
towers stand in stark contrast to the flat surroundings. They create a new
layer and bring new mean-
ing to the landscape (e.g.
harvesting of strong winds
in a barren landscape, see
Frantál et al. [2017] for a
discussion of perceptions
of pictured landscape).
(Photo: Tadej Bevk).

landscape type (Pasqualetti and Stremke 2018, 95). This should be utilised to further the new narrative of the sustainable renewable energy landscape, which could and should become another layer in this complex system (Figure 3.1.6.1).

In the continuous play of rewriting landscapes, they became a palimpsest of current and past land uses, societal organisations, values, and beliefs. With a thoughtful approach through multiple layers of complexity, RE will provide energy and climate benefits, but also add new land uses, meanings, and values which can increase the vitality of a landscape.

DEVELOPING RENEWABLES IN CROWDED LANDSCAPES: IN SEARCH OF INTERNATIONAL SMART PRACTICE

Authors: Bohumil Frantál, Dan Van der Horst, Stanislav Martinát,
Serge Schmitz, Na'ama Teschner, Luís Silva, Mojca Golobic & Michael Roth

3.2.1
Introduction

The growing renewable energy sector has altered landscapes and land use dynamics and brought about new land use conflicts and disconnections between policy-makers and stakeholders (Calvert & Mabee 2015, Frantál & Kunc 2011, Van der Horst & Vermeylen 2012, Warren 2014). Renewable energy is spatially diffused and the desire to harness it at scale creates new productive demands on locations and landscapes that may already be struggling to accommodate different interests in development and conservation.

There have been significant differences between countries in the level of successful deployment and the extent of controversies and public opposition. While some countries have already almost exhausted their realizable potential and the on-land space for new developments in some respects (e.g. for large wind farms or hydropower plants), other countries are far behind, reluctant, or just starting out. That's why there is clearly scope for international comparisons and learning. But learning from comparative analysis is not necessarily straight-forward, given that there are often significant national differences in economic, legal-procedural, socio-political, and cultural-historic contexts.

Focussing on the siting of more renewable energy projects in already crowded diverse landscapes, we attempt to explore what international lessons can be gleaned from specific projects (case studies) that are nationally perceived to be innovative and successful. More concretely, we seek to synthesise wider lessons from a range of nationally perceived 'best', respectively 'smart' practice projects, and examine how these examples can be analysed in order to yield guidance for other countries.

3.2.2
Questing a 'Smart Practice' in Renewable Energy Development

'Best practice' has become a buzzword for procedures, techniques, or methods that, through experience and research, have proven to reliably lead to a desired result (Gulati & Smith 2009). The various definitions of best practices (Kruse, Marot 2018) show that their rationale is based on not only constant learning, feedback, and reflection of what works and why, but also on what does not work. When it comes to the question of how to identify best practice, the experts are somewhat ambiguous (Bretschneider et al. 2005). Bardach (2000) even suggests that the term best practice is misleading in itself. There is an ontological aspect to this; how can we really know what is the best? And, even if at one particular moment in time the number of options are sufficiently limited to help experts reach a consensus, how can we know that this label still sticks when conditions, policies and technologies continue to change?

A 'smart practice' may therefore be a more useful concept to explore. Although smart is also a rather vague and popular word in management, it can be distinguished from the best practice by its greater focus on the processes that produce agreeable outcomes. The task of researchers should be to explore the smartness of a given practice, to verbalise and evaluate it for applicability in the context of the target site (Bardach 2000).

Smart practice studies can be found across disciplines, some have been provided even in the context of renewable energy development—yet mostly with a focus on individual energy production systems and within one or similar regional contexts. For example, Wolsink (2007) and He et al. (2016) have focused on successful measures in the promotion of wind farms; Cabraal et al. (1996) and Tsikalakis et al. (2011) have highlighted smart practices in solar schemes; Dolman and Simmonds (2010) have examined wave and tidal energy, and Ciervo and Schmitz (2017) have turned their attention to smart practices in bioenergy production.

Based on the insights of such previous studies, as well as the above-mentioned definitions, *smart practice* in the planning and siting of renewable energy production systems would at least have to (i) effectively produce energy based on renewable sources; (ii) seek to minimise environmental harm in each stage of its production, operation, and disposal (life cycle); and (iii) seek to decrease potential conflicts among individual users (or groups of users) of the landscape where it is sited, throughout participation, collaboration, and planning.

Fulfilling the above-defined criteria to a very high standard can be quite challenging, due to various geographic, socioeconomic and cultural conditions of individual sites and communities where the projects are located. Yet, given the need for the energy transition to see renewable energy niches becoming the standard, the transferability of solutions for renewable energy development is very important (Raven et al. 2008). Like in other planning-related practices, there may be a need for some 'sustained effort and imaginative adaptation', but collecting and analysing a rich pool of cases, provides opportunities for ideas to be reapplied elsewhere and under somewhat different circumstances (Selman, 2004, 388).

3.2.3
Research Method and Procedure

Inspired by Delphi methods (Miles et al. 2016), we deployed a mix of methods characterised by several rounds of expert engagement, starting with a qualitative phase of expert elicitation of national cases of smart practices, followed by a quantitative exploration of recurring characteristics of case studies, using computer assisted procedures (Frantál et al. 2018). The explorative research included the following steps:

- More than 100 experts from 30 countries working together through the COST Action were invited to identify what they considered to be smart practice in renewable energy development in their own country. We asked them to pick one or more case studies, providing basic description of the project and justifying by what criteria the project should be considered a smart practice.
- The narrative sections of all case studies were collectively reviewed and analysed by authors of this chapter to identify specific characteristics representing criteria of smart practice. The focus was primarily put on the outcome criteria, although criteria related to the process (such as participatory planning) were also mentioned in the descriptions.
- Presenters of national case studies were consulted again to determine which specific criteria each case study meets (using a simple binary coding of 1 /project meets the criterion/ or 0 /project does not meet the criterion/).
- The information from the case studies have been coded, categorised and converted into a data matrix usable for statistical analysis. The data were analysed using SPSS software, providing basic descriptive statistics, principal component analysis, multiple correspondent analysis, and cluster analysis.
- A typology of smart practice projects was designed based on the interpretation of results of statistical analyses. The proposed typology consists of 'empirically grounded types' (Kluge 2000) combining empirical analyses and theoretical knowledge of the experts. Potential use of the results and policy implications were discussed and formulated.

3.2.4
Dataset of Smart Practice Case Studies

The dataset created includes 51 case studies from 20 countries. The complete list of case studies is presented in the paper by Frantál et al. (2018). The basic characteristics of case studies are presented in Table 3.2.4.1 Eight case studies represent innovative policies, plans, methodologies, or tools, four case studies are examples of innovative technologies, procedures or projects, which were implemented multiple times at different locations, and 39 case studies are examples of specific projects realised on a particular site. The majority of projects are located in rural areas; only a small number is found in urban areas. More than half of the projects are located in borderland areas, which can be recognised even from the location map (see Figure 3.2.4.1).
Wind energy constitutes almost one third of all case studies. A quarter of the case studies concerns solar energy (with a majority of ground-mounted and rooftop-mounted photovoltaic systems). One fifth is represented by projects of either or both biomass cultivation and biogas production. Six case studies concern hydropower, including one pumped-storage plant, and five case studies are examples of multifunctional (mixed-energy) projects.

Category		Number	[%]
Type	Specific project realised at a particular site	39	76
	Technology or procedure implemented at different locations	4	7
	Policy document, plan, method, or tool	8	17
Energy	Wind (onshore)	16	31
	PV (ground-mounted)	7	14
	PV (on roof)	5	10
	Solar-thermal (on roof)	1	2
	Biogas	4	8
	Biomass	6	12
	Hydro (small)	2	4
	Hydro (large)	4	8
	Mixed	6	12
Location	Rural area	29	57
	Urban area	10	20
	No specific location	12	23
In total		51	100

Table 3.2.4.1
Basic characteristics of the COST RELY smart practice case studies (Source: survey carried out within the COST RELY TU1401 Action)

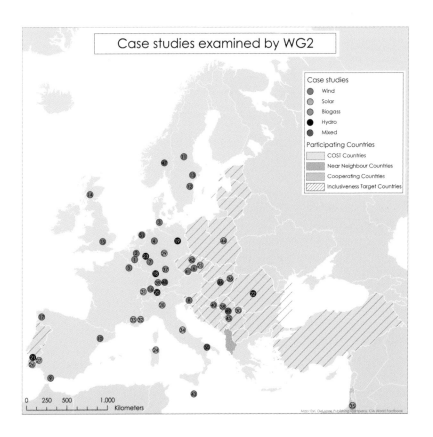

Figure 3.2.4.1
Location of smart practice case studies (Source: elaborated within the COST RELY TU1401 Action)

Criterion	Description	Abs.	Rel. [%]
Rural area	Located in a rural area	29	9.2
Visual impact	Reduces visual impact	26	8.3
Local benefits	Provides economic benefits for local people/community	24	7.6
Border periphery	Located in a border or peripheral area	22	7.0
Low population	Located in areas with low population density	21	6.7
Pilot project	Represents a pilot or experimental project	20	6.4
Local demand	Meets the local demand for energy	16	5.1
Deconcentration	Provides spatial deconcentration of impacts	16	5.1
Land use synergy	Allows multifunctional use of land	15	4.8
Environmental synergy	Compatible with environment, using local sources	14	4.4
No conflict of use	Located on land without other (significant) use	13	4.1
Technological innovation	Represents technological improvement or innovation	13	4.1
Small scale	Consists of small size and/or small number of units	12	3.8
New landmark	Creates a new visual landmark	12	3.8
Co-benefits	Provides by-products and/or co-benefits	10	3.2
Reversibility	Easy removal of technology, thus restoring the area	9	2.9
Demonstration effect	Serves for demonstration and public education	9	2.9
Regulation function	Provides some eco-system regulation function	9	2.9
Heritage synergy	Is compatible with cultural heritage objects	7	2.2
Degraded land	Uses environmentally degraded land (brownfields)	5	1.6
Infrastructure synergy	Utilises existing infrastructure	5	1.6
Improving stigmatised land	Improves image of environmentally stigmatised land	4	1.3
Energy region	Located in an area already used for energy production	4	1.3
In total		**315**	**100**

Table 3.2.4.2
Criteria of smart practice projects: absolute and relative frequencies (Source: survey carried out within the COST RELY TU1401 Action)

3.2.5
Results

Specific Criteria of Smart Practice

During the qualitative analysis of case studies we identified 23 specific criteria by which the case studies can be considered as a smart practice. The description of these criteria and their absolute numbers and relative frequencies (dependent on how much projects meet each specific criterion) are presented in Table 3.2.4.2.

Generic Criteria and Typology of Smart Practice

In order to explore the structure of relations among specific characteristics of smart practice and to find out if they can be divided into groups representing more generic criteria we applied principal component analysis, multiple correspondent analysis, and cluster analysis (Frantál et al. 2018). The result is an identification of eight generic criteria of smart practice, including two criteria related to the location and geographical context of projects, one criterion representing their scale and spatial concentration, and five criteria representing different kinds of syn-

TYPE	CHARACTERISATION
1. Spatial targeting	
1.1 Rural peripheral areas	Low population, less favoured, economically deprived rural areas Borderland or inner peripheries Landscape of no special value, no environmental protection Stronger motivation effect of economic benefits for local people
1.2 Industrial or (post-industrial areas	Using environmentally degraded or derelict land (brownfields) Improving the image of environmentally stigmatised areas Concentration of externalities (already affected energy regions)
2. Synergy provided	
2.1 Infrastructure synergy	Synergy with existing infrastructure (e.g. road or rail networks) No conflict of use Easy reversibility Reducing visual and other impacts
2.2 Local economy synergy	Increasing local energy independence Direct economic benefits for local communities Stimulating public participation and shareholding
2.3 Environmental synergy	Compatible with the environment Using local resources and/or wastes Providing regulation functions Generating co-benefits or by-products
2.4 Land use synergy via technological innovation	Enabling multifunctional use of land Promoting technological innovations Pilot or experimental projects (practice as a laboratory)
2.5 Heritage synergy and education	Synergy with historical-cultural heritage Energy tourism (information centres, observation towers, eco-trails) Demonstration and education effect

Table 3.2.5.1
Typology of smart practices (Source: survey carried out within the COST RELY TU1401 Action)

ergies (infrastructural, economic, environmental, land-use, and heritage) provided by projects.

Based on these generic criteria, we propose a typology of smart practice projects, consisting of two main groups and several sub-types. The first group consists of smart practices which are characterised by spatial targeting of projects to specific geographical areas in order to minimise potential land use and social conflicts. In this group, projects are typically targeted towards either low-population, peripheral, and economically under-developed rural areas or towards industrial or post-industrial areas, which are often environmentally degraded. The second group includes projects which provide some kind of synergies (with the local environment, cultural heritage and education, local economic development, and multiple land uses). A more detailed description of types is in Table 3.2.5.1.

In reality, most of the projects meet several criteria of smart practice and thus fit into several categories simultaneously. Part of the differentiation between the smart practice types of 'targeting' and 'synergies' could be understood in terms of relative scale; spatial targeting relates to broader geographical areas which can be identified at a strategic, national level, whereas synergies are likely to require more locally specific knowledge and joined-up thinking between different sectoral agencies.

3.2.6
Discussion and Policy Implications

One of the key challenges in the transition to a low-carbon society is to find how various renewable energy systems can be deployed in diverse, crowded and ever-changing landscapes. There are various ways in which the findings of this research could potentially be used for international and comparative learning purposes. But it is important to make a very clear distinction between the generic typology and concrete examples. This study did not seek to provide advice at the concrete project level, and indeed academics must be careful not to uncritically extrapolate and over-extend practical approaches branded as smart practice. It has been argued that the nature of projects, and uniqueness of local political-economic conditions, challenges the transferability potential of smart practices of renewable energy siting (e.g. Garcia 2011). In other words, concrete examples of national smart practice cannot be transferred across or even within national borders without caution and critical thinking for reasons of context and scale.

With regard to policy relevance, our findings could be utilised in a number of ways, from setting guiding principles for project design, spatial planning, and consent (for countries that are still in the process of developing these) to evaluating the portfolio of renewable energy projects developed to date (for countries that are already forging ahead with renewables). Such evaluations may yield insight into the extent to which a national (sub)sector has progressed faster or slower when adopting or ignoring these forms of smart practice, whilst subsequent international cross-comparisons can yield insights into the extent to which different borders have been permeable to lessons learned by early adopters (i.e. the extent of international policy learning in specific countries).

With regard to ex-post evaluation, it is furthermore important for future studies to recognise that the more renewables a country has installed, the more difficult it becomes to find 'easy' locations or 'win-win' configurations that are still available for a new project. As a consequence, novel and more specific forms of targeting may need to be developed and new and more contextually specific opportunities for synergy may need to be identified. Since smart practice implies learning by doing, and since our typology is conceptually independent of the extent of technology adoption, there is potential scope for this study to inform a more detailed sectoral analysis of policy learning, adaptive governance, and socio-technical innovation.

3.3

NARRATING THE SMART PRACTICES OF RENEWABLE ENERGY DEVELOPMENT

The typology of smart practices of renewable energy development as presented in the previous chapter consists of two main groups comprising seven sub-types of practices. The first group includes projects which are characterised by spatial targeting of developments to specific geographical areas (either to low-population, peripheral, and economically under-developed rural areas or to industrial or post-industrial, often environmentally degraded areas) in order to minimise potential land use conflicts. The second group includes projects which provide some kind of synergies, whether with the existing infrastructure, local economies, the environment, multiple land uses, cultural heritage and educational activities, or a combination of several of them. The following sub-chapters present more detailed descriptions of selected examples of such smart practices.

Figure 3.3.1.1
The use of solar energy in a household in the Bedouin village of Ovdat, Israel. (Photo: Bohumil Frantál)

3.3.1
Tackling Energy Poverty with Renewables: Solar Energy in Peripheral Rural Areas

Na'ama Teschner

Rural peripheral areas are usually characterised by low population density. Communities in borderland or inner peripheries are often also socially and economically disadvantaged, lacking access to various public services. Despite the fact that completely off-grid communities are becoming increasingly rare, in various places around the world, especially in the global South (but also in other locations including Eastern Europe), there are still large amounts of people living without connection to the electricity grid. According to the World Bank, approximately 28 % of the rural population worldwide has no access to electricity (World Bank 2014).

The case of the Bedouins (seminomadic tribal descendants) in Israel's southern Region, the Negev, is an extreme example of such off-grid population. Across the region, there are about 75,000 residents living off-grid, many of them in temporary structures with little or no state provision of infrastructure (Meallem & Garb 2010). By today's standards, such conditions create high levels of energy poverty. The social and material inability to secure enough electricity for the household adversely affect people's health, education, gender equality, and general well-being.

Advantages	Risks	Table 3.3.1.1 Competitive advantages and risks of PVs at the households level in rural peripheral communities (Source: author´s compilation)
• Reduce energy poverty and improve quality of life • Low environmental impact • Decentralised, self-owned production • Security of supply (depends on geo-climatic conditions) • Gain social acceptance from local communities • Stronger motivation effect of economic benefits for local people	• Price of the installations (solar panels) and high initial costs • Lack of knowledge regarding operation and maintenance of the systems • Lack of safety and building regulations and low level of enforcement in rural areas • Potentialyl unstable surface for new constructions • Low governmental motivation to initiate support schemes	

Small-scale solar PV installations are increasingly considered a smart practice and effective solution for rural electrification, especially in arid regions or other peripheral regions, e.g. mountainous areas. The cost viability of stand-alone PV systems for rural applications have numerous advantages. Studies conducted with relation to Nigeria (Shaaban & Petinrin 2014), Sub-Saharan Africa (Amankwah & Amoah 2015), Yemen (Hadwan & Alkholidi 2016) and South Asia (Palit & Bandyopadhyay 2016) all concluded that the initial high cost may be a barrier, but low maintenance and operation costs are making solar PV attractive, in comparison with both other renewable and non-renewable sources.

3.3.2
From Wasted Land to Megawatts: Developing Renewable Energy on Post-industrial Brownfields

Bohumil Frantál

The concept of 'brightfields', which was introduced few years ago in the USA (EPA 2010), is also emerging as a kind of smart practice of renewable energy development—representing a spatial targeting of projects to industrial or post-industrial areas—within Europe. So-called brightfields are brownfields (underused, abandoned, derelict, and often contaminated lands), which have been converted into a newly usable land by implementation of renewable energy technologies. Brownfields are the result of changing patterns of industry and development in many regions. They are largely regarded as liabilities which degrade also the value of the surrounding land; however, it is often difficult to sell them and municipalities are unable to revitalise them using their own resources. The regeneration of brownfields has become more common during the last two decades since vacant developable land (or so-called greenfields) is less available, more expensive, and more protected in densely populated areas and as a result of emerging policies, economic instruments, and management tools supporting the regeneration processes.

It has been reported (e.g. Adelaja et al. 2010, Kunc et al. 2011) that the technical potential of post-industrial and other brownfields for renewable energy development is of a considerable degree in many regions. Brownfields can be used for the construction of either solar panels or wind turbines, for the capture of landfill gas and conversion to electric energy (LFGE) systems, and also for cultivating crops and trees for bioenergy production, or these uses can be combined with other, non-productive land-uses (e.g. recreational or educational).

Compared to greenfields, brownfields can provide many competitive advantages (Table 3.3.2), even though their successful revitalisation requires complex risk management (Neuman & Hopkins 2009). The brightfield approach meets at once several challenges we need to globally cope with, such as urban sprawl, toxic waste cleanup, environmental restoration, climate change, energy demand, and energy sustainability. Moreover, the brightfields phenomenon can be considered as an example par excellence of the so-called 'third transition', a transition from the fossil fuel-powered age into the post-industrial era, an era characterised by scarcity—of energies, natural resources, and living spaces (Whipple 2011).

Figure 3.3.2.1
Solar park constructed
on a previously used coal
mining dump in Osla-
vany, Czech Republic.
(Photo: Bohumil Frantál)

Advantages	Risks
• Preserve green spaces • Access to existing infrastructure and grid • Available at lower land prices • Abandoned estates get new productive use • Create new job opportunities • Gain social acceptance from local communities • Assist in environmental restoration and improve the image of stigmatised land • Can serve as demonstration and education sites	• Unknown pre-existing conditions (potential or unexpected contamination) • Often complicated ownership relationships • High initial costs for investigation and re-vitalisation (decontamination) • Potentially unstable surface for new constructions

Table 3.3.2
Competitive advantages
and risks of brownfields
for renewable energy
development (Source:
author´s conceptualisation)

155

3.3.3
Energy Roads and Roofs: Providing Synergy with Existing Infrastructures

Sina Röhner & Michael Roth

As pointed out several times in the previous chapters, the installation of renewable energy facilities is often connected to various conflicts concerning the change of the visual landscape. One possibility of smart practice to avoid or at least reduce such conflicts is to combine RE with existing infrastructure, which offers the opportunity to install RE facilities without further conflicts of use and reduce the visual and also other impact significantly (Weiland et al. 2016). Another advantage is that RE built in combination with existing infrastructure can easily be reversed in most cases.

The easiest RE technology to combine with existing infrastructure seems to be solar power. The most obvious and most common example for this smart practice is the installation of solar power or solar thermal plants on roof tops. For flat roofs, where there might be a conflict of use with rooftop greening, newer technologies make it possible to combine these two types of use. On steeper, roughly south-facing roof tops, solar power is without competition.

Another possibility is the installation of PV panels on noise protection walls along traffic routes. There are examples in Germany, where up to several hundred metres of PV modules were installed along motorways (Emden, Lower Saxony, or Vaterstetten near Munich). Several smaller solar power plants can be found along other motorways or also railway tracks (Röpcke 2007).

Bicycle lanes can also be combined with solar power. Instead of using traditional material to pave the

Figure 3.3.3.1
Solar power plants on a noise protection wall along a motorway in Baden-Wuerttemberg, Germany (Photo: Sina Röhner)

Figure 3.3.3.2

Solar power plant on noise protection wall around residential area in Großbettlingen, Germany
(Photo: Michael Roth)

lanes, PV modules which are especially produced for this kind of use can be used to build parts of the bicycle lanes and produce energy while providing safe ground for cyclists (cf. CleanTechnica 2017).

In Germany, field studies are being carried out to test the use of PV modules above agricultural areas to avoid the conflict between agriculture and energy production (see also chapter 3.3.6). The PV modules are built at a certain height above the ground, letting enough sunlight pass through for the plants below and enough space for the farmers to cultivate the land. Results are positive, meaning that the energy production is very good while there is only little loss in harvest. (Fraunhofer Gesellschaft 2017, Fraunhofer-Institut für Solare Energiesysteme ISE 2016)

Wind energy can also be combined with existing structures. In the Ruhr-Region in Germany, formerly a big coal producer, wind turbines have been built on mining dumps. These locations offer great wind conditions and if the turbines are located within industrialised areas, there is less resistance to their installation. Besides this, the combination of old mining dumps with new wind turbines symbolises the change from fossil fuels to renewable energy in this region, stressing that the Ruhr region is still an energy region, despite the decline in the coal industry (Schürmann 2015, Schmeer 2009).

In the field of water energy, old mills can be reused for energy production and to reduce landscape impact by using and preserving the existing traditional building instead of building new facilities, which has been done for several mills in Germany (Böck 2012, Steinhilber 2010). Such projects are also examples of smart practices using a synergy with historical and cultural heritage, which have the potential for development of energy tourism (see Chapter 3.3.7).

Figure 3.3.4.1 An Anaerobic Digestion plant on an Orkney farm. (Photo: Dan van der Horst)

Figure 3.3.4.2 'Wind-to-heat'; Community wind turbines on the island of Westray, providing electric heating to the youth centre and old people's homes right next door. (Photo: Dan van der Horst)

3.3.4
Farming, Fishing and Fermenting: A green Island Economy

Dan van der Horst

Socio-economic arguments are often at the heart of public discourse about the acceptance and appropriateness of developments in renewable energy that are mandated or supported by the state. Economic costs and benefits are easiest to assess at the national scale, where sectoral statistics can capture the transactions involved in building a renewable energy plant (Dvořák et al. 2017). Social costs and benefits should be mostly assessed at the local level, but despite the high level of interest in community energy, there are still significant evidence gaps (Berka, Creamer 2017). The most readily available evidence of local benefit relates to the income that local landowners and communities receive from large commercial wind farms, but this does not address potential indirect economic benefits or economic synergies.

In theory renewables can have negative impacts on tourism, but research to date has found very little empirical evidence of this (Frantál, Kunc 2011; Silva, Delicado 2017). In contrast, the most readily observable synergy between renewable energy development and landscape management is provided by examples of enhanced tourism potential. Many countries have pioneering examples of renewable energy projects that have been designed with picknick places, visitor centres, and observation decks.

Visitor information can be collected through automated visitor counts and the tracking of local spending (e.g. entry tickets, money spent in the shop or café). Some questions can be raised about the scalability of these benefits beyond the handful of leading and iconic projects, and indeed it is useful to consider the tourism potential of non-renewable energy landscapes too (Frantál, Urbánková 2017). Community income from renewable energy plants may also relate to the local landscape by providing funding for local services that can slow down or reverse the threat of rural depopulation (Del Rio, Burguillo 2009). In the absence of this income (i.e. in the absence of the renewable energy plant), it can be envisaged that the future landscape would feature more derelict houses, abandoned farmland, and fewer local shops, schools, and businesses.

In off-grid locations, renewable energy can play an important role in reducing the cost of electricity, displacing noisy, dirty, and expensive diesel generators that were often only run intermittently. Renewable electricity can power modern and convenient cooling and heating systems, typically through ground or air-source heat pumps. Electric heating often provides a displacement of labour-intensive and dirty heating practices based on solid fuels like coal, peat, and woody biomass. This impacts on the landscape in various ways, from the potential disappearance of the sight of black smoke belching from the chimneys of private homes in otherwise beautiful rural locations, to the seizing of traditional but potentially destructive practices of the local harvesting peat or firewood. When local inhabitants are beneficiaries of improved energy

Benefits	Costs and challenges
• Satisfying local energy needs • Better disposal of existing farm waste • Cheaper disposal of shellfish waste • Better local water quality • Income from selling electricity • Job creation • Reputational benefits for the island; more visitors, new residents, more money spent locally, securing local services and jobs	• High up-front investment cost • Limited scalability (farm-size, small islands) • Difficult to cover all costs through private investment when most of the benefits are public • Low energetic value of feedstock • Various permits required • Lack of appropriate expertise in Orkney or UK; learning by doing • Cost of installation is high due to a lack of ready-made components and remote location (transport limitations)

Table 3.3.4.1
Benefits and costs of AD plants. (Source: author´s conceptualisation)

services provided by renewables, they are likely to be supportive of such developments. Finally, smaller renewable energy projects are more likely to be locally and community owned, ensuring stronger levels of public acceptance (Warren, McFadyen 2010; Musall, Kuijk 2011).

The Orkney island of Westray has some 700 inhabitants and is connected to the UK national grid. The key economic activities on the island are farming, tourism, and fishing; the island is home to Orkney's only shell-fish processing plant. Much of the land on Westray is privately owned by farmers who grow crops or keep cattle that has to be housed indoors for some eight months of the year when the weather is cold and often extremely windy. Farmers have to produce and store enough silage (cut grass fermented in large plastic bags) for their cattle to last the whole winter because it is a bulky product that cannot be easily transported. In other words silage is not a tradable commodity but is produced for the purpose of self-sufficiency. Cattle slurry and left-over silage need to be disposed of in a safe manner; just dumping this on the fields (as farmers used to do) leads to nitrification of the fresh water supplies that the island depend on for the production of drinking water. Poor quality drinking water requires extra costs of purification and concerns about drink water quality can put off tourists and visitors and force local inhabitants to buy bottled water which is expensive and adds to the local burden of waste management. The shell-fish processing plant also has a waste management problem; the nearest approved landfill site that is allowed to take this kind of food waste, lies near Inverness; an expensive trip by lorry, involving two ferry rides and some 200 km of driving with the likelihood of requiring an overnight stay.

Cattle slurry and silage are not the best feedstock for an AD plant because they are already half-fermented and have a very low pH which limits the growth of anaerobic bacteria. But they are waste products that the farm needs to deal with, and they have the potential to produce energy that the farm needs in order to operate. The whole-island solution is simple; the shell-fish waste has almost no calorific value but it can be made bacteriologically safe and mixed in with the silage to lower the pH of the feedstock and allow the anaerobic bacteria to flourish and produce more methane. This methane is used in combustion engines to produce both heat and electricity. The heat is used to warm the farm and manage the mesotrophic AD plant, whilst the electricity is sold through the national grid. The winners of this operation are not just the farmer, the local engineer, and the investors in the AD plant, but also the owners and employees of the fish processing plant and all the islanders for having cleaner and more affordable drinking water. Westray's reputational benefits of having this integrated island-waste-energy system and being able to brand itself as a carbon neutral island can translate into more tourist visits and the attraction of new residents, i.e. more children in the local school, more customers in the shop, more volunteers to help out in community activities, including the running of community facilities like the youth centre and the retirement home.

Figure 3.3.5.1
Alqueva Dam in Portugal
(Photo: Bohumil Frantál)

3.3.5
Exploiting Water Courses Sustainably: Seeking a Synergy of Hydropower with the Environment

Luís Silva

Starting to produce electricity in the late 19th century, hydropower has become the main source of renewable energy around the world today (International Hydropower Association 2017). Hydroelectric dams may effectively serve other purposes, such as flood control, water supply, irrigation, and recreational activities. But they are not without environmental and socioeconomic disadvantages and risks (see Table 3.3.5.1), particularly large hydropower projects (e.g. World Commission on Dams 2000).

Yet, smart practices may be adopted to mitigate such risks. The Alqueva Dam in Southern Portugal exemplifies the point. In terms of public participation, an effective and permanent consultation with the inhabitants—about 400 people—mostly affected by the construction of the Alqueva Dam—those living in the old village of Aldeia da Luz—was put into practice. The first surveys on the residents' priorities and needs reach back to 1977–1978 and continued over time. In the process, inhabitants could express their dissatisfaction and negotiate agreements. They not only rejected the proposal to build a wall to protect the old village from the dam's water, the simplest and cheapest solution, but they also made a set of demands in relation to the new village, namely, that it should be constructed close to the old one, that the graves were located before, and that all the houses had a yard. The replication

Advantages	Risks
• Compatible with the environment	• High construction costs
• Uses local resources	• Negative impact on ecosystems and biodiversity
• Provides regulation functions	• Loss of reservoir land
• Generates co-benefits or by-products	• Reservoir siltation
• Reduces CO_2 emissions	• Relocation of adjacent populations
• Provides water supply and irrigation	• Modified water quality and chemistry
	• Emission of greenhouse gases
	• Failure risks
	• Water loss by evaporation
	• Submersion of archaeological and cultural sites

Table 3.3.5.1

Advantages and risks of hydropower dams (as exampled on Alqueva Dam) (Source: author´s conceptualisation)

of the village and the relocation of its inhabitants was accompanied by a multidisciplinary team of experts as well as by professional anthropologists, who designed a museum to preserve the memory of the submerged village and the archaeological artefacts found in the parish.

Despite some limitations, care was also taken with the minimisation of the impacts on vegetation and the environment: centuries old olive trees were moved and planted elsewhere. The company that manages the project (Empresa de Desenvolvimento e Infraestruturas do Alqueva, hereinafter EDIA) also bought the Noudar Natural Park, an area of about 1,000 ha, in order to compensate the loss of habitats in ecosystems of traditional montado, an agroforestry system characteristic of Alentejo. In addition, an elevator was installed in the Pedrógão Dam—built 23 km downstream from the Alqueva Dam in 2006—to allow the circulation of migratory fish species that seek the tributaries of the Guadiana River to spawn.

Moreover, efforts were made to minimise heritage impacts. EDIA funded several archaeological excavations and studies, and more than one hundred archaeological sites were subject to intervention and protection, mainly through their removal and placement elsewhere, as occurred with the Xerez cromlech, but also in situ, as happened with the Lousa Castle, which was safeguarded through a 'shelter' of bags filled with sand and concrete. EDIA also funded a multidisciplinary study of the water mills (about one hundred) submerged by the Alqueva and the Pedrógão lakes, most of which were built to resist to the seasonal floods of the Guadiana River, hence their designation as 'submersion water mills'.

3.3.6
Harvesting Energy and Food: Land Use Synergy via Technological Innovation

Alessandra Scognamiglio

Renewables can be considered a new form of land use, or, even the most important land use of the 21st century (Stremke et al. 2013). It is largely acknowledged that among renewable energies, a feasible short and long-term solution against climate change is the generation of electricity through photovoltaic systems (PV), providing significant environmental benefits in comparison to the conventional (fossil) energy production (De Marco et al. 2014). With the increase of the number of installations new concerns about land use issues are arising (Sacchelli et al. 2016), especially the risk of conflicts of PV with agriculture (Zanon, Verones 2013). Arable lands are ideal areas for PV, because they offer an optimal solar potential, and moreover,

their flat or terraced morphology favours both the easiness and the economy of installation (in areas with a complex morphology, expansive supporting structures and increased design costs would be required).

New land use synergies can contribute to the best possible use of land through technological innovations, also when strict regulations and guidelines that limit the use of arable lands constitute a barrier to the implementation of PV. 'Agrivoltaic' systems are combinations of food crops (at ground level) and PV modules (above layer) on the same land unit (Dupraz et al. 2011; Marrou et al. 2013; Dinesh & Pearce 2016; Majumdar, Pasqualetti, 2018, Malu et al. 2017). In the following, an Italian example is presented.

In Italy, in the early 2000s, a very good feed in tariff (Decreto legislativo 387/2003; D.M: 19.02.2007-Secondo Conto Energia); (D.M. 06.08.2010-Terzo Conto Energia; D.M. 05.05.2011-Quarto Conto Energia) favoured a fast increase of the installations of PV. In an early im-

Figures 3.3.6.1 and 3.3.6.2
Agrovoltaico® system at Monticelli d'Ongina (PC), Italy. On the right: soybean harvesting; on the left: wheat crop in the tillering phase. Images courtesy of REM.

plementation phase, in the absence of specific regulations and guidelines aimed at the preservation of arable lands, many PV systems were located in agricultural areas, with a consequent strong opposition of the public (landscape changes). Afterwards, measures were taken at national, regional and local level: i.e. the supporting incentives for PV systems were stopped in the case of on-ground PV whose power is 1 MW or more (L. 24.032012, n. 27); systems whose nominal capacity is 1 MW or less are allowed, but with some restrictions. For example, the system should cover not more than 10 % of the total available surface; the distance between two different systems has to be at least 2 km if the land plot belongs to one owner. These restrictions are not valid if the agricultural land is abandoned for over five years. Regional and local regulations can be more restrictive than the national ones.

In such context, the Italian company REM, together with the Università del Sacro Cuore, developed an innovative agrivoltaic system, patented as Agrovoltaico® (Figure 3.3.6.1).

The PV modules are installed on a suspended metal structure, placed at the height of 5 m from the ground, and they are anchored to a double axis sun tracking system. The nominal power of the system is 3.3 MWp (about 2.3 ha of modules area), on a land area of 20 ha. The land area occupation ratio (LAOR, the ratio between the area of the modules and the area of land that they occupy, expressed in percentage) is quite low (about 10 %), compared to a standard PV arrays, which is normally about 30–40 % (Dijkman, Benders 2010), but there is an advantage due to the food production (Scognamiglio 2016).

The influence of the dynamic shading on the crops has been recently investigated through a long term simulation (Agrovoltaico vs. open field crops). Results show that when water is a limiting factor for the crop growth the yield is higher for Agrovoltaico systems than in open fields. This indicates that agrivoltaic systems, supporting crop yield, can play a significant role at the energy-food-water nexus (Amaducci et al. 2018).

Figure 3.3.7.1
Educational trail among
wind turbines, Gilboa
Mountain, Northern Israel
(Photo: Bohumil Frantál)

3.3.7
Energy Tourism: Embracing the Visibility of Renewables as an Asset in the Contemporary Place Competition

Bohumil Frantál

An important evaluative criterion for perceiving the landscape impact of different technical objects are the symbolic associations attached to them. Devine-Wright (2005) illustrated this symbolic dimension of perceptions by an example of small-scale hydropower stations in the Peak District National Park (UK), which were perceived in a highly positive manner when associated with the historic water mills found in many places in the park. This example shows how innovation in technology can be perceived positively when representing a continuity between the past and the present. In this sense, a proper spatial targeting and branding strategy may lead to more positive perceptions of wind turbines if they symbolically emphasise a continuity with historic windmills,

which can be found in many regions across Europe.

Many projects have implemented ab initio plans for their use as educational centres or exhibition venues (e.g. Sotavento wind farm, Spain or Whitelee wind farm, UK), they become part of nature trails (e.g. wind farm in Kotka, Finland or biogas plant in Boretice, Czech Republic), and wind turbines serve as observation towers (Westerholt, Germany; Vancouver, Canada), with the aim of utilising their tourist potential. The 'Energy Farm' in a village of Brandbu (Norway) incorporated newer facilities into historical farm buildings originating from the mid of the 18th century, allowing visitors to experience alternative ways and forms of small-scale bioenergy production and consumption. For some municipalities wind turbines or solar plants have become icons, which contribute to the creation of their place brand. Such projects can be considered examples of embracing the visibility of renewable energy facilities not as a problem but as an asset in the contemporary place competition.

These projects are examples of tourist attractions for the emerging 'energy tourism' which became

a new niche of special-interest tourism (Frantál, Urbánková 2017). This involves visits to former, retired, or regenerated sites (e.g. old coal mines or hydropower plants), as well as to still-operational energy sites where some facilities, services, or activities have been provided specifically for tourists' use. Energy tourism overlaps with other types of special-interest tourism, such as cultural and heritage tourism, agricultural tourism, and even adventure tourism.

It has been suggested that energy tourism can play a more important role than being just another segment of industrial tourism contributing to higher levels of brand loyalty or place branding. Energy tourism seems to have a potential to improve people's energy literacy, and environmental awareness, and to change people's energy behaviour towards more sustainable energy citizenship. New forms of energy tourism combining environmental education, the presentation of novel technologies, interactive science experiments, and various outdoor activities (e.g. boat tours, climbing, kiting, biking, or camping) have been designed to attract not just a technologically curious tourism segment but also

family-oriented or adventure-seeking tourists (Jiricka et al. 2010).

There is, of course, a question of cumulative effects and possible thresholds; how many local projects are required to give the local area a positive place brand and when does the increase in projects result in a negative effect on the overal place image. One thing that energy tourism certainly has in its favour is the novelty factor, as it can attract people who want to spend time away from the usual places, to see and to do something different (Frantál & Urbánková 2017). In Iceland, several tour operators offer tourists a mixture of nature and spa tourism with energy education, including visits to geothermal power plants and greenhouse cultivation centres. This kind of tour packages, where visits to energy sites are combined with other tourism activities, seem to have a huge development potential. It is not likely that every new project will become a tourist attraction, yet every project at least symbolically represents a material reconnection to the energy that supports all of us (Pasqualetti 2000).

Figure 3.3.7.2
Small restaurant in a greenhouse using geothermal energy, Friðheimar, Iceland
(Photo: Bohumil Frantál)

3.4

THE POTENTIAL AND VULNERABILITY OF LANDSCAPES FOR SPECIFIC RENEWABLE ENERGY PRODUCTION SYSTEMS

Authors: Adolfo Mejia-Montero, Michael Roth & Bohumil Frantál

3.4.1
The Bond between Landscapes and Renewable Energy

Traditionally renewable energy production systems (REPS) have been spatially assessed solely on the basis of resource distribution. This positivist approach has systematically ignored the inherent geographic bond of REPS with social, environmental and cultural elements of pre-existent landscapes.

Multifunctional landscapes are able to provide a variety of functions, resources, and options over different land uses. However, most of the times land uses are interconnected, creating competitiveness among stakeholders (local residents, farmers, tourists, nature conservationists, etc.), or between certain landscape functions or features (such as e.g. cultural or aesthetic). Prone to conflict are, in this sense, primarily landscapes characterised by heterogeneity, fragmentation, dynamics, and competition of potential users for limited resources and space —typically peri-urban landscapes at the interface of urban and rural spaces (see e.g., von der Dunk et al. 2011).

Landscape changes are therefore, to a large extent, a by-product of market forces and sectorial policies (Mann & Jeanneaux 2009), whose impacts often have the form of unintended consequences (Röhring & Gailing 2005). However, while some landscape functions are regulated (e.g. laws on nature and landscape protection, laws on the protec-

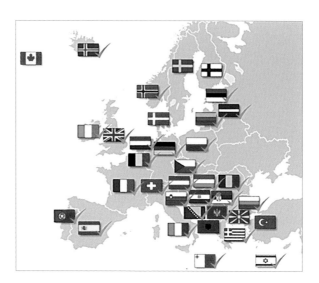

Figure 3.4.2.1
Countries from COST action which participated in this pan-European assessment.

tion and exploitation of mineral resources, etc.), other landscape functions (aesthetic or cultural) usually lack standardisation and are controlled by informal institutions in the form of cultural traditions and norms (Mann & Jeanneaux 2009).

The multifunctionality and heterogeneity of landscapes makes it impossible to create a comprehensive institutional system to regulate all areas, or to reach general consensus on visual fit and compatibility between specific REPS with specific types of landscapes. This is also due to perceptions and evaluations varying in geographical, cultural, and socioeconomic contexts, traditions, and personal experiences.

This is why the present work comprises the first highly participative pan-European expert assessment (99 experts from 28 different European countries) looking at compatibility between different European landscapes, represented by 44 Corine Land Cover (CLC) classes, and different REPS. CLC was chosen as a proxy for landscapes due to its great resolution and the potential homogeneity to asses different European physical landscapes.

The results of this work will hopefully support downscaling to policy-making and guidelines at the national level in Europe, to assist the European energy transition without jeopardising its landscape quality, maximising the multifunctionality of synergies between landscapes and REPS, and improving the overall long-term quality of energy landscapes.

3.4.2
The How and Who of the Participative pan-European Expert Assessment

In order to assess the compatibility between CLC and REPS an assessment matrix and questionnaire were developed, with both categories displayed as rows and columns respectively, using as inspiration the work of Burkhard et al. (2009, 2012) assessing compatibility between Ecosystem Services (ESS) and CLC. This questionnaire was then shared with the RELY network of experts, for them to rank each combination of CLC and REPS depending on their professional perception of compatibility, including options for respondents who didn't feel confident enough to asses specific values (0 = Not relevant, 1 = Completely compatible, 2 = Rather compatible, 3 = Neutral, 4 = Rather conflicting, 5 = Absolutely conflicting, 9 = I don't know/can't judge).

In a parallel fashion and attached to this document, a second matrix to assess compatibility between Ecosystem Services (ESS) and REPS included.

To enrich the analysis of results each respondent was assigned a number, country of procedence, and area of knowledge: technology (engineers, physicist, etc.), people (sociologists, human geographers, etc.), landscape (landscape architects, landscape managers, etc.), and multi (geographers, energy planning, etc.). A total of 99 expert responses (64,251 data cells), from 28 European countries,

Classification	Key color	Description
High consensus	White ☐	50 % of respondents agrees it's compatible (1 and 2), conflicting (4 and 5), or neutral (3) and the modal is in the same range of values, even taking into account non-meaningful responses (0, 9, or blank space).
High consensus of meaningful responses	Yellow ☐	50 % of respondents agrees it's compatible (1 and 2), conflicting (4 and 5), or neutral (3) and modal is in the same range of values. Discarding non-meaningful responses (9, 0, or blank space).
Consensus around neutral effects	Light red ☐	A significant mMajority choose neutral (3) and a leveled ratio exists between compatible (1, 2) and conflicting (4, 5) responses.
Contentious majority	Mars red ■	The majority of significant responses. This category also included the case where two conflictive responses had the same number of responses (e.g. where 1 and 4, 1 and 5, and vice versa had the same number of responses); for these highly contentious cases where no apparent consensus criteria seemed to apply, the data cell wsd marked as -99.

Table 3.4.2.1
Classification, key colour, and description for assessing agreement among the expert community around compatibility between landscapes and REPS

Israel, the USA, and China (See Figure 3.4.2.1) were collected and transcribed into two different excel documents (one for the CLC-REPS and another for the ESS-REPS assessment), comparing around 1 % of the data cells to give certainty about the transcription quality.

Once transcribed, a SPSS frequency analysis was carried out as a first phase to later use the results on a GIS map showing the compatibility between REPS and CLC on a spatial setting. The SPSS analysis was focused on the frequency values of the (ordinal) responses instead of mean values to provide a more detailed analysis on the level of agreement among experts through a key colour classification, shown in Table 3.4.2.1.

These different categories for assessing agreement based on frequency values open future opportunities to quantitatively assess how expert agreement fluctuates around different combinations of CLC and REPS.

3.4.3
A Spatial Representation of the Experts' Assessment

The final result of this assessment consists of the matrix initially sent to experts now containing the mean values of compatibility between CLC and REPS from the SPSS frequency analysis of all 99 expert responses (Table 3.4.3.1). Moreover, each data cell includes information regarding the level of agreement among the expert community, represented by the key colour classification.

Finally, in order to create the 11 ArcGIS layers which spatially represent the expert-based compatibility assessment among the 11 RESP and 44 different CLC classes GIS, the following colour code shown on Table 3.4.3.2 was created.

The resulting maps are shown in the next images. However, it's worth noticing that the purpose of these maps is only to show a general overview of results at a pan-European scale; a more in depth analysis can be carried out with the GIS tool.

Table 3.4.3.1
Final matrix for assessing compatibility between CLC and REPS, filled in with results from the SPSS analysis and key colours for agreement criteria

Landscape type			Renewable energy production system										
CORINE nomenclature		Code	11	12	21	31	32	41	42	45	51	61	62
Level 1	Level 2	Level 3	Large wind power on-shore	Large wind power off-shore	Marine energy	Small and micro hydropower	Large hydropower	Large solar PV ground-mounted	Lage solar PV on-roof	Large solar PV thermoelectric	Large geothermal	Biomass production	Biogas station (AD plants)
1. Artificial surfaces	1.1. Urban fabric	1.1.1. Continuous urban fabric	5	0	0	3	5	5	1	1	4	2	4
		1.1.2. Discountinuous fabric	4	0	0	2	4	4	1	1	2	4	4
	1.2. Industrial, commercial and transport units	1.2.1. Industrial or commercial units	1	0	0	2	2	2	1	1	1	2	1
		1.2.2. Road and rail networks and associated land	2	0	0	2	5	2	1	1	2	1	1
		1.2.3. Port areas	1	0	0	1	5	1	1	1	1	1	1
		1.2.4. Airports	5	0	0	5	5	1	1	1	5	1	3
	1.3. Mine, dump and construction sites	1.3.1. Mineral extraction sites	1	0	0	1	1	1	1	1	1	1	1
		1.3.2. Dump sites	1	0	0	1	5	1	1	1	1	1	1
		1.3.3. Construction sites	1	0	0	3	5	3	1	1	1	3	5
	1.4. Artificial non-agriculture	1.4.1. Green urban areas	5	0	0	2	5	5	1	5	5	4	5
		1.4.2. Sport and leisure facilities	5	0	0	1	5	5	1	5	5	5	5
2. Agricultural areas	2.1. Arable land	2.1.1. Non-irrigated arable land	1	0	0	2	5	5	1	1	4	1	1
		2.1.2. Permanently irrigated land	4	0	0	1	4	5	1	5	4	1	1
		2.1.3. Rice fields	4	0	0	1	5	5	1	4	5	5	1
	2.2. Permannet crops	2.2.1. Vineyards	4	0	0	3	5	5	1	2	5	5	-99
		2.2.2. Fruit trees and berry plantations	4	0	0	2	5	5	1	4	4	1	2
		2.2.3. Olive groves	4	0	0	3	5	5	1	4	4	1	1
	2.3. Pastures	2.3.1. Pastures	1	0	0	1	3	5	1	-99	1	1	2
	2.4. Heterogeneous agricultural areas	2.4.1. Annual crops associated with permanet crops	2	0	0	1	5	5	1	1	4	1	1
		2.4.2. Complex cultivation patterns	4	0	0	2	4	5	1	-99	4	1	2
		2.4.3. Land principally occupied by agriculture, with significant areas of natural vegetation	4	0	0	1	4	5	1	4	4	1	2
		2.4.4. Agro-forestry areas	4	0	0	1	4	5	1	5	4	1	1
3. Forests and semi-natural forests	3.1. Forests	3.1.1. Broad-leaved forest	5	0	0	1	5	5	1	5	4	1	5
		3.1.2. Coniferous forest	5	0	0	2	5	5	1	5	4	1	5
		3.1.3. Mixed forest	5	0	0	2	5	5	1	5	4	1	5
	3.2. Shrub and/or herbaceous vegetation association	3.2.1. Natural grassland	2	0	0	1	3	5	1	3	5	2	4
		3.2.2. Moors and heathland	2	0	0	1	5	5	1	3	5	5	5
		3.2.3. Sclerophyllous vegetation	2	0	0	1	4	5	1	2	5	2	5
		3.2.4. Transitional woodland shrub	2	0	0	1	4	5	1	4	3	2	4
	3.3. Open spaces with little or no vegetation	3.3.1. Beaches, dunes, and sand plains	5	0	0	5	5	5	1	1	5	5	5
		3.3.2. Bare rock	1	0	0	1	2	5	1	3	3	5	5
		3.3.3. Sparsely vegetated areas	2	0	0	2	3	2	1	2	2	5	5
		3.3.4. Burnt areas	2	0	0	1	2	2	1	1	1	5	5
		3.3.5. Glaciers and perpetual snow	5	0	0	3	5	5	1	5	1	5	5
4. Wetlands	4.1 Inland wetlands	4.1.1. Inland marshes	5	0	0	4	5	5	1	5	1	5	5
		4.1.2. Peat bogs	5	0	0	5	5	5	1	5	5	5	5
	4.2. Coastal wetlands	4.2.1. Salt marshes	5	0	0	5	5	5	-99	5	5	5	5
		4.2.2. Salines	5	0	0	5	5	5	5	5	5	5	5
		4.2.3. Intertidal flats	5	0	0	2	5	5	5	5	5	5	5
5. Water bodies	5.1. Continental waters	5.1.1. Water courses	0	4	2	0	0	0	0	0	0	0	0
		5.1.2. Water bodies	0	4	1	0	0	0	0	0	0	0	0
	5.2. Marine waters	5.2.1. Coastal lagoons	0	4	4	0	0	0	0	0	0	0	0
		5.2.2. Estuaries	0	4	2	0	0	0	0	0	0	0	0
		5.2.3. Sea and ocean	0	1	1	0	0	0	0	0	0	0	0

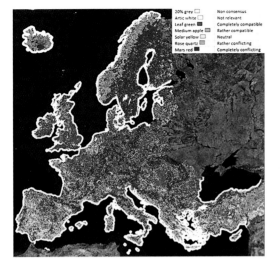

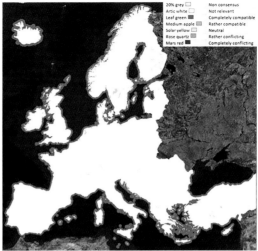

Figure 3.4.3.1
Map showing the compatibility between large wind power onshore and landscape (represented by CLC classes) according to the results from the expert compatibility assessment

Figure 3.4.3.2
Map showing the compatibility between large wind power offshore and landscape (represented by CLC classes) according to the results from the expert compatibility assessment

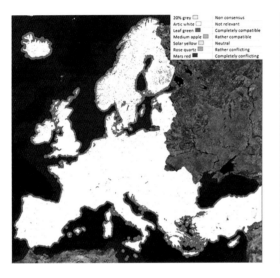

Figure 3.4.3.3
Map showing the compatibility between marine energy and landscape (represented by CLC classes) according to the results from the expert compatibility assessment

Figure 3.4.3.4
Map showing the compatibility between small and micro hydropower and landscape (represented by CLC classes) according to the results from the expert compatibility assessment

Number	Color	Criteria
-99	20% grey	Non-consensus
0	Artic white	Not relevant
1	Leaf green	Completely compatible
2	Medium apple	Rather compatible
3	Solar yellow	Neutral
4	Rose quartz	Rather conflicting
5	Mars red	Completely conflicting

Table 3.4.3.2
GIS colour code for the CLC and REPS compatibility assessment

Figure 3.4.3.5
Map showing the compatibility between large hydropower and landscape (represented by CLC classes) according to the results from the expert compatibility assessment

Figure 3.4.3.6
Map showing the compatibility between large solar PV ground-mounted and landscape (represented by CLC classes) according to the results from the expert compatibility assessment

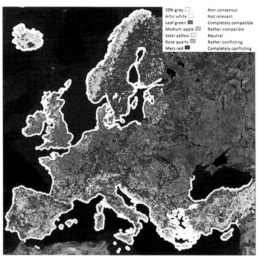

Figure 3.4.3.7
Map showing the compatibility between large solar PV on roof and landscape (represented by CLC classes) according to the results from the expert compatibility assessment

Figure 3.4.3.8
Map showing the compatibility between solar thermoelectric and landscape (represented by CLC classes) according to the results from the expert compatibility assessment

171

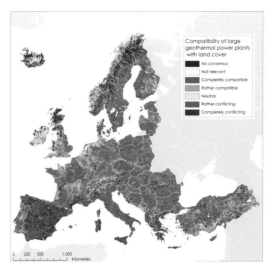

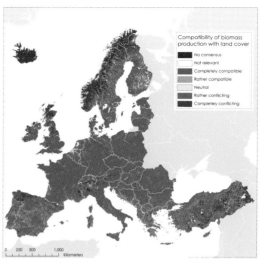

Figure 3.4.3.9
Map showing the compatibility between large geothermal and landscape (represented by CLC classes) according to the results from the expert compatibility assessment
Figure 3.4.3.10
Map showing the compatibility between biomass production and landscape (represented by CLC classes) according to the results from the expert compatibility assessment

Figure 3.4.3.11
Map showing the compatibility between Biogas stations (AD plants) and landscape (represented by CLC classes) according to the results from the expert compatibility assessment

3.4.4
Discussing Limitations and Future Improvements for this Assessment

Naturally an assessment at the European level has its limitations, shortcomings, and potentials criticisms, many of them coming from respondents or the group responsible for the research themselves. For instance, it could be worth questioning if individual experts possess the capabilities to assess compatibility from such a wide diversity of land covers, renewable energy technologies, and ESS, or if this assessment creates 'artificial knowledge'. This was pointed by some of the members of the COST RELY participants when receiving the question-naire, even after the questionnaire was modified to include an option for self-assessment as unfit for assessing specific values of the matrix.

As it was previously stated, one of the objectives for this assessment is to provide a tool for downscaling policy-making and supporting REPS on the European level. However, it is necessary to take into account the existing differences among national guidelines, benchmarks, and policies from different member countries or regions.

On the same note, the assessment does not take into account renewable energy resources, availability. This would be recommendable for planning and/or policy-making to underpin the potential best areas for project development, based on both resource availability (e.g. wind speed distribution, geothermal potential, or insolation) and landscape compatibility with RESP.

Lastly, it remains still open to discussion whether the results of these maps need to be evened out to avoid the salt-and-pepper effect, and to identify larger areas of study for future research or implementation of REPS on a European regional scale.

3.4.5
Next Steps towards
Research and Publications

The data collected from this expert assessment lays the path for a number of future publications related to the reciprocative analysis of CLC, REPS, and ESS, respectively. Three different ideas and working plans for future research with the data presented have been already highlighted:

A REPS and CLC compatibility (spatial and cluster analysis) study departing from a theoretical framework around REPS, land use, conflict, and compatibility due to the spatial impacts of REPS, deepening on the descriptions of materials and methods (survey, SPSS, and GIS analysis and data) and focusing mainly analyzing maps of REPS and CLC compatibility, different levels of consensus among experts, and potential influence that groups such as technologies, land cover classes, region, or area of knowledge might have over the results.

A RESP and ESS compatibility (cluster analysis) study focused on the second matrix of the questionnaire between REPS and ESS. The analysis would depart from a theoretical framework around REPS, ESS, conflict and compatibility drawn from both empirical evidence and theory. The consensus matrix will be used, as in the REPS-CLC previous paper, to explore levels of agreement among different groups.

A triangulation study between RESP, ESS, and CLC to inform good practice on REPS by using the previous papers on compatibility between REPS-CLC and RESP-ESS, in combination with existent research (Burkhard et al. 2009, 2012) on compatibility between ESS-CLC. The aim would be to triangulate results to analyse how these three different concepts could better complement on a spatial setting, using practical examples of smart practices for REPS to explore planning strategies to strengthen compatibility or overcoming incompatibility among REPS, CLC, and ESS.

This assessment could also be repeated in the future in order to analyse landscape dynamics, comparing this with the results of the current assessment by using a comparative approach to study both the physical (update of CLC) and psychological (repetition of the survey) landscape dynamics.

4

SOCIO-CULTURAL ASPECTS OF RENEWABLE ENERGY PRODUCTION

Deploying renewable energy infrastructure not only affects landscapes, but also people and communities living close to its sites. Humans tend to establish a sense of belonging to the community of people with which they directly interact, but also to the landscape that represents that community. Changes of this local landscape or lifeworld through people or institutions from outside, such as RE projects, are therefore felt by them as an intrusion if this intervention is not adequately introduced in the local context. This sense of intrusion is often expressed with concerns about place disruption, landscape encroachment, or disesteem of local autonomy, which often lead to local opposition against such projects. It is therefore, as recent studies (Devine-Wright 2010, Wolsink 2006) have shown, not a simple 'not in my backyard' reflex that makes locals oppose RE projects, but often a neglect of socio-cultural aspects in the planning of RE projects. Our goal was to understand how renewable energy projects can be optimally integrated in local contexts, not only striving for a minimum of local resistance but rather for a maximum gain of social support. The principal outcome was to outline the requirements of a tool-box for effective RE planning. This outcome was achieved pursueing the following procedure.

- The first step was to carry out a literature review to identify the range of factors that affect local acceptance of or local supportfor renewable energy projects. A concise summary of the literature review which introduces the research results is presented in Chapter 4.1.
- The second step was to create an inventory of innovative practices of participatory RE planning across Europe to under-

stand the state of planning practice. The analysis of innovative practices enabled the identification of challenges in renewable energy planning. This underpinned the planning of the content of the surveys reported in Chapter 4.2.

- To provide an improved understanding of RE planning in different contexts, evidence was collected of existing guidelines and tool-boxes of participatory RE planning. The process of evidence-gathering included analysis of legal systems of RE planning in Switzerland compared to those in Poland, undertaken by a short-term scientific mission.
- To produce an overview of participatory RE planning in Europe, a first online semi-standardised questionnaire was developed and promoted, to which responses were obtained from experts from 36 countries. Based upon the results of the first online survey, and findings of recent research studies on innovative participatory planning approaches, a second online survey was developed which focused on expert assessments of the quality of wind energy planning in European countries. This second survey addressed issues of public acceptance, the added-value of innovative planning tools, and their compatibility with the national planning culture. The second online questionnaire was completed by the members of the COST Action RELY, their national contacts, and members of the organisation Wind Energy Europe. Findings from this survey are presented in Chapter 4.1.
- A synopsis of the findings from the analysis of the two online surveys, the inventory of innovative practices, the collection of tool-

boxes, and the literature reviews enabled the design of a tool-box for participatory RE planning which is described in Chapter 4.3.

- The knowledge-base for the inventory can be expanded upon. A more robust update of this tool-box will require extensive and systematic empirical research on the acceptance of renewable energy projects in different European contexts, and more extensive expert surveys on RE planning in Europe which can expand on the prototype published in this section.

The output has provided a good platform for the identification of gaps in the research field and for the setting-up of a pan-European database of recent research findings, practices, and trends. Comparisons between countries in relation to the state-of-the-art and innovations in the field of socio-cultural aspects of renewables were undertaken at meetings held by RELY. The outcome has been the identification of new research questions which also address issues of methods and networking. The relationship between technology, society, and landscape or natural assets are not easily generalised but have to be viewed simultaneously on different (spatial) levels and across diverse historical political background.

Matthias Buchecker & Dina Stober

4.1

PUBLIC ACCEPTANCE OF RENEWABLE ENERGY PROJECTS: A FOCUS ON WIND ENERGY

Monika Suškevičs, Matthias Buchecker, Sebastian Eiter, Dina Stober, Igor Kuvač, Berthe Jongejan, Stanislav Martinat & Cheryl de Boer

4.1.1
Introduction

Different countries in Europe have adopted policies which aim to increase the share of renewable energy sources. At the pan-European and EU level, coordinated efforts in this regard exist as well. In its roadmap for moving to a competitive low-carbon economy in 2050 (COM 2011), the EU has set several ambitious goals for achieving a resource-efficient Europe. For instance, in February 2011 the 'European Council reconfirmed the EU objective of reducing greenhouse gas emissions by 80–95 % by 2050 compared to 1990' (COM 2011). This means that innovative approaches are needed for the transition to RE systems.

In addition to technological innovations, the transition to renewable energy sources requires efforts from multiple stakeholders. One of the challenges here is the *acceptance of new technologies* by different parties, such as businesses, local communities, or the public at large (Haas et al. 2004, van Rijnsoever et al. 2015). In research literature, acceptance of different types of RE technologies has been discussed, such as hydropower (Cohen et al., 2014), wind (Ellis and Ferraro 2016, Fournis and Fortin 2017), solar and photovoltaic (Carlisle et al. 2014, Heras-Saizarbitoria et al. 2011), tidal sources (Devine-Wright 2011), and biomass (Emmann et al. 2013). This research has concluded that the general public acceptance of renewable energy systems has increased during the past decades. However, the concerns and factors behind acceptance are still debated.

In Europe, private and public interest to increase the wind energy sector is growing. In 2016, nearly 9 % of Europe's electricity production was from wind—a third of the total renewable energy production (ENTSO-E 2016). At the same time, wind energy is among the most controversial sources of renewable energies, most probably due to its highly perceivable visual impacts on landscape and other environmental concerns (Leung and Yang 2012, Pasqualetti 2011).

There is as of yet no certainty in the scientific community regarding what the most important or common concerns are related to acceptance or non-acceptance of wind energy. Currently, we have a rather good overview of wind energy acceptance and related issues across different European countries, focusing either at (sub-)national or local level, e.g. Ellis and Ferraro (2016). However, a broader European perspective is largely missing, especially regarding Southern and Eastern Europe.

In this chapter, we first briefly discuss the *role of different acceptance concerns*, particularly in relation to wind energy, and then *illustrate them with examples from a recent expert survey in Europe*, as part of the COST Action RELY. We conclude with a *set of general recommendations for planning practice* of wind energy.

4.1.2
Concerns behind Non-acceptance

One can think of the notion of 'acceptance' related to renewables in many ways, e.g. Wolsink, (2010),Wüstenhagen et al. (2007). Usually, acceptance is understood mostly with regard to local residents' acceptance in terms of a given renewable energy project (Wolsink 2010). For instance, acceptance has been defined as 'a favourable response related to the proposed or *in situ* technology by members of a given social unit' (Upham et al. 2015). In acceptance studies, though, often the opposite, i.e. non-acceptance, 'opposition' or 'resistance' against wind energy is more frequently explored (Fournis and Fortin 2017). The empirical part of this article also focuses on non-acceptance. Concerns about renewable energy systems or projects acceptance are embedded in the wider socio-economic and biophysical context. Acceptance problems thus have many facets. These may be con-

cerns about income generation, beliefs in economic benefits (for others) or losses (for oneself), controversial siting aspects, depending on factors such as the type of technology used, type of landscape involved, or how the process of siting is carried out (Carlisle et al. 2014, Cowell et al. 2011, Wolsink 2007, Wüstenhagen et al. 2007).

Based on a literature review, we divide wind energy acceptance concerns into three broad groups: landscape and identity, perceived impacts, and governance.

Landscape and Identity

Aesthetic and visual aspects of the landscape are the most commonly voiced and also the most studied concerns regarding non-acceptance of wind energy projects (Devine-Wright 2005, Firestone et al. 2015, Pasqualetti 2011). People tend to develop bonds with their local socio-physical environment—also known as *place identity* (Devine-Wright 2011). Changes in this environment are often perceived as negative disruptions. The change not only depends on the objective spatial proximity of developments but also on how such visibility is perceived by people, e.g. 'out of sight' developments are more acceptable (Jones and Richard Eiser 2010).

Environmental Impacts

Environmental and other *impacts* are often referred to as concerns when it comes to wind energy acceptance (Dai et al. 2015, Leung and Yang 2012, Mann and Teilmann 2013, Saidur et al. 2011). The impacts may, for instance, concern biodiversity, such as the collision of birds or bats with wind turbines, or other land uses, such as agricultural land (e.g. dissection of fields) or forests (e.g. fragmentation) (Steinhäußer et al. 2015). Local residents are also often only concerned about the potential negative health effects of wind farms, such as noise pollution, shadow flicker, or electromagnetic fields (Ellis and Ferraro 2016).

Governance

Procedural aspects have been found to be highly important in affecting acceptance (Zoellner et al. 2008). This means that the fairness perceived by stakeholders and the public is important for acceptance, and procedural aspects need to be taken into account when planning wind energy projects. It also means trust-building with the stakeholders who are most affected by the projects, or who are important for some other reason, such as having the potential to provide local knowledge (Ellis and Ferraro 2016).

Social justice, i.e. how the costs and benefits of a decision, e.g. a policy or plan, are divided among the affected groups and persons, is another important predictor for acceptance. For example, people tend to support renewable energy projects which are believed to have community economic benefits (Bidwell 2013).

Wind energy projects are often implemented via spatial planning procedures. We assume that these procedures are of key importance in affecting how wind energy is perceived (Stead 2013). Potential differences in how wind energy acceptance concerns are perceived may be rooted not only in administrative and legal contexts but also in the ways these formal aspects are implemented (Inglehart and Welzel 2010, Jauhiainen 2014, Othengrafen 2010). We do not analyse responses from individual countries, but take a wider perspective in Europe, by grouping Europe into four large supra-national regions (Figure 4.1.3.2). We elaborate on participatory planning of RE and innovative examples from different countries in more detail in Chapter 4.2.

4.1.3.
Wind Energy Acceptance across Europe: What Do Experts Think?

Survey

We initiated, designed, and conducted an online expert survey in summer 2017 among the members of the COST Action. The participants were also been encouraged to spread the survey link among their national personal networks. The Action consists of scientists and practitioners with a background in relevant fields, such as (applied) geography, landscape planning, architecture, environmental sciences, and renewable technologies. Therefore, we assumed to address a set of knowledgeable persons who could give an informed opinion about the situation in their countries. The survey consisted of different topics around planning and acceptance, such as strategic planning, community initiatives, and local involvement and acceptance.

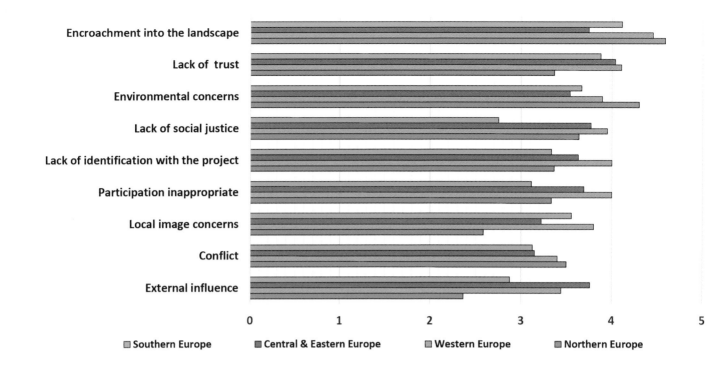

The questionnaire included of mainly closed-ended questions, where the experts had been asked to assess items on a 5-point Likert scale (1 being the lowest and 5 the highest value). We pre-tested the questionnaire in spring 2017 and revised it accordingly. Altogether, we received 108 responses from 33 countries, which represent EU-27 but also EU-candidate and adjacent countries across Europe (plus Israel). We did not aim for the sample to be a strictly representative for the countries, as we did not aim to compare results between single countries. We conducted an analysis of expert responses based on descriptive statistics (ANOVA-tests in Statistical Program for Social Science [SPSS]), focusing on acceptance or non-acceptance concerns.

European Regions: A Framework for Analysis

As elaborated above, we hypothesised that the national socio-political contexts and especially the spatial planning aspects may affect wind energy acceptance. To analyse wind energy acceptance concerns we grouped the countries that have received responses from our survey into *four European supra-national regions* (Figure 4.1.3.2), based on the spatial planning literature, e.g. Knieling and Othengrafen (2015), Othengrafen (2010).

Reasons for Non-acceptance of Wind Energy in Europe

The ways in which reasons for acceptance issues are perceived share several similarities across Europe. The most relevant overall reasons for non-

Figure 4.1.3.1
Perceived reasons for non-acceptance of wind energy development in Europe. Mean values from the questionnaire, on a scale from 1 (least relevant) to 5 (most relevant).

Socio-cultural Aspects of Renewable Energy Production

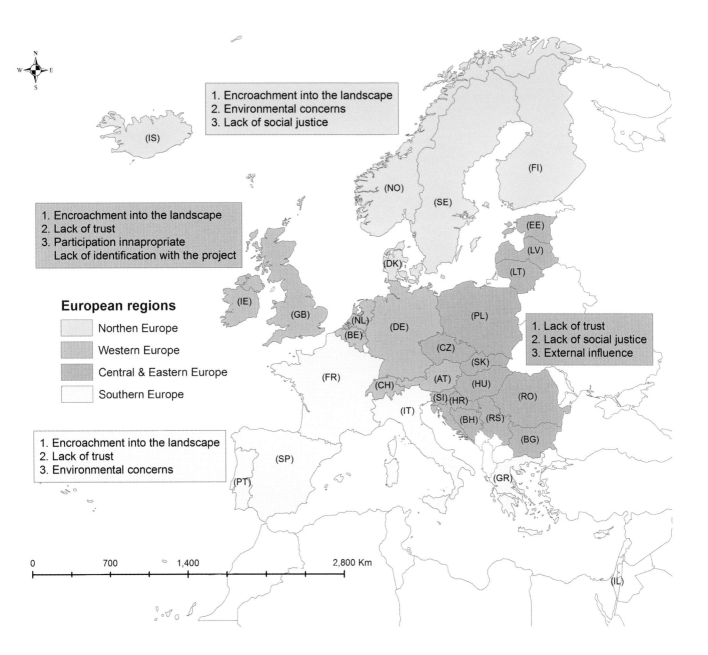

The following labels and legend appear on the map:

1. Encroachment into the landscape
2. Environmental concerns
3. Lack of social justice

1. Encroachment into the landscape
2. Lack of trust
3. Participation innapropriate
   Lack of identification with the project

European regions

- Northen Europe
- Western Europe
- Central & Eastern Europe
- Southern Europe

1. Encroachment into the landscape
2. Lack of trust
3. Environmental concerns

1. Lack of trust
2. Lack of social justice
3. External influence

0   700   1,400   2,800 Km

acceptance of wind energy development are believed to be *encroachment into the landscape, lack of trust* in the process or decision-makers, and *environmental concerns,* followed by *concerns over social justice* and *lack of identification with the project* (Figure 4.1.3.1).

There are however some slight differences also in various parts of Europe. These differences concern most especially how *external influence, local image,* and *social justice* are perceived (in terms of statistically significant differences, at p<0.1[1]). The Central and Eastern European (CEE) region stands somewhat out here, as issues of social and procedural justice (e.g. trust, participation inappropriate) tend to outweigh landscape concerns (Figure 4.1.3.2). The reasons for these trends may be partly ground-

ed in the socio-economic and political contexts of these countries, i.e. differences in planning cultures (Othengrafen 2010): for some further details, see Chapter 4.2.

However, when interpreting these results, in addition to the ranking of reasons (Figure 4.1.3.2), one has to bear in mind also the absolute mean values and their differences among regions. For example, environmental concerns seem to be important in Northern and Southern Europe; however, the absolute mean values expert assigned to them differ almost by one point (Northern Europe: mean = 4.30; Southern Europe: mean = 3.67). We can observe similar patterns across other acceptance concerns too, e.g. *lack of social justice* tends to be problematic all over Europe, although more im-

Figure 4.1.3.2
Highest ranked reasons for non-acceptance of wind energy: some differences across Europe, map prepared by S. Martinat

portantly in the CEE region (mean = 3.77) and in Western Europe (mean = 3.95). This partly aligns with earlier research suggesting that concerns about inappropriate participation are specific to Eastern and some Southern European countries, possibly due to their relatively young democracies (Palo-niemi et al. 2015). However, it also adds to this body of evidence by suggesting that aspects concerning decision-making of wind energy projects, e.g. related to trust, and to the distribution of costs and benefits of wind energy projects, are important in other parts of Europe as well.

We further asked the experts to give an estimate about the groups who usually do not accept wind energy projects. The responses indicate that mostly *local* groups are opposed, who are sometimes organised specifically against specific wind energy projects. We can broadly see two trends why such opposition occurs: a) specific concerns about *environment or landscape*, e.g. nature conservation or heritage protection groups, or b) *specific interests represented by local stakeholder groups*, e.g. estate owners' associations, or tourism developers.

4.1.4.
Conclusions and Implications

We know that a wider range of concerns relate to wind energy acceptance. The literature has however paid little attention to cross-country comparative or supra-national perspectives when analysing acceptance issues in Europe. The COST RELY survey addressed a wider, pan-European supra-national view. The results suggest that in wind energy development, landscape concerns are important all over Europe, but socio-economic problems are significant too, especially in eastern and southern parts of Europe.

This survey is however limited to experts' opinions and thus cannot reflect the views of other stakeholders or the wider European public. As we took a broad-scale approach in our study, based on our data, we cannot suggest specific measures for addressing acceptance problems in particular regions. Instead, we can suggest some general aspects which planners and policy-makers should be aware of when designing wind energy policies and implementing different planning approaches.

- **Be aware of diversity**. In *all regions* and different planning contexts, there is a *diversity of acceptance concerns* present, i.e. never only just one problem. Thus, planners should be aware of such diversity and not (over)focus on one concern.

- **Adapt, don't adopt**. Planners should consider *problem* and *context* in a particular location. This is intuitively understandable, but still often not practiced (Reimer and Blotevogel 2012). This should be part of mutual learning between planners and stakeholders.

- **Planning process design matters**. Planning should not be too much object-focused, i.e. on landscape concerns or renewable energy technology. Although landscape was the most important concern in all regions, procedural aspects of wind energy, e.g. trust, allocation of costs and benefits, and meaningful participation matter strongly overall across Europe as well.

- **Use knowledge at detailed scales**. Our analysis was based on a broad-scale view of acceptance problems. If specific measures for addressing the problems are to be built, *more detailed analyses on country, sub-national, or local scales* are needed.

For a more detailed discussion on different measures (tool-boxes) for improving renewable energy acceptance, see Chapter 4.3.

---

1   Due to the small sample size in this survey the p-value was agreed to be higher than conventionally.

PARTICIPATORY PLANNING OF RENEWABLE ENERGY WITH A FOCUS ON BEST PRACTICE

Dina Stober, Matthias Buchecker, Berthe Jongejan, Sebastian Eiter, Monika Suškevičs, Stanislav Martinat & Igor Kuvač

4.2.1
Introduction

Motives behind local resistance to the implementation of renewable energy projects have often been attributed to 'not in my backyard' (Van der Horst 2007, Burningham et al. 2015), explained as being egoistic protectionist attitudes against a local contribution to societal goals. However, recent studies have shown that local resistance often reflects a reaction to an intrusion into the local life space/world experienced as a disruption to place attachment (Devine-Wright 2017). These negative reactions were found to be mainly influenced by procedural aspects, that is a perceived lack of power of the local public in project planning (Wolsink 2007, Devine-Wright 2011). Concerns about landscape change and perceived loss of landscape quality regarding renewable energy infrastructure have featured in opposition campaigns in many European countries (Upreti and Van der Horst 2004, Burningham et al. 2006, Newig and Fritsch 2009, Wolsink 2010, Devine-Wright 2017, Kienast et al. 2017), even though renewable energy can facilitate sustainable development (Stremke and van den Dobbelsteen 2012, Dale et al. 2016, Pasqualetti and Stremke 2017).

For the last thirty-five years, the concept of 'public participation' as a key element of societal acceptance has been widely used (see Kruse et al. 2018). The concept of participation from the very beginning was perceived as a 'good thing' (Arnstein 1969), although one can find opposing opinions like Newig & Fritsch (2009) claiming that public participation can be counterproductive in achieving expected goals.

Stakeholder and public participation is considered to be an essential part of decision-making and managing issues of public concern (Koirala et al. 2018). It includes aspects such as investing in relationships and channels for the sharing of information, transparent communication, consensus building, and informed decision-making. There are many different levels of public participation, from basically informing the public to effective public involvement in decision-making (Langer et al. 2017).

In Europe, public involvement in environmental planning has been secured in specific situations by the Aarhus Convention (United Nations Economic Commission for Europe 1998). This Convention states that the affected public shall be informed at an early stage about a planned activity, including expected impacts on the environment, actions to mitigate these impacts, and an overview of alternatives considered (Art. 6, Parag. 6). People are to be given the opportunity to submit objections in a written form or express them in a public hearing (Art. 6, Parag. 7). In addition to the Arhus Convention, the EIA directive and SEA directive are highly relevant for RE planning, and they include a strong mandate for public participation, as well as addressing landscape, environmental, and cultural impacts of RE projects.

4.2.2
Standard Participatory Planning in Europe

Findings of studies about participatory planning RE in some European contexts suggest that the level of involvement in this domain is rather low, being mainly consultative, limited to local level decisions, and consigned to the second phase of spatial planning processes (Devine-Wright 2011, Späth et al. 2014, Castan Broto 2017). Furthermore, comparisons of different spatial planning practices across Europe have been published, introducing the term 'planning culture' (Reimer and Blotevogel 2012). We established a research framework for evidence-based, qualitative research of participatory planning of renewables in Europe following the concept of 'planning culture'.

In order to reveal the state of participatory planning in Europe in renewable energy production, we conducted an online survey about participatory planning of RE projects involving the COST Action RELY members between May and August 2016. Sixty members responded (N=60) and the geographical coverage was rather high. Responses came from 30 out of 36 COST Action RELY member countries. The survey focused on the national planning situation in the different European countries. It included questions on the forms of communication about the projects, options for the public to provide feedback, the involvement of local stakeholders in the planning process, and implications for the implementation of the project,

	Stakeholder participation	Local authority approval	Financial participation	Compensation measures	Countries
Type A	•	Ø	•	Ø	BE, CH, DK, DE, IE, EL, HU, NL, NO, AT, PL, PT, FI, SE, UK
Type B	Ø	•	•	Ø	FR, IS, IT, SP
Type C	•	Ø	Ø	•	CZ, EE, HR, LT, ME, SI
Type D	Ø	Ø	Ø	Ø	AL, BA, MT, RS, TR

Table 4.2.2.1
Strategy types for integrating RE projects in local context

opportunities for the local population to influence decisions, and options available to local inhabitants to participate in RE projects financially.

The survey revealed that:
- In most European countries, inhabitants are informed about planned RE projects at least in a written form, with public hearings conducted depending on the developer and the size of the project.
- Feedback from the public can usually have direct consequences for the implementation or design of RE projects.
- In most Western European countries, and a few in Eastern Europe, stakeholders are involved in reasoned discussion regarding the planning of larger RE projects.
- In nearly all Western European countries, and several in Eastern Europe, formal stakeholder involvement has considerable implications for the implementation of RE projects.
- Only two European countries (Switzerland and Poland) have a mandatory voting mechanism for approving larger RE projects, but in a considerable number of Western and in some Eastern European countries, petitions and strong resistance can provoke voting.
- In many European countries local governments have a power of veto. In some countries, this local veto can compensate for the lack of stakeholder involvement.
- Individual financial participation is identified in a few European countries such as Denmark, Greece, Germany, Poland, and Portugal. This appears to be the most significant instrument of participation in Denmark, Greece, and Portugal.
- Municipalities are often involved in the financial arrangements; hydropower plants are widespread examples of such practices.
- Compensation measures are the most common form of sharing benefits.
- The promotion of energy regions, in particular in countries with powerful regional authorities, is another widely used strategic instrument to foster local implementation of RE projects.

In summary, the findings from the survey show that each European country has introduced a mix of measures to control (through public participation) and share the benefits of RE projects, and in so doing increase the acceptance of the projects. These practices can be classified as follows (Table 4.2.2.1):

Type A: Using a mix of measures, including both aspects, participation within decision-making process and sharing benefits
Type B: Participation provided by local authority approval
Type C: Share benefits through compensation measures
Type D: Neither control through public participation nor measures to share the benefit available

Very few countries have either a relevant control or an appropriation system for sharing benefits. In all Western European countries, a combination of both systems has been adopted (type A). In some European countries, participation is guaranteed by the local government (type B). In Eastern Europe, nearly half of the countries adopted the double strategy (type A), and the rest of these countries achieve a system of sharing benefits through compensation measures (type C).

4.2.3
Innovative Practices on
Participatory RE Planning in Europe

The planning of infrastructure for the production of new renewable energies such as wind and solar energy is a new challenge in many European countries. One means of developing a basis for suggesting improvements to the participatory planning in RE is to study innovations in planning approaches being trialled in Europe. Therefore, we conducted an inventory of best or innovative practices in participatory RE planning in Europe.

According to Veselý (2011) 'best practice research' is based on the concept that instead of formulating an abstract ideal state we want to reach, we should develop what has been implemented and is proven to work somewhere else. Continuous improvement and iteration are implicit in the concept of 'good practice'. In the preface of the survey, innovative participatory planning was defined as non-ordinary practice if compared to the national context, regarding design of the process, tools used, stakeholders involved, technologies applied, etc. for public participation as well as case studies that are representative of RE projects and landscape compatibility. Twenty-five case studies that represent

innovative practices were provided from COST RELY members.

The systematic analysis of case studies can make an important contribution to the debate by going beyond local-level experiments and looking at opportunities for citizen participation across a diversity of contexts. Context refers to the historical, cultural, institutional, social, economic, political, and geographical settings that shape, and are shaped by, the technology in focus.

Three research questions guided the inventory of innovative practice:

What types of innovations in participatory planning have been implemented in different European contexts?

Are certain types of innovations associated with particular geographic regions?

What characterises innovative participatory practices in RE planning with respect to the goals, stakeholders involved, techniques used, and lessons learnt?

In order to identify innovative practices of participatory RE planning in Europe, a semi-structured questionnaire was designed and experts from 36 European countries involved in the COST Action RELY were asked to complete the questionnaire. This approach was to provide systematic information about innovative case studies of participation processes regarding renewable energy systems and facilities in these countries.

The semi-structured questionnaire included open and closed questions, and comprised three parts: 1) story telling about RE projects, 2) questions on contextual issues, and 3) questions about participa-

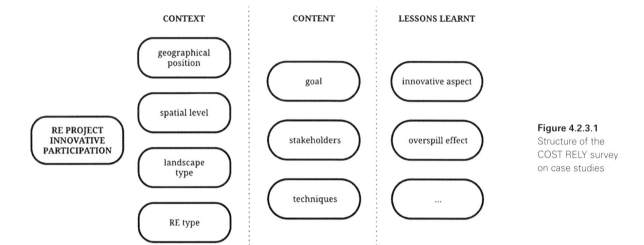

Figure 4.2.3.1
Structure of the COST RELY survey on case studies

tion processes and outcomes with an emphasis on innovation and lessons learnt (see Figure 4.3.2.1). Twenty-five case studies were collected from COST Action RELY members during the first half of 2016 which 1) involved renewable energy system development or planning, and 2) included public participation. These were reviewed with respect to selected characteristics with the aim of identifying any patterns that emerge. The form of participation used in the participatory planning described in the case studies was assessed by the expert group according to three dimensions or criteria:

- goal (identified intention proceed to typology according to Wessenlink et al. 2011);
- inclusiveness (early involvement of stakeholders, level of participation and one-way or two-way flow of information used), and
- level of participation (according to the ladder of citizen participation, after Arnstein 1969; International Association of Public Participation 2 2006).

A framework was used to describe the goal or rationale of the participation, following a classification introduced by Fiorino (1990) and subsequent work by Stirling (2006) and Wesselink et al. (2011). The categories of typology are defined as follows:

- Substantive: the main goal or rationale of this type is to improve the project and to stimulate a learning process.
- Normative: the rationale has the primary purpose of stakeholder involvement that respects stakeholder interests and their rights to be involved in the project.

- Instrumental: the rationale is to persuade the stakeholders of the benefits of the project and ensure its efficient implementation.
- Legalistic rationale: participation is only organised to meet formal legal requirements with no conditional effort to improve its outcome.

The provided case studies are situated in low, middle and high-income countries with varied democratic traditions. Case studies considered in our inventory stem from twelve countries, seven of them belonging to Southern and Western, and five to Central and Eastern European countries (Figure 4.2.3.2). Fifteen case studies were linked to one RE technology, while the other examples included two or more RE types of studies. Beside technology types, the considered RE projects differ also in terms of size, landscape context (Corine Land Cover typology applied), and landscape functions which have been embedded. Most of the case studies could not be specifically assigned to just one type of landscape, but most of the projects present innovative practices of participation in RES situated in agricultural landscapes, followed by urban spatial contexts including industrial zones and sites.

Only four out of 25 cases dealt with specific site or geographic contexts or features, e.g. the island landscape of Krk, HR; the Tisza ecological corridor, HU; the seismic area around L'Aquila, IT; and the flooded settlement of Alqueva, PT. Using the interpretations and explanations of experts, the size of the areas and of the RES [classification IRENA (International Renewable Energy Agency) 2012], and ancillary documentation, the level of the case study

COST RELY countries

Case studies
Source of energy

- Wind
- Hydro
- Solar
- Bio
- Mix

Figure 4.2.3.2
Map of locations of the case studies (provided by S. Martinat)

was categorised as local, regional, or national. Thirteen case studies refer to the local level, twelve to the regional, and one to the national. Most case studies present projects that have been implemented, only one is still conducted as an experiment (NL), and some reached the planning phase and then stopped due to political changes (HR, BIH), but the participation processes in the early phase represent innovative examples in the context of those countries.

4.2.4
Type of Participation in the Planning Processes: Case Studies

Although the case studies represent best and most innovative practices, the results of the qualitative analysis suggest that most of the cases are assessed as being of the instrumental type of participation focused on efficiently implementing projects (Table 4.2.5.1). At the same time, most cases were identified as being at the medium level of participation—classified as involvement or deliberation. Most of the cases, however, were rated to be highly inclusive. Interestingly, just four out of 25 cases included a one-way flow of information without early involvement of all stakeholders. Broadly open processes with early and transparent communication channels, but focusing on efficient implementation, seem to be the most common aspect across all practices described. Only few cases, however, presented processes that hand over the control to stakeholders who are not authorities. Just three of

them presented processes where the improvement of projects was envisaged and where policies were exposed to being designed and controlled by particular interest parties or the community in general.

4.2.5
Characteristics of the Innovative Practices and Regional Differences

Some regional differences of innovative participatory practices were identified by clustering case studies according to geographical regions of Europe, as introduced in Chapter 4.1 (Suškevičs et al.). Practices considered as innovative in a Western European context manifest as being at least of instrumental or substantive types of participation, providing consultative levels of participation. Practices identified as being innovative in the contexts of Eastern and Southern Europe include legalistic and normative types of participation, and information levels of participation. In Eastern Europe, the innovative practices described only reach a level of involvement.

	Project name	County, country	T	I	L	Innovative aspect
AU	Energiekultur Kulmland	Oststeiermark, Austria	O	↔	🏠	innovative actor—energy manager; innovative approach—building a regional energy vision, building trust
CH	Solar modules on avalanche barriers	St. Antönien, Grisons, Switzerland	🔧	↔	☉	innovative approach—bottom up initiative, multiple use of resources
CH	Linthal 2015	Linthal, Grosstal, Canton of Glarus, Switzerland	🔧	↔	🏠	innovative approach—well-designed involvement process, building trust and progress in level of participation
CH	Solarpark La Boverie	Payerne, Yverdon Switzerland	🔧	↔	🏠	innovative approach—building vision on energy region, energy city, multiple use of resources
DE	Energy strategy/ policy Zellertal	Arnbruck and Drachsels-ried (county: Regen), Bavaria, Germany	🔧	↔	☉	innovative visualisation techniques—scenario simulation, interactive mapping
DE	GIS-based and partic-ipative visual land-scape assessment	Mecklenburg-Western Pomerania, Germany	🔧	↔	☉	innovative visualisation techniques—GIS, 3D analysis
DE	Interactive visual landscape assess-ment as a basis for the geodesign of wind parks	Saarland, Germany	🔧	↔	☉	innovative visualisation techniques—GIS, scenario simulation
DE	Dezent Zivil	Schopfheim, Baden-Wuerttem-berg, Germany	🔧	↔	🏠	innovative visualisation techniques 3D simulations, innovative approach—bottom up and high proactive initiative
DE	Energy village Wildpoldsried	Wildpoldsried, Bavaria, southern Germany	O	↔	🌐	progress in participation level, innovative approach—building a local energy vision
NL	Energiewerkplaats Fryslân The Energy Atelier Friesland	Province of Friesland, Netherlands	O	↔	☉	innovative approach—locally tailored approach, innovative techniques—interactive mapping, 3D sce-narios, new actors—local energy cooperation
BIH	Micro hydropower plant Čajdraš	Cajdras, Zenica, Bosnia and Herzegovina	⚙	↔	☉	innovative approach—multiple use of resources as a goal
BIH	Solar power plant Kalesija	Kalesija, Bosnia and Herzegovina	⚙	↔	👤	innovative actor—financial private participation
CZ	Biogas station in Pustějov	Pustejov, Moravi-an Silesian Region, Czech Republic	🔧	↔	👤	innovative technique—study trip

Table 4.2.5.1
Assessment results of the case studies on innovative participation practices

	Project name	County, country	T	I	L	Innovative aspect
HR	Island Krk—Energy Independent Island	Island Krk, Primorsko—Goranska County, Croatia	🔧	↔	▮	progress in participation level; innovative approach—financial participation by inhabitants and learning by established models
HR	Biogas Gundinci	Municipality of Gundinci, Brod-Posavina County, Croatia	🔧	⋈	☉	progress in in participation level; innovative actor—UNDP innovative approach—multiple use of resources
HU	Coach-BioEnergy	Szada, Hungary	≜	↔	☉	innovative approach—multiple use of resources; innovative techniques for information—letters, posters, bringing together—forums and study trips
HU	Csaba Vaszkó—bioenergy feed stock production	Tiszatarjan, Borsod-Abauj-Zemplen County, Hungary	O	⋈	☯	innovative approach—integration of local symbols, meanings and economy context
SR	Energy efficient Kindergartens in Belgrade	Beograd	🔧	↔	☉	progress of the participation level—broad public information and call for consultation
SR	Small biomass power plant	Dragacica	🔧	⋈	▮	innovative approach—financial participation by farmers
SR	Ecoremediation of degraded areas by energy crops production	Sadzak, Municipality Sremska Mitrovica	🔧	↔	☉	innovative approach—financial participation by farmers multiple use of resources
FR	Ailes des Crêtes wind farm	Chagny and Bouvellemont, Ardennes, France	O	↔	☯	innovative approach—shared benefits and learning by established models
FR	Energ'Ethique 04	Digne-les-Bains, Alpes de Haute Provence, France	O	↔	☉	progress in participation level—democratic governance, innovative approach—building regional vision, innovative actor—energy social enterprise
IT	L'Aquila Progetto C.A.S.E.	L'Aquila, Abruzzo, Italy	⚑	⋈	▪	innovative approach—multiple use of resources
PT	Barragem de Alqueva	Alqueva, parish of Alqueva, municipality of Portel, Portugal.	🔧	↔	☉	progress in participation level—high inclusiveness and multidisciplinary approach in early stage of project
PT	Central Solar da Amareleja	Amareleja	≜	↔	☉	progress in participation level but without external obligation; innovative approach—building trust

TYPOLOGY ⚑ legalistic ≜ normative 🔧 instrumental O substantive
INCLUSIVENESS ⋈ narrow ↔ broad
LEVEL ▪ information ☉ consultation ▮ involvement ☯ collaboration ○ empowerment

Figure 4.2.6.1
Public workshop
in Friesland
(© Sandra van Assen/
energiewerkplaats.frl)

4.2.6
Illustrative Case Studies of
Innovative Practice

In order to illustrate the broad range of innovative practices in European participatory RE planning, we will present four case studies at the end of this chapter that were selected according to innovation themes. These themes were identified by qualitative analysis of the answers to the open-ended questions about the indication of the innovation in the case study. Four innovative themes were identified: approach to innovation, level of participation progress, innovative actors, and innovative participation techniques.

The Energy Atelier Friesland (NL)—Innovative Approach
provided by Berthe Jongejan

The Aim of the Project
The main aim of the project was to develop a framework for communities to design their own process of participation, with practical methods and a localised approach.

Since 2015, the regional authority in Friesland (one of the northern provinces of the Netherlands) has been organising villages in so-called communi-

ties of practice. To accelerate the implementation of renewable energy the organisation has brought together people in groups referred to as 'Energie werkplaats' (Energy Atelier). The 'Atelier' helps inhabitants to exchange knowledge about energy efficiency, energy production (e.g. the use of solar panels), and entirely energy-neutral villages.

Innovation Aspect: Innovative Approach
The Frisians like to refer to their strong sense of community, while demonstrating their desire to do things together. This is not a formal collaboration, but rather partnerships that involve the entire community. At the end of 2015, Friesland, with a population of less than 650,000 inhabitants, counted 37 energy cooperatives. Consequently, the province of Friesland has become the most innovative collaborative cooperation nationally. People establish energy cooperatives to jointly insulate their homes or to purchase or produce renewable energy.

In many cases, initiators appear to lack a clear picture of the steps they need to take to bring their plans into action. Therefore, the Energy Atelier Friesland designed a set of techniques for fostering participation.

The approach was experimentally tested through designed workshops (Figure 4.2.6.1 and Figure 4.2.6.2) in three communities and by now has benn implemented in eight. All were evaluated as

Figure 4.2.6.2
Direct participation in planning and design process in Friesland (© Sandra van Assen/ energiewerkplaats.frl)

successful (by the project leader, mail communication, 2018). Workshops were directed to support planning strategy with several foci:

- Explore spatial implications when integrating renewable energy technologies (RETs) in local landscapes and the conceptualisation of local energy transition perspectives by enabling local communities to visualise the relevant spatial implications
- Enable policy-makers to evaluate whether the spatial implications of the energy transition, facilitated on a local scale, would contribute to the energy transition on a regional scale against the strategic horizon of 2050 consistently, harmoniously, and in an integral way

- Advance local energy initiatives' development and growth through supporting the development of comprehensive energy visions wherein the visual and land-use effects of these renewable energy technologies can be easily appreciated by local communities
- Indicate first elements of a strategy the provincial authority could adopt to address the possible spatial implications that follow the progresses of local energy initiatives in time and environment

However, despite the strong sense of community, the energy transition continues to lag in Friesland. Many local initiatives are being obstructed due to intractable political obstacles. The cooperation

Figure 4.2.6.3
Sunpark Griene Greide (Friesland) owned by energy cooperation (Author: Alexander Kooistra)

and the Energy workplace are now discussing with politicians about new ways of democratic decision-making by affording cooperation more decision power. This way of working is still in an experimental phase. Meanwhile the Province of Drenthe has also started to work with communities of practice to increase the success rate of its solar power cooperatives (Figure 4.2.6.3).

Outcomes

The outcomes of presented case study are numerous: trust built, future vision designed, consensus on ambitions and goals, increasing the amount of participants, involvement of stakeholders, awareness of the visual and spatial impact, awareness of important spatial and historic values and features, inspiration, consensus on the allocation of RE facilities and minimisation of negative impacts, consensus about further going ambitions on creating positive spatial impact.

Spillover effect can also be identified as keeping money in the community and improving sociocultural connectivity while encouraging local ownership.

The results of Energy Ateliers are presented by Dijkman (2015): Towards the Energy Transition in Fryslân. In: Investigating the Spatial Impacts of the Energy Transition at the Local Scale, H. Dijkman, 2015 (http://edepot.wur.nl/390140).

Figure 4.2.6.4
Inauguration of photo
voltaic installations at
the school Gaubert in
December 2013, run by
the local people. (Source:
Energ'Éthique 04)

Energ'Ethique 04 (FR) — Levels of Participation Progress
Provided by Alain Nadai

The Aim of the Project

The case study named after social cooperation in energy, Energ'éthique 04, originates from the south of France, Digne les Bains, and presents activities which were initiated bottom-up by the citizens. The leitmotiv was a high involvement of local participants, financially and socially, presenting new active roles in society.

Innovation Aspect: Levels of Participation Progress

Energ'éthique 04 presents social enterprise and crucial stakeholder that was the initiator and main leader of further networking around the idea of renewables in the community of Digne les Bains. This social enterprise was created through a citizens' initiative of approximately fifty citizens who were willing to share the benefits of the production of renewable energies. Energ'éthique declared their mission was to develop financial participation, to provide knowledge, and support changes in behaviour towards the reduction of energy consumption. The most innovative idea being developed is to organise the energy solidarity and reduce risks to energy security.

Although energy solidarity is a well-known term, it is most widely used on higher levels, such as EU (Herranz-Surrallés 2016), and only in several cases it has been used to investigate on the local level (Walker and Devine-Wright 2008, Bomberg and McEwan 2012). The enterprise acts as a democratic governance, since it is open to all and functions on the basis of 'one person one voice' while being locally connected through long-term partnerships with local actors. Regulated by the statute of the cooperative, an equitable distribution of the activity's benefits between the employees, the enterprise, and the shareholders are distributed. In this case, public participation did not have to be initiated by external stakeholders such as local, regional, or national governments.

Outcomes

Several projects were developed including different technologies and financial models. Energ'éthique 04 sees its role in providing assistance and know-how so that citizen groups can realise their own projects. Public participation presented by this project gives a story on the activities of the general public and interested citizens that have initiated social enterprise, following a bottom-up approach and the participation of various stakeholders (private and public) (Figure 4.2.6.4). No techniques were highlighted to be innovative in by themselves, but it is essential that stakeholders' activities and very clear social and financial goals and operative rules are in place from the beginning of the project. The whole initiative can be scored as an approach innovation with regard to the fact that in France still more than 70 % of the electricity is provided by nuclear power and that the French state is highly centralised. Therefore, local initiatives carry even more weight.

Island Krk — Energy Independent Island (HR) — Innovative Actors

Provided by Dina Stober

Aim of the Project

A group of municipalities on the Island of Krk, in the northern part of the Adriatic Sea, are working together with the community infrastructure operator Ponikve d.o.o. They have become the main beneficiaries of environmental and climate protection of the island area. The stated vision is to become an 'energy-independent island', an aspiration which was created during an international win-win networking project. The newly established energy cooperatives developed a strategy and vision that promotes Krk as 'the island with 0 % CO_2 emissions'.

Innovation Aspect: Innovative Actors

The actor that started initiative of the RE development and the establishment of an energy cooperative was recognised as an innovative aspect of participation in national context. An energy cooperative was set up by nineteen individuals, experts in RE technologies, environmentally conscious individuals, representatives of administrations, local government and local services such as the company Ponikve d.o.o. To achieve the vision, an energy cooperative launched a series of workshops for residents (Figure 4.2.6.5). They developed knowledge of renewable technologies and promoted its dissemination to aid their implementation and construction, such as that of solar panels. At the level of the island area, the scope and means of developing solar farms and bio-energy facilities were promoted. Technical information and assistance were provided to the local population on the procurement and deployment of equipment, obtaining permits, loans, and grants.

To enable participation of the residents, a questionnaire and leaflets were sent to residents from which more than 300 responses showed significant interest in installing solar power. The cooperative offers help with obtaining documentation, reduced costs of equipment, renting roofs to set up photovoltaic power plants, and loans from the Croatian Bank for Reconstruction and Development.

Outcomes

The innovation of the study mainly reflects the step change in the level of participation, which was depicted as broad, instrumental, and trying to achieve defined goals. Spillover effects can be found in the promotion of an overarching concept of sustainability and CO_2 reduction, implemented in the school curriculum and art competitions which were organised by the communal operator.

Figure 4.2.6.6
Water buffaloes on the floodplain forest in Tiszatarjan, Author: Csaba Vazsko

WWF Tiszatarjan Bioenergy Feed Stock Production (HU)—Innovative Participation Techniques
Provided by Csaba Vaszko

Aim of the Project

WWF Hungary has initiated a pilot project next to the Tisza River in north-eastern Hungary with the goal to develop and test innovative payment for ecosystem services schemes to restore the area's natural floodplains while increasing and diversifying local income streams.

Innovation Aspect: Innovative Participation Techniques

WWF used a participatory approach to implement a landscape-based bioenergy project in Tiszatarjan village. The local community started to remove wild bushes of the highly invasive Amorpha species, which is partly used as a source of biomass for heating the local public buildings, and the extra biomass is being sold to an energy company. Additional project 'mechanisms' included the introduction of grazing animals such as grey cattle and water buffalo (Figure 4.2.6.6) to prevent the return of invasive species, and to assist with grassland management. As a result, a new local biomass supply chain has been developed providing biomass for heating the public buildings. Removing and grazing invasive shrubs have increased the floodwater retention capacity of the floodplain and increased the resilience of the local community. The flood risk has been significantly reduced across 100 hectares of floodplain areas contributing to better flood security. The resultant improvements to the landscapes and biodiversity also makes the area more attractive to tourists.

Outcomes

This case study was identified as well rooted in the local economy, society, and history, the success of which was evaluated as being successful due to its building of new partnerships and the involvement of a diversity of actors. During the participation process, farmers, public workers, community members, and local decision-makers had the opportunity to visit and get to know the key economic drivers of the RES, which was seen as an opportunity to establish networks and open communication channels. This also provided an opportunity to build trust between stakeholders based upon the sharing of experiences by stakeholders with similar interests. Some key conclusions were drawn based on the participation evaluation: clear goals have to be communicated from the outset of the project, and the main drivers of the local and regional socio-economic processes have to be engaged.

4.3

TOOL-BOX FOR EFFECTIVE RENEWABLE ENERGY PLANNING

Matthias Buchecker, Sebastian Eiter, Dina Stober, Monika Suškevičs, Cheryl de Boer & Berthe Jongejan

4.3.1
Introduction

In most European countries, the deployment of renewable energy projects as a cornerstone of climate change policies lags behind the plans related to energy transition (Flacke and de Boer 2017). Sovacool and Ratan (2012) identify nine factors in their international comparison of renewable energy development that foster or block the diffusion of renewable energy: institutional capacity, political commitment, favourable legal or regulatory frameworks, competitive production costs, mechanisms for information and feedback, access to financing, individual or collective ownership, participatory project siting and recognition of externalities, and

positive public image. In these factors the three dimensions of social acceptance as suggested by Wüstenhagen et al. (2007) are represented in equal proportions:

- Socio-political acceptance referring to key stakeholders' acceptance and support of energy policies and technologies
- Market acceptance involving consumers', investors', and energy firms' adoption of renewable energy innovations
- Community acceptance relating to residents' and local authorities' acceptance of specific renewable energy projects and respective siting decisions

Policy-makers and the mainstream of research literature have so far mainly focused on problems of community acceptance of renewable energy, ne-

Figure 4.3.1.1
The three pillars of acceptance of a local change, such as an installation for renewable energy production

glecting the fact that the underlying policy goals and investors' objectives have also been contested (Wolsink 2010). Some studies even see the main societal acceptance problem of renewable energy deployment in the lack of trust in the energy sector, companies and regulators (Mumford and Gray 2010), and, more fundamentally, in the lack of institutional capacities for learning, in particular for enhancing collaborative planning (Breukers and Wolsink 2007).

Accordingly, renewable energy projects have to be considered like other non-local infrastructure projects as external interventions in local contexts—often with limited local benefits—that raise mainly three concerns (Figure 4.3.1.1):

- Why is the intervention necessary and why exactly here? This concern refers to consistent national and regional policies and strategies that justify the project.
- How is the project legitimised locally? How is the project communicated and adopted in the local lifeworld (Habermas 1981)? This concern refers to procedural justice.
- Who benefits from the project and who bears the costs, i.e. who is mainly affected by the project? This concern also refers to distributive justice.

The role of national or regional policies and strategies on the acceptance and efficient realisation of renewable energy projects has so far been rarely considered within the research literature (Wolsink et al. 2010). Sovacool and Ratan (2012) have shown that the withdrawal of political commitment for wind energy in Denmark in 2001, including lower feed-in tariffs, drastically limited further growth of wind power capacity. A comparative study in three European regions on core beliefs related to effective implementation of wind power (Wolsink & Breukers 2010) found that hierarchical and technocratic planning approaches showed the least chance of successful implementations whereas approaches facilitating local ownership, early participation, and local designation of areas for wind power reduced opposition to new wind power projects. Our expert surveys within COST Action RELY (see Chapters 4.1 and 4.2) revealed that a) characteristics of strategic planning differ significantly between European regions and b) fostering participatory wind energy planning increases acceptance related to specific resistance issues.

The role of procedural justice for social acceptance of renewable energy projects is often emphasised (Gross 2007, Haggett 2008, Pasqualetti 2011), but very few studies have considered the respective practice in Europe or the contribution of local participation to successful project implementation (Langer et al. 2017). Recent literature highlights that public involvement in local renewable energy planning in Europe normally takes place at a late stage—in the spatial planning phase or even the permitting phase, and only seldom in the need-determinaton phase. It is often limited to one-directional information and pursues primarily instrumental goals, such as convincing the public of a project or a site, rather than open discussions about the project design (Polatidis and Harlambopoulos, 2004, Devine-Wright 2011, Höppner et al. 2012, Aitken et al. 2016, Späth et al. 2016). This

Figure 4.3.1.2
Community workshop in Stilfs (South Tirol). (Photo: Matthias Buchecker)

approach of 'decide-announce-defend' (Bell et al. 2005) can be considered as undemocratic (Simcock 2016). Walker et al. (2010) suspect the avoidance of more substantial and interactive forms of public involvement as a consequence of decision-makers' conceptions of the public assuming opposition based on selfish 'not in my backyard (NIMBY)' motives. This lack of trust in the public might in many cases be self-fulfilling, increasing local opposition—at least in contexts with high expectations in opportunities to participate (Bickerstaff 2012). In fact, more deliberative planning approaches enhance acceptance of renewable energy projects (Breukers and Wolsink 2007a, Devine-Wright 2005). A systematic analysis of the relationship between different modes of participation and acceptance of wind energy projects revealed that citizens' assessment of wind energy projects is considerably influenced by the degree of participation (Langer et al. 2017). Interestingly, information appeared to have the strongest effect, followed by cooperation, whereas consultation, which represents the highest level of stakeholder involvement in many European countries, showed the lowest effect. In fact, Simcock (2016) found inadequate information and lack of influence on the substance of the project (e.g. technology, site, size) to be key concerns for lacks of perceived procedural justice in England.

The role of distributional justice has mainly been addressed in research on community energy projects and financial participation in renewable energy projects. Local ownership of renewable energy projects may enhance their acceptance (Geissler et al. 2013, Jami and Walsh 2014). Community energy is associated with project schemes that are developed through broadly collective decision processes or ones that distribute their benefits locally and collectively (Walker and Devine-Wright 2008). Financial participation, i.e. providing options for citizens to acquire a share of local energy projects and its benefits—in contrast to community projects—focuses on individual participation. So far, very few studies have provided evidence that community projects or financial participation really enhance public support of renewable energy projects (Devine-Wright 2005, Warren and Mc Fadyen 2010). Other studies, however, have reported of community projects that encountered opposition (Simcock 2014, Walker et al, 2010), and Simcock (2016) revealed in a case study in the UK that a community wind energy initiative was only supported by those citizens who considered the decision process as being impartial and just. Accordingly, a choice experiment in Germany concluded that financial participation alone contributed only marginally to the acceptance of a wind energy project (Langer et al. 2017). Neither solely allowing communities to determine if a wind project proceeds, nor solely options for financial participation are sufficient to secure local support for an energy project. An additional bottom-up integration process into community life is necessary, as a successful case study from Denmark suggests (Sperling 2017). From this perspective, the potential financial profit seems to be less important to citizens than the right to have a democratic say in the process (Langer et al. 2017). Comparisons between European countries show that opportunities for financial participation are strongly related to success in the growth of wind power development (Sovacool and Ratan 2012) but nevertheless also suggest, that locally shared profit is also a relevant acceptance factor.

Overall, we conclude that three possibly interlinked aspects need to be considered or even kept in balance in order to achieve an effective planning of renewable energy deployment:
1) national or regional strategic planning,
2) local control through involvement, and
3) securing of local benefits.

Balancing these aspects requires considering the interconnections between the demand and supply side of renewable energy, as well as integrating national and local levels of planning (Devine-Wright 2011). Furthermore, optimal solutions cannot be generalised, as they depend on the local context in terms of earlier conflicts, vulnerability or discourses (Hostmann et al. 2005, Simcock 2016), especially expressed in the main acceptance issues, and on the wider cultural context, in particular norms of justice (Bickerstaff 2012, Vermeylen and Walker 2011). Before we suggest a tool-kit based on these considerations, we present the basic recommendations of existing tool-boxes.

4.3.2
Literature and Tool-boxes on Effective Participatory Planning

Although the relevance of procedural justice for the support of renewable energy projects is widely acknowledged, there is relatively little internationally acknowledged literature focusing on effective participatory planning of renewable energy projects. Three recent studies focus on effective renewable energy planning from the perspectives of developers and project managers, respectively.

A comprehensive analysis has been elaborated in the EU-Project INSPIRE-Grid focusing on stakeholder participation in power line planning (Battaglini et al. 2012). Based on evaluations of planning processes in different European contexts, Späth et al. (2014) suggest a tool-box for 'best' planning practices. Key elements suggested as evaluation criteria are

- **Early involvement**: the need for the project should be discussed prior to detailed planning, preferably in the need determination phase.
- **Representation of stakeholder groups**: if possible based on a systematic stakeholder mapping.
- **Task definition**: the expected tasks of the stakeholders should be clearly formulated.
- **Participatory decision-making methods**: the decision mechanisms should be clearly stated (e.g. multi-criteria analysis).
- **Influence on the outcome**: stakeholders should have substantial influence on project design, but also on procedural design.
- **Independence of the key participants**: the organiser, and in particular the moderator should be independent of the project.

Another European study analysed the planning of 27 European renewable energy projects, which were very different in terms of technology and geographical context, with the goal of adapting a tool for managing innovations for acceptance building (Raven et al. 2009). The study identifies challenges in the integration of a project into a specific context and stresses the importance of reflecting, articulating, and negotiating expectations and visions of the planned project. The suggested tool-kit is largely in agreement with the criteria of INSPIRE-Grid, but focuses more specifically on problem structuring (Gregory et al. 2013) and negotiating project variations based on minimised interaction. The evaluation of the approach, however, revealed the key role of the interactive elements.

In accordance with this finding, recent research literature in natural resource management stresses the key role of interactive problem structuring for facilitating social learning and shared solutions (Pahl-Wostl et al. 2007, Mostert et al. 2008). This social learning approach, which has not been considered in the context of renewable energy planning so far, requires a more collaborative and thematically more open approach than ordinary participatory planning.

A third study on renewable energy planning suggests a participatory multi-criteria approach (Polatidis and Harlambopoulos 2004). Based on public consultation a multi-criteria evaluation of project alternatives should be defined. The participatory multi-criteria evaluation will help to achieve a consensus on the preferred alternative, including a "'zero-action alternative'. However, such a highly structured procedure can only be successful if a) the involvement takes place at an early stage and b) the participatory multi-criteria evaluation has a discursive character rather than being a numeric procedure. And of course, even if zero action may be seen as a necessary alternative for fundamental participation, it cannot be a ubiquitous option as RE development has to take place somewhere, when EU and national goals shall be achieved.

Tool-boxes on Participatory (RE) Planning

An inventory among COST RELY experts revealed a considerable number of guidelines and tool-boxes for participatory (RE) planning in Europe that can be grouped in three categories, whether they describe

- generic steps of good planning of renewable energy projects,
- generic steps of good participatory planning, or
- methods of public involvement.

The guidelines and tool-boxes referring to renewable energy mainly focus on wind energy projects. A prominent exception is a guideline *Advocating for Sustainable Energy in Central and Eastern Europe* (Pagan and Vollmer 2017). The guidelines from different countries are characterised by a considerable discrepancy between the standardised (minimum) involvement and a recommended involvement. The standardised procedure normally includes the following steps:

- Early information on the planned project.
- Assessment of the impact of the project by authorities. If the impact is considered substantial, an environmental impact assessment (EIA) and public involvement are required.
- Pre-application public consultation involving key stakeholders or an existing local contact group in order to define the content of the EIA, identify community benefit measures and possibly project alternatives.
- EIA according to guidelines, if required.
- Formal public consultation to submit objections, etc.

Best practice engagement for communities additionally recommend to

- Proactively set out the energy aspirations for the community or region and designate priority areas or no-go zones
- Establish a local contact group or an actor forum

- Conduct open public deliberations on project alternatives and community benefits prior to the submission of a detailed plan.
- Form a community engagement plan

Guidelines of some European regions define further recommended or mandatory measures to reduce conflicts and mitigate local impacts of renewable energy projects:

- In Denmark, wind turbine areas have to be designated in the municipal plan and the use of these areas has to be described in a guideline before projects can be submitted. In Switzerland and some German regions, wind park areas have to be defined in regional strategic plans.
- In Denmark and some regions of Germany, promoters of wind energy projects have to offer local residents options to purchase shares of the wind turbines.
- In Germany and Switzerland, ecological losses induced by renewable energy projects have to be compensated by the promoter, if possible nearby the site.
- In Denmark, the federal state finances within 'green scheme' (according to the energy produced) measures to enhance the scenery and recreation opportunities within the site municipalities to enhance the acceptance of wind energy projects.
- In Denmark, the federal state offers local initiatives a guarantee fund in order to more easily obtain commercial loans for financing preliminary investigations. Local initiatives may also apply for a guarantee to take out a loan.
- In Denmark, the promoter of a wind energy project has to pay for any loss of property value.
- In Holland, the department of cultural heritage encourages local governments to utilise the qualities of historical landscape and heritage in the production of sustainable energy.

General guidelines for generic participatory planning stress the reflection of the context and the purpose of participation in a specific project (Hostmann et al., 2005; Höppner et al. 2012), and the importance of the quality of the participatory process. The main quality criteria for a participatory process are

- Clear scope: the purpose and limitation of the process, but also the liability of the outcomes should be clarified at the beginning.

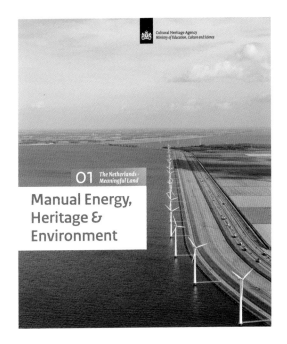

Figure 4.3.3
This is an example of a guideline of the Netherlands for how to develop strategies for integrating sustainable energy production in the landscape (Cultural Heritage Agency 2017)

- Fairness: all participants should have the same options to influence the outcome.
- Transparency: the interests of the participants have to be clarified, and the decision process should be reproducible.
- Mutual learning options: all arguments should be respected and the legitimacy of a diversity of perspectives should be acknowledged.
- Early involvement: a participatory process should already include the definition of the problem and the goal.
- Direct and complete information: comprehensive information from the responsible agencies may raise more conflicts, but residents will feel respected and conflicts based on misunderstandings will be avoided.
- Competence: participants should only participate in discussions on issues about which they are sufficiently informed.
- Inclusion of non-organised interests: efforts should be taken to involve weakly organised individuals.
- Avoidance of losers: discursive procedures are successful if all parties leave the process as winners. Concessions are needed from all sides (Renn et al. 1998, Buchecker et al. 2013).
- Institutional integration: participatory processes should have practical relevance and therefore be integrated into superior decision-making processes.

Tools suggested in Recent Research Projects
- **Integrated regional natural resource management**: participatory processes involving regional actors to discuss visions of the future use of natural resources tend to extend the solution space and make it possible to find shared solutions, also regarding the deployment or extension of renewable energy projects (Buchecker et al. 2013, Gaus et al. 2016). The acceptance of such projects is enhanced through a better mutual understanding among the actors, social learning, and a cross-sectoral approach that allows win-win situations for all actors involved.
- **GIS-based interactive planning tool to deliberate about ideal renewable energy sites**: in a European project, a planning tool was developed and tested encouraging workshop participants to agree on an optimal set of technologies and sites for renewable energy production within a certain region (Flacke and de Boer 2017). This bottom-up process is expected to increase local acceptance of projects. The GIS-based tool also has an awareness-building effect, as it calculates the potential energy production on the site.
- **Scenario workshop to reveal or deliberate about the preferred size or sites of renewable energy plants**: a visualisation tool based on land use and land cover data and scenario techniques (Wang et al. 2016) helps to reveal different stakeholder groups' preferred project designs. Such visualisation tools have also been applied

successfully using low-budget techniques such as real-time-illustrations (Tobias et al. 2015).

- **Concept mapping in an early planning phase to agree on a shared problem understanding**: this tool, which has been developed for process moderation mainly in devlopment contexts, strives for shared understanding among stakeholders about consequences of a project within a local or regional system using a graphical method (Heeb and Hindenlang 2008). The tool helps to bring ideological conflicts down to factual processes and allows for finding shared solutions that might include compensation or improve distributional justice.

4.3.3.
Towards a Tool-box

The COST RELY online survey on participatory wind energy planning revealed that strategic and local planning as well as local capacities influence residents' acceptance of projects. Interestingly, however, the relevance and even direction of the influence factors depends on the specific non acceptance problem (Table 4.3.3.1). This means that different measures have to be chosen to increase local acceptance, dependent on local issues. In case of non-acceptance problems related to lack of procedural justice, a stronger involvement of local

stakeholders and more incentives for community initiatives help, in particular if local self-organisation is well developed. In case of non-acceptance problems related to landscape issues, the definition of national or regional priority areas and early communication can reduce resistance, in particular if local social capital is high. Accordingly, the main local acceptance issue and the socio-cultural context matter when choosing an effective tool.

Accordingly, depending on the relevance of non-acceptance issues, experts rated specific measures and tools as particularly effective to achieve acceptance for wind energy projects (Table 4.3.3.2). In the context of resistance to wind energy projects due to lack of trust, deliberation about the size of wind parks, options for individual financial participation, and prospect for the provision of local jobs, as well as the negotiation of wind energy projects within processes of integrated natural resource management on regional level, are considered necessary to promote implementation of projects. In the case of resistance due to landscape issues, however, deliberation on the site and on the benefits of planned projects are considered to be most effective. Compensation measures are only considered to be effective in contexts with non-acceptance due to lack of participation and place image issues, and even counterproductive in contexts with resistance due to landscape or environment issues.

Selecting and applying efficient measures and tools to increase the acceptance of renewable en-

	Reasons for non-acceptance								
Context situation	Social injustice	Lack of participation	Threat of local image	Landscape encroachment	Environment encroachment	Social conflicts	Exterior intervention	Lack of identification	Lack of trust
Priority areas defined				+		−			
Financial participation possible		−							
Comprehensive strategic planning	+			−	+	+			
Comprehensive communication of strategy	−			+	−				
Timely local communication			+	+					
Comprehensive local communication	+			−	+	+		+	
High deliberation quality				−					
High influence of local actors	+	+	−	−		+	−	−	
High local social capital				−			+		
Good local communication culture		+							
Sufficient financial resources	+			+	+	−			
National incentives for community initiatives		+			−	+			+
High degree of local self-organisation		−							

Table 4.3.3.1
Relationship between planning quality as well as context situation and reasons for non-acceptance based on regression analyses

	Social injustice	Lack of participation	Threat of local Image	Landscape encroach-ment	Environmental encroachment	Social conflict	Exterior inter-vention	Lack of Identifi-cation	Lack of trust
Deliberation on the site		x*		x*					
Deliberation on the size						x*			x*
Deliberation on allocating benefits	x*		x(*)	*					
Individual financial participation	x*		x*			x*			x*
Financial participa-tion municipality									
Environmental compensation		x*			x(*)				
Financial com-pensation			x*	-x*	-x*				
Integrated ressource management									x(*)
Transformation in community project	x*	x*	x*						
Justification with local needs							x*		
Justification with providing jobs									x*
Justification with national strategy								x*	

ergy projects is only realistic if these instruments are compatible with the planning culture and planning practice in their respective planning contexts. Our expert survey on wind energy planning suggests that most measures and tools considered to be promising in making renewable energy planning more effective are not compatible with specific planning cultures across parts of Europe (Table 4.3.3.3). Substantial forms of participatory planning such as deliberation on the site or early and transparent communication are considered to be only reconcilable within the planning culture of Eastern Europe, where the practice of involving the public has only been tentatively reintroduced after the socialist period. In Southern Europe, measures of individual financial participation and community initiatives seem to be problematic in their implementation, whereas in Western Europe measures of national strategic planning seem to have limited value in gaining acceptance. Interestingly, reservations to the widest set of measures were found in Northern Europe, where in particular measures to promote distributional justice are assessed as little compatible with the regional planning culture, possibly because distributional injustice is not considered to be a major societal issue. Unexpectedly, as compared to the high expectations in the literature (Sperling 2017, Simcock 2016) the transition of projects to involve community input is overall rated as the least compatible within the planning cultures across Europe.

Table 4.3.3.2
Relationship between experts' assessed effectiveness of instruments and the relevance of non-acceptance issues based on regression analyses (*: $p < 0.05$; (*): $p < 0.1$)

Tools / Measures	Total	Northern	Western	Central and Eastern	Southern	p-value (ANOVA)
Early and transparent communication of project	3.5	4.18	3.83	2.88	3.67	0.09
Public deliberation on the site	3.27	3.73	3.39	2.88	3.56	
Public deliberation on the size	3.11	3.36	3.44	2.71	3.22	
Public deliberation on site selection criteria	3	3.18	3.17	2.65	3.33	
Public deliberation on allocation of benefits	3.1	2.55	3.5	3	3.22	
Offering opportunities for individual financial participation	3.08	3	3.67	2.63	3.25	0.05
Financial participation in the project by the municipality	3.15	2.9	3.33	3.21	2.88	
Granting environmental compensation measures	3.28	3.4	3.44	3.13	3.25	
Granting financial compensation for public measures in the municipality	3.33	2.9	3.39	3.36	3.63	
Including the project in a integrated natural resource management	3.11	2.9	3.17	3.08	3.38	
Transforming the project into a community project	2.59	2.3	3	2.48	2.38	
Justifying the project with local energy needs	3.23	2.9	3.22	3.48	2.88	
Provision of new jobs	3.3	2.94	3.92	3.92	3.63	0.059
Propagating the project with national strategic energy planning	3.39	3.6	2.94	3.64	3.38	
Justifying the project with inclusive national strategic energy planning	3.35	3.3	3.06	3.63	3.25	

Cultural contexts and planning practices will evolve and develop towards an alignment within Europe, in particular if supported by future EU directives on environmental planning, but the poor overall adoption of compatible planning measures and tools suggest this will take place only very gradually over the next years and maybe decades, as unique cultural patterns are normally rather persistent. Seeing the different effectiveness of planning measures and tools depending on the cultural context, we suggest a tool-kit that recommends

- A best practice general planning design for effective implementation of renewable energy projects and

- Specific measures and tools depending on the acceptance situation and planning culture (Table 4.3.3.4).

Our recommendation of specific planning tools is based on an expert survey focusing only on wind energy. As the acceptance situation and the adoption of planning measures is supposed to mainly depend on the local or national context rather than on types of technology, our recommendations probably also apply to other renewable energy sources.

Table 4.3.3.3
Experts' assessments (n=110) of the compatibility of tools or measures with the planning culture in the context according to European regions (mean values: 1= very little; 5= very high)

Best Practice General Planning Design for Effective Implementation of Renewable Energy Projects:

- Communication of national and regional policies and strategic planning approaches as justification of the project, e.g. priority areas.
- Context analysis: earlier conflicts, state of planning, plans and visions, local and regional energy policy, history of regional energy production (Llewellyn et al. 2017) as a basis for project planning and a communication strategy.
- Early communication about first project ideas (who, what, where) and the envisioned planning procedure including options for public involvement: avoid misunderstandings.
- Identification of relevant regional and local stakeholders.
- Dialogue with regional and local stakeholders on the envisioned project.
- Early stakeholder involvement (interactive workshop) on project impact criteria and preferred alternatives or accompanying measures (e.g. relevant for EIA).
- Public information on detailed project planning using different channels and providing options for feedback.
- Stakeholder involvement in detailed project planning, with options to negotiate amendments locally.
- Information about the decision process, e.g. alterations of the project or accompanying measures.
- Formal approval by the local government or population: local democratic legitimation of the project.

Conclusions

The implementation of RE projects means, very similarly to that of other supra-local infrastructure projects, an intervention into a local context. There are, however, three challenges that are specific for RE projects: a) the key justification for them is on a global level, to mitigate the climate change, which might not be consistently supported by national policies; b) RE-projects provide not just costs, but also benefits, the distribution of which need to be negotiated; and c) potential sites of these projects are relatively ubiquitous. Planning RE projects therefore need to be more comprehensive in order to tackle these additional challenges. These are normally not considered in the actual RE planning and not taken into account in existing tool-boxes on participatory planning. In recent years, new approaches and tools have been developed in research and practice that help integrate these challenges in planning. Planning, however, must also take the cultural contexts, in which it takes place into account. In these contexts new challenges raise different (acceptance) issues, and the accepted ways to solve them are specific. Planners of RE projects must on the one hand optimise their approaches by learning from best practice in other fields of environmental planning that have a longer history of adaptively improving public participation. On the other hand, they have to assess the situation in the specific planning context to select appropriate additional tools to overcome the specific challenges of integrating RE projects in local contexts. According to our findings, the main contextual aspects they have to consider are the main acceptance concerns towards RE projects and the local planning culture that limits—at least in short term—the acceptance of planning tools. Accordingly, our tool-box for an effective implementation of RE projects suggests a best practice general planning procedure and specific additional planning tools tailored to contextual acceptance situations and regional compatibilities with planning approaches.

	Local involvement difficult (east)	Strategic planning difficult (west)	Individual financial inclusion difficult (east)	Decentralisation difficult (north, south)	Local compensation difficult (north)
Social Injustice	Participation on benefits Individual financial participation	Participation on benefits Individual financial participation Community project	Participation on benefits Overall strategic planning	Participation on benefits Individual financial participation	Individual financial participation Overall strategic planning
Lack of participation	Environmental compensation	Participation on site Community project	Participation on site Environmental compensation	Participation on site Actor involvement	Participation on site Actor involvement
Lack of trust	Integrated resource management Provision of Jobs	Integrated resource management Provision of jobs Participation on size	Integrated resource management Provision of jobs Participation on size	Provision of jobs Participation on size	Integrated resource management Participation on size
Threat of local image	Financial compensation	Community project participation on benefits	Financial compensation Participation on benefits	Financial compensation Participation on benefits	Community project Participation on benefits
Landscape encroachment	Participation on benefits Defining priority areas	Participation on site Early communication	Participation on site Early communication Defining priority areas	Participation on site Early communication Defining priority areas	Participation on site Early communication Defining priority areas
Intervention from outside	Justify with local energy needs (Financial participation municipality)	Justify with local energy needs (Financial participation municipality)	Justify with local energy needs (Financial participation municipality)	Integrated resource management (Individual financial Participation)	Justify with local energy needs (Financial participation municipality)
Environmental encroachment	Environmental compensation Strategic planning (priority areas)	Environmental compensation Integrated resource management	Environmental compensation Strategic planning (priority areas)	Environmental compensation Strategic planning (priority areas)	Environmental compensation Strategic planning (priority areas)
Social conflicts	Individual financial participation Strategic planning Integrated resource management	Individual financial participation Deliberation on size Integrated resource management	Deliberation on size Strategic planning Integrated resource management	Individual financial participation Deliberation on size Strategic planning Integrated resource management	Individual financial participation Deliberation on size Strategic planning
Lack of identification	Justification with national enery strategy Integrated resource management	Early communication Integrated resource management	Justification with national enery strategy Early communication Integrated resource management	Justification with national enery strategy Early communication Integrated resource management	Justification with national enery strategy Early communication

Table 4.3.3.4
Appropriate measures and tools to be included in RE planning depending on the acceptance situation and the compatibility with the planning culture of the context

5

OUTREACH OF COST RELY AND REFLECTION ON FUTURE STRATEGIES

5.1

TRANSFORMATIONS IN EUROPEAN ENERGY LANDSCAPES: TOWARDS 2030 TARGETS

David Miller

5.1.1
Policy Context

The European Commission EU 2030 Energy Strategy set a target of at least 27 % of energy consumption from renewable sources by 2030, a 40 % cut in greenhouse gas emissions (GHG) compared to 1990 levels, and at least 27 % energy savings compared with business-as-usual (European Commission 2014). The international policy context for the 2030 Energy Strategy is the UN Framework Convention on Climate Change and its Paris Agreement on Climate Change (United Nations 2015). Its central aim is to 'strengthen the global response to the threat of climate change', which will involve new financial, technological, and capacity building frameworks. The ambitious cycle set out in the Agreement comprises a technical phase of a global stocktake in 2023, followed by a political phase between 2023 and 2025 for domestic (and EU) policy processes, culminating in a meeting of the Parties to the Paris Agreement in 2025. Therefore, there is scope for new or revised policy initiatives to influence the trajectory of renewable energy development through to 2030, and beyond.

The European Union policy framework for climate and energy 2020 to 2030 (European Commission 2014) recognises that existing measures of Member States are expected to achieve a 32 % reduction in GHG emissions compared to 1990 levels. The updated target of a 40 % reduction will

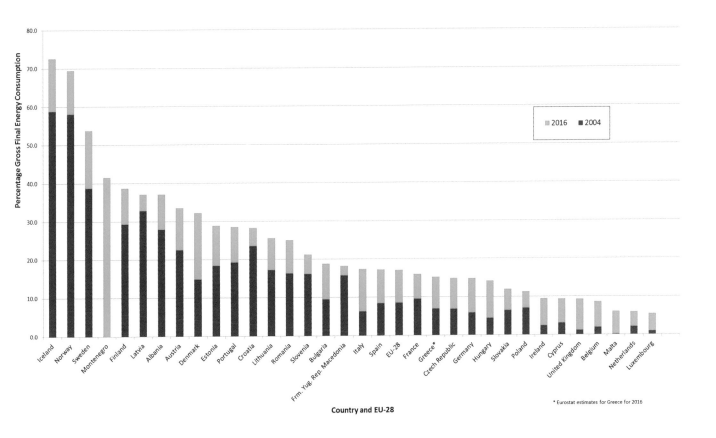

Figure 5.1.1.1

Share of energy from renewables and 2020 targets in the EU, 2004, 2016 (% of gross energy consumption; source: Eurostat 2017)

include agriculture, land-use, land-use change, and forestry (European Commission 2013), but will require an acceleration in development of renewable energy.

Figure 1(a) shows the proportions of renewable energy consumed in the European Union in 2004 and 2016. This is an indicator of the progress of the EU and Member States towards the 2020 targets, representing the changing extent to which renewable energy have substituted fossil or nuclear fuels. By 2016, eleven countries had reached their targets for 2020 (Sweden, Finland, Denmark, Croatia, Estonia, Lithuania, Romania, Bulgaria, Italy, Czech Republic, and Hungary). Targets for 2030 will be established in the strategy and national plans for the global stocktake by 2024, updated by 2028 (European Union 2017).

Figure 1(b) shows the contribution of different types of renewable energy to total production between 1990 and 2016. This expansion has been achieved by technical improvements (e.g. more efficient wind turbines), differential development of renewable energy systems (e.g. wind energy increased from 22.2 TWh in 2011 to 302.9 TWh in 2016; solar increased from 0.1 TWh in 2000 to 110.8 TWh in 2016), and innovation in design and siting (e.g. urban solar, 11.6 % of gross electricity generation from renewable sources in the EU in 2016; Eurostat 2018). These trajectories in production of renewable energy will need to be maintained to achieve the 2030 targets.

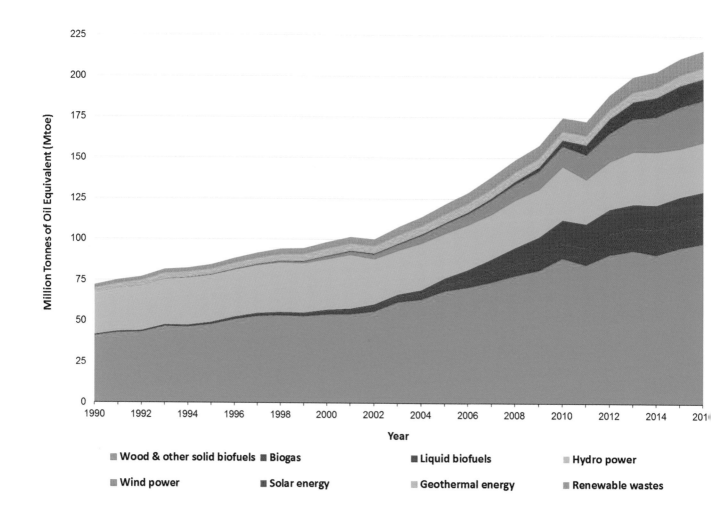

Figure 5.1.2
Primary production of
renewable energy in the
EU: 1990 to 2016 (Mtoe)
(Source: Eurostat, 2018).

5.1.2
Landscape Change

The 2030 Energy Strategy aims for 'greater transparency, enhanced policy coherence and improved coordination across the EU' (European Commission 2014). It will require consistency with the 7th Environment Action Programme to 2020 (European Commission 2013a) which seeks to protect, conserve, enhance, and value the natural capital of the EU, which Europe's landscapes link to the associated cultural and social capital. The quality and global significance of many of those landscapes are articulated in the European Landscape Convention (Council of Europe 2000). This recognises

that landscapes evolve, accelerated by improvements in primary production (which increasingly will include renewable energy), transport (the fuel of which will change radically from hydrocarbon based to electric), and changes in regulation and planning (e.g. updated EU Environmental Impact Assessment Directive), all delivering to global economic and political factors. Understanding landscape change requires understanding of the context of its wider geography and related dynamics (Antrop 2004) of which political responses to climate change are the highest profile.

5.1.3
Renewables and Landscape Change

A framework and visual concepts of landscape characteristics (stewardship, coherence, naturalness, complexity, disturbance, historicity, imageability, scale and ephemera, [Tveit et al. 2007]) is used to discuss potential changes or transformations in landscapes that can be reasonably expected with evolution of renewable energy in line with the EU 2030 Energy Strategy. The focus is on the renewable energy systems which are likely to be significant or can be expect to increase their production most by 2030.

5.1.3.1 Hydropower

Without significant changes in public strategies, it will be biophysical and economic constraints that limit extensive new development of renewable energy in certain types of landscapes, such as large scale hydropower in mountainous regions. Mid-20th century investment in large-scale hydropower schemes (reservoirs, pumped storage, and run-of river) has enabled the provision of 14.2 % of total primary energy production and 36.8 % of gross electricity generated from renewable sources in the European Union in 2016 (Eurostat 2018). In several countries, hydropower is the most significant source of electricity production in 2016, providing 97 % in Norway and 40 % in Sweden, but it represents only 1 % in Poland (International Energy Agency 2017). Further development of large scale hydropower schemes by 2030 is likely to be very limited. In most of Europe, a balance has been reached between environmental benefits and disbenefits of large scale hydropower, with attention directed to medium and small scale developments. Although there may be no landscape scale transformations due to hydropower, plans for its management and development require it to be understood in the wider context of the water-energy nexus, a priority of the European Innovation Partnership on Water. This partnership aims for an increase in the contribution of base-load renewable energy produced from water in the European energy mix, alongside development and application of efficient and smart technologies for water use, treatment, and recovery of energy from waste water (e.g. household water and sewage for producing biogas which is used in turn to generate heat and electricity, in Aarhus, City, Denmark). The cumulative effects of such developments on physical landscapes will not be transformative but will produce social benefits that stimulate new configurations of use and character.

5.1.3.2 Biomass

The European Commission (2014) recognises the need for biomass policy that covers sustainable use of land and forests, and addresses indirect land use effects, delivering multiple benefits to biodiversity and energy policies. The 2030 Energy Strategy reinforces the value of forest biomass to the renewable energy mix, projected to provide 42 % of the total renewable energy target for 2020, equivalent to the total wood harvest in 2013.

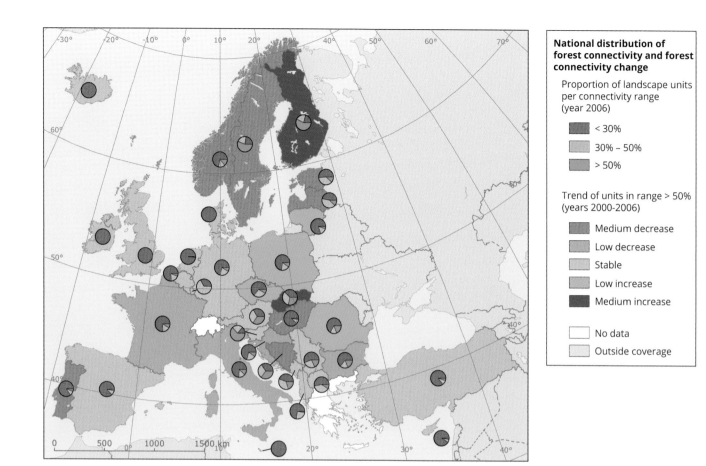

Figure 5.1.3.1
Proportion of landscape
units per connectivity
range reported by country
for 2006. The trend in the
proportion of units in a
high connectivity range
(>50 %) is for 2000-2006
per country (Source:
European Environ-
ment Agency, 2013).

With over 40 % of Europe's land area in trees or silvicultural management, forestry makes a significant contribution to its landscapes through direct and indirect effects. These include multiple uses offered by different habitats, textures, colours, and cultural aspects that contribute to sense of place. The significance of forest landscapes over the longer term will reflect their status, alongside moorland and peatlands, as key carbon sinks and 'repositories for biological diversity' (European Commission 2013b), but some of these are threatened by plant disease accentuated by climate change. They can also have negative impacts on biodiversity, replacing native with invasive species of tree and depleting soil resources. Therefore, there are trade-offs to be assessed and decisions made regarding the eco-nomic return of biomass production for renewable energy, the benefits of some aspects of natural capital, and the carbon payback balance between wood biomass for energy and the release of carbon from soils and peatland. The area of forestry in Europe is increasing at approximately 0.4 % per year (European Commission 2013) so overall the effects will not transform Europe's landscapes by 2030. However, public policy in many regions with relatively low proportions of forestry promotes its expansion (e.g. Scotland: to 25 % from 17 % land area) which, if successful, and depending upon species, will have significant local impacts on openness, stewardship, naturalness, and complexity. For example, in Hungary, there has been significant expansion in forest biomass through planting of *Robinia pseudoacacia* (Black Locust), in monocultures, with consequent impacts on landscapes in terms of their perceived naturalness, openness, and complexity.

Analysis of changes in patterns of woodland cover, using metrics of connectivity and fragmentation (Figure 2), highlighted differences across Europe. The highest levels of woodland landscape connectivity were in Slovenia, Finland, and Sweden, and the most poorly connected were in the United Kingdom, Denmark, and Ireland, as can be expected in those countries with the lowest proportions of woodland cover. There is no evidence that expanding wood biomass production is changing those patterns, but it does create opportunities.

5.1.3.3 Photovoltaic and Solar Power

Since 2000, there has been a dramatic increase in medium and large-scale above-ground photovoltaic farms in rural areas in Europe, although not across all countries. These range in size from small (less than 1 ha) to large scale, the largest in Europe (2017) at Cestas, south-west France, which occupies 2.5 km^2 for an installed capacity of 300 MW. These landscapes are transformed in use but not character. The developments are introducing new, geometric patterns into landscapes, in most cases replacing one form of production (agriculture) with another (energy), both coherent and evidence of stewardship, but energy being potentially less ephemeral. Multiple developments in close proximity are creating a cumulative effect of land use which is predominantly for renewable energy pro-duction, and disturbances which are human-made rather than managed.

Photovoltaic energy is being designed into urban landscapes. Examples of integration are roofs of industrial and agricultural buildings, openspace in transport infrastructure (e.g. motorway junctions, Figure 3), sound barriers, and recreation facilities (Figure 3b). Such installations provide added benefits to the prime function of the host structures, but their lifespan may be dependent upon the evolution or removal of those structures.

5.1.3.4 Wind Power

The largest increase in source of renewable energy between 2011 and 2015 was in wind power, a trend which is likely to continue. The versatility of wind farm design has enabled onshore development to be associated with most types of land use in Europe (e.g. crop and livestock agriculture, forestry, moorland, peatland, built environment), and thus a very broad range of landscape types. The availability of turbines with generating capacity of a few KW to 6.2 MW (as of 2016), ranging in height to tip of blade from under 10 m to 246 m (Gaildorf, Germany) have powered domestic to large-scale commercial developments, with the largest onshore wind farm in Europe (2016), the 240 turbine, 600 MW capacity site at Fantanele-Cogealac, Romania.

Plans for larger developments, greater capacity, and taller turbines are reflected in scenarios of installed capacity of wind energy by 2030 (Wind Europe 2017) of 323 GW of wind energy capacity, comprising 253 GW onshore and 70 GW offshore.

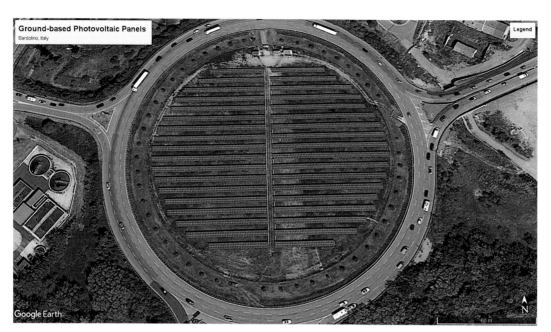

Ground-based Photovoltaic Panels
Bardolino, Italy

Legend

Google Earth

60 m

N

Figure 5.1.3.2
Above-ground solar
energy; integrated into
transport infrastructure;
Bardolino, Italy (Source:
Google Earth Pro)

Figure 5.1.3.3
Above-ground energy;
integrated into sensory
urban park; Tirol, Austria
(Source: D Miller)

Outreach of COST RELY and Reflection on Future Strategies

This is over double the combined figure at the end of 2016 of 160 GW. To accommodate such expansion, future siting can be expected in almost any landscape type, subject to wind resource and planning priorities. Some landscapes are highly likely to be transformed from their current character, the significance and desirability of which will depend on local context and contemporary perceptions of landscapes. Such changes can be assessed (Figure 4a) and cumulative impacts monitored.

The development of onshore wind energy has been encouraged by policy-makers due to its proven technology, and scalability to suit the wind and land resource available. This technology has enabled many countries to stay on track to meet targets for renewable energy production whilst other technologies (e.g. solar, marine renewables) are rolled out.

Over the lifetime of a wind energy development, the experience gained of siting, reliability, and productivity inform investment decisions linked to replacement of the turbine nacelle (10 years) or renewal of planning permission (25 years). These moments offer opportunities to change aspects of design, significant change through returbining, or decommissioning. As more developments go through these phases, they provide feature-specific dynamics of landscapes, and means of remaining contemporary to societal expectations, including the possibility of landscapes that retransform to post-renewable energy uses.

Figure 5.1.3.4
View of new wind farm development on north west Norwegian coast from cruise ship (Source: D Miller)

The visual influence of onshore wind development includes coastal areas, where views of wind turbines are influenced by the shapes, uses, and colours of land and sea. Seascapes in which there have been considerable increases in wind energy development include ports and harbours, in which turbines are not changing landscape characteristics significantly. In other areas, wind turbines may change local landscapes by introducing the first, visually identifiable, built structures (e.g. Figure 4b). As coastal areas usually have good wind resources, these areas will continue to be attractive for development, and so cumulative landscape impacts may increase.

Wind farms at sea can be described as inshore and offshore, both of which will make a signifi-cant contribution to achieving the 2030 targets. Some of the largest developments being built, including floating structures, arguably form new seascapes visible and experienced only from the sea. Those closer to the shore have direct impacts on seascapes, extending the area of ostensibly permanent, built structures out from the coast. The nature of such impacts will be influenced by the context of the shapes of local coastlines and topography and coastal uses of land and sea. Some such seascapes will be transformed from predominantly natural and minimal human influence, whereas in others wind energy will be part of a mix of primary and secondary industry focused on aquaculture, fishing, energy (oil and gas), or transport.

Figure 5.1.3.5
Number of wind turbines by Landscape Character Assessment areas, for Scotland, 2012 (Datasets: SNH, 2013; SNH, 2015)

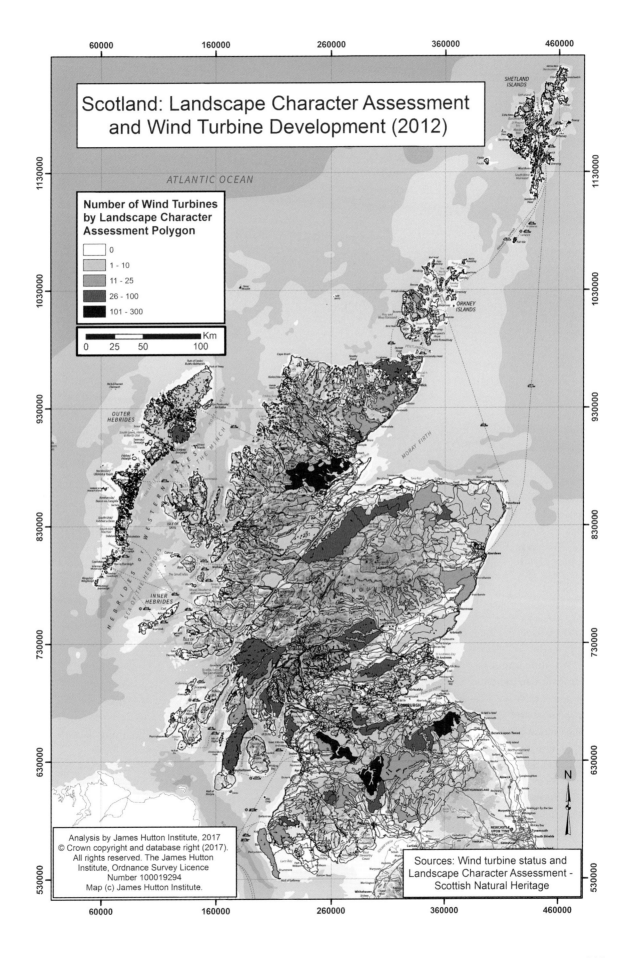

Scotland: Landscape Character Assessment
and Wind Turbine Development (2012)

**Number of Wind Turbines
by Landscape Character
Assessment Polygon**

	0
	1 - 10
	11 - 25
	26 - 100
	101 - 300

Km
0   25   50   100

ATLANTIC OCEAN

SHETLAND
ISLANDS

ORKNEY
ISLANDS

OUTER
HEBRIDES

MORAY FIRTH

INNER
HEBRIDES

EDINBURGH

Aberdeen

NEWCASTLE
UPON TYNE

N

Sources: Wind turbine status and
Landscape Character Assessment -
Scottish Natural Heritage

223

5.1.3.5 Tidal and Wave Power

The development of tidal and wave renewable systems through to 2030 will impact on coastal landscapes and seascapes. Projections of installed capacity, from almost zero in 2005 and up to 2014, is to increase to 84 MW in 2018 and 100 GW by 2050. The geography of such developments is focused on the Atlantic coast, entrances of estuaries, and where water flow creates the greatest potential tidal or wave energy (e.g. between islands).

Previous to such development, these areas may have had no permanent or long-term built structures. The effects on the seascapes will be primarily visual and ecological. The introduction of energy systems will be visual evidence of human intervention and production from natural resources, representing disturbance with potential reductions in perceived naturalness associated with those seascapes. However, social and economic implications for some coastal communities (e.g. allocation of financial benefits dedicated to local communities) may reduce risks of community abandonment with indirect consequences for the seascapes of those areas (e.g. adverse impacts on tourism resources).

5.1.3.6 Geothermal Power

Increased uptake of geothermal energy, in some places linked to former industrial mining areas, is likely to support economic developments on site. Implementation of these systems is rapid (less than a year). Significant direct or cumulative change in physical landscape is unlikely. However, geothermal energy systems are being used to create opportunities for economic development which may be transformational in social terms. In the medium term, such change may lead to new associations between activity or outputs and place.

5.1.4
Integrated Landscape Change

Changes in spatial development and implementation of renewable energy systems will differ by technology type, with increasingly decentralised locations of solar, wind, and micro hydro systems. The existing distribution of renewable energy systems across the diversity of social and environmental conditions in Europe is evidence of how they can be designed for consistency with landscape quality, to stimulate business, and to be used as a catalyst to increase social engagement and innovation. Alongside technology, social innovation in communities is driving development of smallscale renewables schemes and landscape change.

However, the lifespan of such landscapes is unknown. As new and significant sources of renewable energy come online, notably tidal and offshore wind systems, combined with less favourable economic incentives for other forms (e.g. onshore wind), the geographic focus of renewables may change. An implication could be reduced interest in existing developments in locations which have proven inappropriate economically, socially, or for landscapes. That is, opportunities will be created for reversing, or modifying, decisions relating to renewable energy developments in light of experience of its operation (e.g. lower financial return due to poor siting of turbines, or landscape sensitivity to cumulative visual impacts which can be ameliorated by decommissioning of particular developments).

Transitions in landscapes through to 2030, and beyond, can be expected to occur as disturbance or incoherence in landscape characteristics due to renewable energy systems evolving into new combinations of patterns, structures, and relative positions of features. In these landscapes, new proportions and configurations of characteristics will emerge.

5.1.5
Conclusions

European and international policy is set to 2030, with indications to 2050. As details of national strategies and plans evolve to achieve contemporary national targets, so will the prospects of new patterns emerging in landscapes in response to the 2030 Energy Strategy. Specific landscape change will depend upon allocation of land with potential for renewable energy production, relative priorities for such areas, political will, and social attitudes. Some landscapes may be transformed in the short term, but with technology that is reversible and replaced by a new or different mix of renewable energy. A challenge for signatories to the European Landscape Convention is how to ensure that such transformations of landscapes take place consistent with the principles of public participation in decision-making (Council of Europe 2000).

Achieving the aims of the 2030 Energy Strategy will require increasing the rate of renewable energy development. The global stocktakes will be opportunities for revising strategies for reducing future GHGs, and the role of renewable energy. The transformations implied for some landscapes or areas of Europe will need public acceptance. So, of greatest significance for achieving the aims of the energy strategy to 2030 and beyond will be the attitudes of new generations of citizens towards energy use, generation, and landscapes. Those land managers, residents, and tourists will be the agents of the European Landscape Convention statement that 'protection, management and planning entail rights and responsibilities for everyone'.

5.2

ADAPTIVE MANAGEMENT STRATEGIES FOR RENEWABLE ENERGY LANDSCAPES

David Miller, Alexandra Kruse & Michael Roth

5.2.1
Introduction

The pressures of climate change are leading to policy, economic, social, and biophysical responses to mitigate its effects, or adapt to new circumstances, both opportunities and threats. Adaptation to climate change is the process of adjusting to actual or expected climate change and its effects (EU Court of Auditors 2017). The basis for EU action in the area of adaptation is its Adaptation Strategy (2013), which encourages Member States to take action but does not make it mandatory. This Strategy sets a key objective to 'climate proof' action at a European level, promoting adaptation in vulnerable

sectors, including agriculture and cohesion policy. The Strategy identifies key principles for adaptation that include reducing risks and realising opportunities associated with climate change. Whereas the increased rollout of renewable energies, and the energy transition in general, is considered the main focus of the climate change mitigation strategy in order to reduce greenhouse gas emissions (GHG) (IPCC 2013, 1458), renewable energies are also a growing part of adaptation strategies. Examples are seeking diversification of income for farmers to be resilient against income losses associated with climate change, or a decentralisation of the energy production system to be less vulnerable against damage induced by the effects of climate change. These goals are also reflected in the EU 2030 Energy Strategy in terms of its aims of reducing green-house gas emissions (GHG) by 40 % compared to 1990 levels, and generating 27 % of energy consumption from renewable sources by 2030 (European Commission 2014).

Alongside enabling its targets being achieved, the EU 2030 Energy Strategy recognises the need to ensure the sustainable use of land, and the resource efficient use of biomass, and to addres indirect land use effects (e.g. with biofuels). The challenge for policy, industry, and citizens is to achieve the aims of renewable energy production, whilst managing the associated changes in uses and functions of land and understanding the implications for landscapes, and the people who live in work in, or are beneficiaries of them.

The United Nations Convention on Biological Diversity (United Nations 2000) sets out principles and approaches for the integrated management of land, water, and living resources to promote conservation and sustainable use in an equitable way. The Convention advocates for the ecosystem approach and adaptive management as a strategy for management recognising that change is inevitable. Applied to land management, this principle invites an approach that has an understanding of the dynamics of the local land systems, learning lessons from past changes (direct and indirect), consideration of possible adaptation strategies, and planning approaches to trajectories of future change.

The promotion, support, or uptake of renewable energy systems has been an important element in institutional and individual adaptation to climate change. Primarily these are examples of 'passive

adaptation' (Prato 2017). The incorporation of renewable energy into landscapes across Europe is reflecting strategies of adaptive management, over different geographic and temporal scales. Biophysical, economic and social contexts provide constraints and opportunities for different types of renewable energy, leading to changes in some landscapes that are consistent with their current or recent characteristics, and changing those of others. The change of the physical characteristics of landscapes is taking place contemporaneously with changes in societal attitudes and responses to landscape change as well as energy use and provision. Both types of adaptation (physical landscapes and cognitive landscapes) are closely linked, but not necessarily synchronised, and therefore can lead to either acceptance or rejection by the public of renewable energies in the changing landscape.

5.2.2
Landscape Change and
Adaptive Management

An adaptive management strategy is one which implements a management action or decision (e.g. to generate renewable energy), monitors the system to identify the response (e.g. peer and societal reactions), and then adjusts the action in a continual cycle of feedback based upon monitoring and learning (e.g. efficiency of development, economic return, price regimes) about how the system responds to

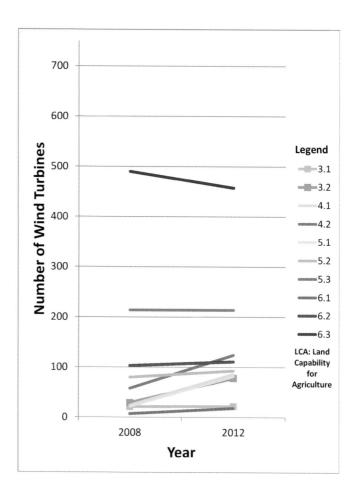

Applications for Wind Turbines
Figure 5.2.2.1
Applications for planning permission for wind turbines, by Land Capability for Agriculture Class, Scotland, 2008 to 2012.

management decisions and actions (e.g. expansion or decommissioning of a development).

An example of such adaptive management strategies in relation to renewable energies (and indirectly the landscape change induced by them) is the so-called 'floating cap' for the new installations of wind turbines in Germany. The 'floating cap' was introduced as a reaction to the fact that, in 2014 and 2015, more wind turbines had been built than planned. These wind turbines could not all be connected to the electricity grid and/or, the energy transmission system in general could not accommodate all of the energy they produced. With the reform of the Renewable Energy Sources Act in 2017 in Germany, a flexible instrument of reducing feed-in tariffs was introduced, mainly to control the economic efficiency of the system. Although principally an economically motivated adaptive management strategy, this is an illustration of how, even with national policy, flexible approaches are possible.

Fazey et al. (2009, 416) argue that many adaptation strategies focus on improvising short-term capacity to deal with environmental change, but can increase vulnerability to unforeseen changes in the future. Landscapes in which renewable energy systems are being introduced often have characteristics of providing capacity (renewable energy) in response to environmental change (climate change), and in some cases are exposed to risks that are new to an area. For example, the conversion of land use to the production of woodland biomass energy

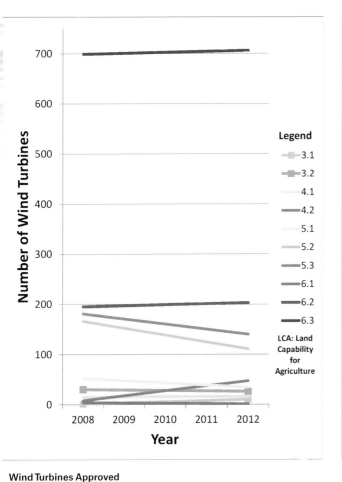

Wind Turbines Approved
Figure 5.2.2.2
Applications for wind turbines
approved, by Land Capa-
bility for Agriculture Class,
Scotland, 2008 to 2012

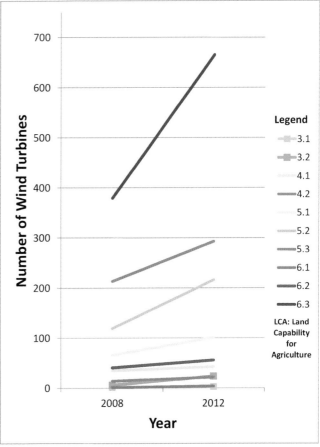

Wind Turbines Installed
Figure 5.2.2.3
Wind turbines installed,
by Land Capability
for Agriculture Class,
Scotland, 2008 to 2012

has an associated increase in the risk of forest fire.
The magnitude of the risk may remain small, but
the exposure and vulnerability may be significant.
Therefore, over a long term the potential for 'disruptive' events increases.

Human responses to drivers for renewable energy or
territorial-related policies and socio-economic conditions include investments in financial and social
capital. Actors responding to opportunities for the
development of renewable energy are taking advantage of technologies which are new, or as they emerge
in different places at different times. The consequences for landscapes have been changes in their
characteristics, the types and rates of which varies
across Europe reflecting differences in biophysical,
economic, and social opportunities for such change.

Landscapes can be managed to adapt to climate
change, led by changing societal priorities. Such
societal priorities are amongst the driving forces
that continuously modify the 'state' of a landscape,
making it neither steady nor constant. They may
undergo a process of development that can be
chaotic and autonomous (Antrop 2005, 31), with
more intense dynamics of change leading to it
being faster and more extensive (Antrop and Van
Eetvelde (2017, 142), i.e. cumulatively large-scale.
Precisely which processes of landscape change are
dominant, and the consequences of these changes,
can vary under different geographic and climatic
conditions. Such influences are exerted mainly
through the policy-induced acceleration of processes of intensification and extensification, and

directly or indirectly influencing the character of the landscape.

Most types of renewable energy are consistent with the intensification of land use, but operate over different timescales. The emergence of renewable energy systems implemented by early adopters may create incoherence in landscapes as features are introduced in places where they are unfamiliar.

Wind, Solar, and Hydro Power

A strategy involving wind, solar, or micro-hydro-power requires changes in land use that can be significant. The physical construction of renewable energy by these sources is relatively rapid (i.e. time periods of months), as can be the direct impacts on characteristics of landscapes (e.g. visual, experiential). Return on investment may also be rapid. However, the overall timescale for development requires consideration of the planning process, which is significant in terms of the strategy of the developer (e.g. farmer, energy company) and associated stakeholders. Timescales for physical change away from renewable energy, i.e. decommissioning, can be equally rapid, with the likelihood that it would revert to previous uses (e.g. crops, livestock).

The increase in use of wind power in Europe, from 22.2 TWh in 2011 to 301.9 TWh in 2015, has been the most rapid of all forms of renewable energy. Increases have been across scales of development from small (e.g. 10 KW, single turbines for domestic production) to large (e.g. 240 turbine, 600 MW capacity, Fantanele-Cogealac, Romania). Wind energy is an example of adaptation strategies which have had significant impact on landscapes.

Figures 1 (a) to (c) show the change in wind turbine development in Scotland, UK, between 2008 and 2012 (Scottish Natural Heritage 2013) with respect to the Land Capability for Agriculture (Bibby et al. 1991). The patterns of adaptation appear to have followed opportunity, demand, and financial incentives. Phases of development which can be interpreted evident are those of medium and large-scale energy companies identifying and developing large geographic sites, generally on land which is less flexible for agricultural production such as exposed peatland, hill sides, and plateaux of moorland habitats (e.g. LCA Class 6.1 to 6.3), and land previously occupied by forestry. Between 2008 and 2012 the number of applications for such areas was already falling due to issues of visual impacts, cumulative environmental impacts, and increasing difficulties and costs for linking to the electricity grid, whereas the number of installations was increasing dramatically, which reflected the approval of significant numbers of earlier applications. All such areas are subject to constraints such as land designations, access to the electricity grid and with the support of the land owner (sometimes the state). Most such areas are now developed, under development, or have had proposals rejected.

At a smaller scale, development has often been farmer led. For example, in north east Scotland, the adaptation strategies of land managers was to diversify their enterprises, invest financial capital and land in small scale wind energy and associated infrastructure. Within the same constraints as for larger developments, the focus of attention was on windier, higher, lower-quality land (grazing). This interest is reflected in the increase of applications for wind turbines on better quality agricultural land (LCA Class 3.1 and 3.2) between 2008 and 2012 (Figure 5.2.2.1). No applications, approvals, or installations were recorded for the best agricultural land of LCA Classes 1 or 2. The number approved reduced slightly (Figure 5.2.2.2), and those installed increased (Figure 5.2.2.3).

Successful development appeared to have two significant implications with respect to other developments. One was to stimulate interest and further applications for development through demonstrable success and financial returns, and the second was to increase the baseline for cumulative impacts of wind power developments, leading to an increasing number of proposals being rejected.

One implication of the pattern of development has been to increase development on lower land, including that used for arable agriculture, progressively introducing visual evidence of renewable energy production in association with agriculture. Feedback from early adopters has influenced uptake by other land managers, with formal (e.g. training) and informal (e.g. observation) learning about wind energy systems, leading to its evolution and scaling out. Arguably, this has led to landscapes which are characterised by renewable energy developments. This has demonstrated the success of the adaptation strategy at the level of individual enterprises, reinforced by public policy of an Agri-renewables Strategy (Scottish Government 2014).

Biofuels

Figure 5.2.2.4 shows the dramatic increase in the production of liquid biofuels between 1990 and 2015 (Eurostat 2017), in particular the rise increase from < 1 Mtoe to approximately 14 Mtoe since 2000. In 2015, the European Commission introduced rules to restrict the conversion of land use into the production of biofuels (i.e. reduce indirect land use change). These included limiting the share of biofuels from crops grown on agricultural land that can be counted towards the 2020 renewable energy targets to 7 % (European Commission 2012).

To comply with the regulations, increases in the land area in Member States dedicated to biofuel production can be expected to be restricted to changes in arable crops. Conversion of other land uses, in particular forestry, to biofuel crops would not be permitted. However, demand for biofuels in Member States could lead to land use change in other European countries including conversion from forest land in the short term.

Leaving aside important issues of supply chains and accessibility to land, the most rapid changes in uptake of renewable energy is biofuels, most often swapping one arable crop for another, maintaining a general land use of agricultural production albeit fuel rather than food. Their growth is in terms of months, and production associated with seasons. The land manager makes the decision as an individual or enterprise, probably adapting management practices to emerging economic opportunities (e.g. market prices). Return on investment is also likely to be rapid. Change in land use

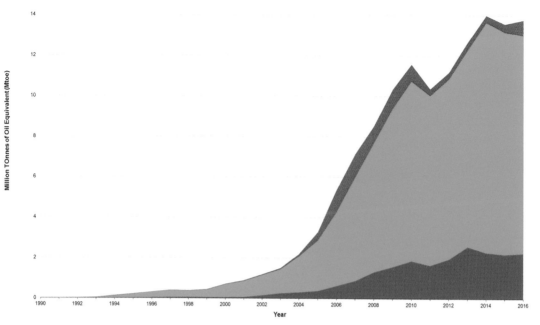

Figure 5.2.2.4.
Primary production of
liquid biofuels, European
Union, 1990–2016
(Source: Eurostat 2017)

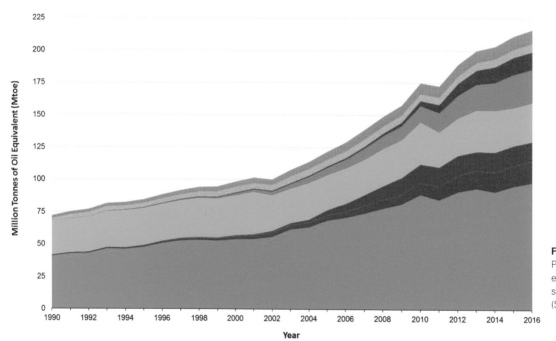

Figure 5.2.2.5.
Primary production of
energy from renewable
sources, EU-28, 1990–2016
(Source: Eurostat 2017)

Outreach of COST RELY and Reflection on Future Strategies

away from renewable energy can be equally rapid, with the adoption of alternative crops. A strategy involving biofuels may involve changes in land use that are significant, but the changes in landscape, almost always agricultural, may be negligible.

Biomass

Primary production from biomass and solid biofuels renewable energy sources increased from 40.6 Mtoe in 1990 to 90.4 Mtoe in 2015 (Figure 2b, Eurostat 2017), with the proportion of renewable energy from biomass projected to increase to 42 % by 2020. No estimate is available of the area of land required for the production of biomass in Europe. However, over the timescales required for the sustainable production of biomass for energy the suitability of land could change in response to changes in climate leading to some areas becoming less suitable for production.

If starting from the planting of a new resource, then the realisation of increased production of biomass as a source of renewable energy is a slower process to achieve its production potential (i.e. years). Increasing the allocation of land to this type of renewable energy can be expected to be consistent with the slow, but cumulatively large-scale, change of a landscape. In some places, biomass production might be combined with the local introduction of new plant species such as miscanthus (e.g. *Miscanthus × giganteus*), various species of Eucalyptus, or a changed in use (e.g. corn and rape seed), but also the important increase of fast growing timbers (*Salix spp, Populous spp., or Robinia pseudoacacia*). The increase of energy plant cultivation will have

consequences (positive or negative) for the local fauna or ecosystems in general, including soil quality and health. Transferring the use of existing woodland to the provision of energy from other products (e.g. timber for construction) may have minimal apparent impacts on the land use or landscape. However, it may have required significant changes in silvicultural practices, species of tree planted, be sensitive to changes in biophysical conditions (e.g. climate), influence land use patterns (e.g. more rectangular, homogeneous in size), and increase monoculture cultivation. In turn, such changes are likely to lead to an intensification of land use, and a loss in landscape structure and diversity (e.g. by a reduction in the number of smaller field units and a decrease in the heterogeneity of landscapes).

5.2.3
Pathways of Landscape Change

Path dependencies in place and time can be embedded in spatial strategies, leading to development in one era evolving, maturing, and requiring replacement or reinvestment during subsequent eras. A strategy can be expected to direct development (e.g. wind, forest, solar) to particular geographic areas or types of landscapes, and away from others. Under this form of spatial planning, the timescales for different types of renewable energy systems (planning, development, operation, and decommissioning or reinvestment) will strongly influence the types of uses of land in areas for different time periods: long-term (e.g. large scale hydropower, forest biomass), medium-term (e.g. onshore wind, above-ground solar) and short-term (e.g. biofuels). Some such uses will preclude others in the same place for the period of their operation (e.g. large-scale hydro prevents forest biomass). Other forms of renewable energy can co-exist such as in spatial units of the same field (e.g. wind power with biofuels, forest biomass or animal waste for anaerobic digestion), or management units (e.g. one farmer with several forms of energy production in the same ownership unit).

Grin et al. (2010, 55) describe a typology of transition pathways, outlining consequences of pressures building on landscapes in a particular direction, gradually becoming more disruptive. Arguably, the uptake and development of agri-renewables has been bottom-up, and largely a response to the promotion of renewable energy technologies and their economic and environmental benefits, in turn leading to evolutions in the character of agricultural landscapes in Europe. Where one technology has become sufficiently disruptive (e.g. cumulative visual impacts of wind turbines) opportunities are taken to invest in other technologies that can be developed in similar land management units. An example in some agricultural landscapes is the increase in investments in solar energy farms instead of wind energy systems, or even the combination of solar energy farms with agriculture or horticulture (agro-photovoltaics).

Linked to the development of the energy resource transitions, pathways will begin to include relevant infrastructure (e.g. electricity transmission), supply chains (e.g. biomass plants and associated transportation), and knowledge (e.g. environmental, legal, and financial opportunities and constraints). Combined, such investment in renewable energy provides a basis for pathways of expansion and diversification in terms of renewable energy system (e.g. uptake of solar). This can lead to the emergence of characteristics of landscapes which are predominantly associated with energy. In due course such pathways may diverge away from renewable energy due to emerging issues such as reactions to cumulative impacts on landscapes. Such divergence would reflect the adaptation of a land management strategy in response to emerging conditions directly associated with the evolution of renewable energy landscapes.

One advantage of renewable energies is their spatio-temporal flexibility as compared to nuclear or fossil energy production systems. With economic break-even periods of less than 10 years (for some photovoltaic systems), expected lifetimes of approximately 20 years, and decommissioning bonds required (for wind turbines), as well as guaranteed feed-in tariffs that are legal binding for 20 years (e.g. in the German Renewable Energy Sources Act), renewable energies offer flexibility for people to adjust their short or medium-term learning cycles in adaptive management strategies. These are not possible for nuclear power systems (which may take several decades to complete all the phases of planning and construction, as well as safely managing waste products for thousands of years), or fossil fuel systems such as open cast lignite mining (which may also require many years going through planning processes and permanent changes in topography and long-term changes in landscape ecological systems).

5.2.4
Conclusions

Plans and policies are being developed at international to local levels to direct and manage land to tackle the challenges of climate change. In several such plans and policies, renewable energy has a significant role in futures that have qualities and characteristics designed to meet social, economic and environmental objectives. The modulation of landscapes to the provision of renewable energy reflects formal strategies for their adaptation to new contexts, as well as biological and human responses, and their cumulative effects.

Increasingly, alignment is integrated across strategies providing multiple functions and benefits. For example, the adaptation may be part of planned interventions to produce energy from a river or reservoir. Indirectly, the principal drivers may be economic as a reaction to environmental climate change and subsequently renewable energy goals, although small or medium scale hydropower can be introduced as part of a broader plan for water management, including flood control.

As drivers evolve so landscapes are changing, biophysical changes such as those linked to climate change are setting constantly evolving contexts (e.g. constraints) for the types of renewable energy systems that are central to the reforming of landscapes. Humans have been responding to opportunities to develop contemporary land systems which have a key function of tackling climate change, which also reflects technical and social learning. Now, attention has to be directed towards avoiding 'litter pollution' through old or non-operative renewable energy installations, such as oversized and unprofitable biogas plants due to insufficient local or regional substrate production capacity, out-dated wind turbines or non-functioning solar panels (ground or roof-mounted). Policies are required across Europe to define standards and rules on visual impacts (e.g. uncontrolled use of roof installations) as well as landscape quality impact assessments. These could have associated spatial plans at all levels of planning (European, national, regional, local) that indicate those areas which are suitable for different types of renewable energy production, and those which are not suitable, with clear explanations for the normative and scientific basis of such plans.

In some areas, the adaptation of landscapes to renewable energy systems is a return to those which existed previously, in others it is a pathway to systems new to those areas. In both cases, in the short or longer term, they represent phases in evolution of the landscapes which will continue to change in response to new, currently unforeseen pressures.

OPEN WINDOWS TO EUROPEAN ENERGY LANDSCAPES

Juan José González, Nieves Mestre & María-José Prados

These days many parts of Europe are characterised by renewable energy landscapes. European and national policies have been jointly developed to advance renewable energies. The results are sparking major technological, environmental, and social changes that have a significant impact on European territory. This has inspired RELY country members to construct a photo database to document geographical differences with the creation of an extensive repository of photographs compiled between 2015 and 2018 offering a visual account of energy systems, installations, and landscapes from all over Europe.

The initial goal was to provide a graphic platform for RELY transfer activities, representing both an instrument for participants to share graphic information and an open-access repository that can be visited by experts or any other interested parties. From the very beginning, the photo database was conceived as a powerful tool to illustrate the wealth and diversity of renewable energy landscape implementations in European landscapes whilst simultaneously helping to enhance public acceptance of renewable energy systems. Information can be accessed at: http://cost-rely.eu/resources/photo-database.

5.3.1
The Photograph Database Project

All country members were asked to provide photographs of renewable energy landscapes to illustrate ongoing research projects in their regions or countries. Country members also submitted photographs taken during fieldwork for the various activities that are part of the COST Action RELY. These photographs are regarded as an intrinsic part of disseminating results and a visual output to reinforce the relationship between renewable energies and landscape quality. Each photograph includes a metadata file with detailed information about the photographer and the photograph itself, including name, location (country and region, also city for renewable energy systems installed on roofs), date, motive for taking the photograph, and keywords selected from the COST RELY Glossary (see chapter 5.4).

Over a hundred photographs taken in twenty-one European countries portray best renewable energy and landscape quality practices. The selection of photographs included here show how extraordinarily widespread renewable energy landscapes have become as a result of technological advances. Real-world renewable energy landscapes display a variety of energy system typologies, installation patterns, best locations, and associated land uses. Five main energy system typologies and twenty-six sub-typologies combining different physical plat-

Figure 5.3.1.1
New PV panels alongside old windmills on Milos, Greece, are an interesting source of conversation about energy typologies and technologies in the village's history (Source: A. Kruse 2015).

237

HYDRO ENERGY

WIND ENERGY

BIO ENERGY

SUN ENERGY

GEO-THERMAL ENERGY

forms and scales were recorded (Figure 5.3.2.2). The sources of renewable power resources illustrated in the photographs are biological, geo-thermal heat, sun; water, and wind. The dominant rural land use for land-based installations is agriculture and forestry. Installations are omnipresent and can even be found in remote, mountain areas, where photographs show the presence not only of hydroelectric power stations, but also of roof and ground-mounted photovoltaic systems. Finally, the database also includes examples of power infrastructure, renewable resources, new siting practices and lastly, but no less importantly, renewable energy landscape bad practices.

5.3.2
Mapping the RELY Photographs

The photo database includes photographs submitted by members from countries ranging from Portugal to Romania and from Greece to Iceland. The map of the photo database sets out renewable energy landscape practices in twenty-one European countries with ample regional representation (Figure 5.3.2.1). Energy systems are well represented by a wider set of resource and installation typologies defined in the COST RELY Glossary (http://cost-rely.eu/resources/glossary). As the map shows and the table proves, solar landscapes are the most common, followed by wind landscapes. Examples of the latter were submitted from thirteen coun-

Figure 5.3.2.1
COST RELY Photograph Database Map. The map shows a predominance of wind and solar farms, both of which are usually associated with scenic landscapes (Source: J.J. González 2017, developed for COST RELY)

Figure 5.3.2.2
COST RELY Photograph Database Chart. The chart shows a preference for both wind and solar farms scenarios—overcoming regional differences. (N. Mestre 2018, developed for COST RELY)

TYPE	SUB-TYPE	COUNTRY	AUTHORS
hydro energy	large	Austria	Csaba Centeri, María- José Prados
	small	Bosnia Herzegovina	Igor Kuvač
	large	Iceland	Csaba Centeri
	large	Norway	Sebastian Eiter
	large	Portugal	Naja Marot, Csaba Centeri, Maria Bostenaru
	micro	Romania	Maria Bostenaru
	large	Slovenia	Csaba Centeri, María- José Prados
wind energy	on-shore	Austria	Marton Havas, Rebecca Krieger
	on-shore	Bulgaria	Georgi Hinkov
	on-shore	Denmark	Naja Marot
	old mill	Denmark	Naja Marot
	on-shore	Francia	Naja Marot
	on-shore	Germany	Olaf Schroth, Jochen Muelder
	off-shore	Germany	Olaf Schroth
	small	Germany	Jochen Muelder, Andre Berger
	on-shore	Hungary	Marton Havas
	on-shore	Iceland	Csaba Centeri
	on-shore	Latvia	K. Reinis
	on-shore	Netherlands	Berthe Jongegan
	on-shore	Portugal	Filipa Soares, Maria JoaoNunes, Naja Marot, Csaba Centeri
	on-shore	Romania	M. Sbarcea
	off-shore	United Kingdom	Olaf Schroth
	on-shore	United Kingdom	Olaf Schroth, Naja Marot
	on-shore	Iceland	Csaba Centeri
	on-shore	Spain	María- José Prados, Juan José González
bio	bio-mass	Germany	Alexandra Kruse
	bio-gas	Croatia	Romulic & Stojcic studio
	bio-mass	Netherlands	Berthe Jongegan
solar energy	PV on-Roof	Austria	Marton Havas, Christina Nöbauer
	PV-on ground	Austria	Sebastian Eiter, Csaba Centeri, María- José Prados
		Austria	Gerald Leindecker
		Czech Republic	Bohumil Frantál.
		Croatia	Romulic & Stojcic studio
	Pv on groud	Germany	Sebastian Eiter, Olaf Schroth
	PV on-Roof	Germany	Jochen Muelder
	PV on-Roof	Greece	Nikos Papamanolis
	Pv on groud	Greece	Alexandra Kruse
		Hungary	Robert Kabai
	Pv on ground	Hungary	Balint Kiss
	on-roof	Netherlands	Berthe Jongegan
	on- Ground	Norway	Sebastian Eiter
	on-roof	Slovenia	Csaba Centeri
	on-ground	Slovenia	Naja Marot, Csaba Centeri, María- José Prados
		Spain	Manuel Perujo
	PV on-Roof	Italy	Maria Bostenaru
	Pv on ground	Portugal	Luis Junqueira, Luis Silva, Naja Marot, Csaba Centeri
	PV on-Roof	Romania	Maria Bostenaru
	Thermo Solar	Spaim	María- José Prados, Juan José González
	PV on ground	Spain	Juan José González
geo-thermal	power plant	Iceland	Csaba Centeri, David Ostman
	natural	Iceland	Csaba Centeri
	greenhouse	Iceland	Csaba Centeri
	greenhouse	Slovenia	Csaba Centeri, Naja Marot

Figure 5.3.2.3
Urban landscapes:
a) Solar balcony railings on a residential building in the village of Sant Elia (L'Aquila province) in the Abruzzo region of Italy. Rebuilding the village opened the door to renewable sources (M. Bostenaru 2010).
b) Three wind turbines spinning on the roof of the office of Greenpeace Germany in Hamburg/Germany (A. Berger 2016).
c) PV car park canopy between Innsbruck and Seefeld/Austria (M. Havas 2016).

tries, which can be seen in the map. The geographical diversity of these examples is a clear illustration of the pervasiveness of this resource and, above all, of the fact that it can be adapted for mounting on both buildings and on the ground. Most of these landscapes are therefore of photovoltaic installations on roofs or on land, although there are also some photographs of thermo-solar plants in the south of Spain, and photovoltaic and wind plants along the Spanish-Portuguese border.

Wind landscapes are also widely represented in territorial terms. The photographs show a selection of twenty examples in thirteen countries: Germany, the United Kingdom, Iceland, Denmark, Austria, Bulgaria, France, Hungary, Latvia, the Netherlands, Romania, Spain, and Portugal. The first three countries provide interesting examples of offshore wind farms in the North and Baltic Seas. Wind farms have also proliferated in inland wind corridors in these two countries, as they have in other countries in the continent, including the transalpine valleys in Austria and the mountains in the south of Portugal and Spain.

Hydroelectric landscapes can also be found in a large number of countries. The photo database includes examples of hydro-energy from seven countries ranging from Romania to Portugal and including Iceland, Norway, Austria, Slovenia, and Bosnia Herzegovina. The wide variety of sites includes water landscapes in fjords and river basins and on lakes, and even along river courses with the construction of tiered artificial channels. Finally, the map shows geo-thermal and bioenergy landscapes, well illustrated by geo-thermal power plants in Iceland. Bioenergy plants in Germany, Croatia, and the Netherlands are good examples of the reuse of a resource that is changing not only the landscape, but also the economic profitability of areas of forestry and agricultural land throughout Europe.

Figure 5.3.3.1
Agricultural landscapes.
a) Amareleja PV plant surrounded by dairy cattle in Moura, Portugal (L. Junqueira 2013).
b) The highest growth rates in energy production in the coming years are expected to come from bioenergy plants: bioenergy corn field in Bergisches Land, Siegkreis, Germany (A. Kruse 2016).
c) The Solucar thermosolar plant outside Sanlúcar la Mayor, Spain; the largest solar complex in Europe. Over a thousand mirrors follow the sun's path across the sky, preventing the emission of about 12,000 tons of CO_2 into the atmosphere per year (J. J. González 2013).

5.3.3
Views of Renewable Energy
Landscape Sites

The photographs cover a wide array of renewable energy landscapes, documenting their regional diversity as part of the integration of energy systems into the territory. They also show renewable energy landscape practices. In line with the criteria of territorial coverage and good practices, this chapter contains eleven photographs that provide an overview of the way in which renewable energy systems interact with their sites and locations. The photographs have been systematised according to a number of criteria, including type, location, type of space, integration into the setting, associated uses, visibility, symbolism, and scale. The result is three sets of photographs of urban, agricultural, and natural landscapes that portray the reality of European energy landscapes today.

The examples labelled urban landscapes show the amalgamation of the urban fabric and energy supply devices. Roof-mounted photovoltaic installations supply energy in a wide range of daily situations. Two instances are given of their efficient or successful integration into residential buildings. The first is a solution in the form of solar panels mounted on the balconies of a residential building in L'Aquila (Italy), part of a new development in the wake of the 2009 earthquake, the CASA project: Complessi Antisismici Sostenibili ed Eco-

campatibili (Figure 5.3.3.1). The second shows an intelligent solution in HafenCity, an urban redevelopment project in Hamburg's port area, where the headquarter of Greenpeace is located (Figure 5.3.3.1). In the third photograph we see a covered car park in Hartberg, Austria, a multifunctional format that can transform the urban landscape of cities by simultaneously providing parking space and generating energy to recharge electric batteries (Figure 5.3.3.1).

However, the best-known renewable energy landscapes are of land installations in the countryside. A differentiation has to be made between energy landscapes in agricultural landscapes and in natural landscapes. The former preserve their traditional links between energy landscapes and the territo-

ry, as they can contribute to economic activity in many rural areas while also providing the opportunity to improve domestic comfort and convenience. And last but not least sometimes provide even a further income. A new look is taken at these relationships between the rural environment and energy with a set of macro-installations that co-exist with agricultural, grazing, and forestry land uses (Figure 5.3.3.2a). The photographs show wind and PV installations in the south of Europe, where they are just as likely to be found on cropland as woodland pastures or river meadows, all of which are economically optimised through the installation of large-scale heliostats, wind turbines, and thermo-solar plants. Cattle grazing on scrubland and grassland, as seen in the photograph of the

Figure 5.3.3.2
Natural landscapes
a) Portugal's extensive network of hydropower facilities provides half of all renewable energy production in the country. The Alqueva Dam in Portugal (N. Marot 2015).
b) Krafla geo-thermal power station in Iceland. Geo-thermal power currently provides 25 % of all Iceland's electricity (D. Ostman, 2017, winter).
c) The Krafla geo-thermal power station in north-east Iceland. The large number of volcanoes in the area is an advantage for geo-thermal energy production (C. Centeri 2016, summer).

Amareleja PV plant in Moura, could be regarded as the epitome of an energy agro-landscape (Figure 5.3.3.2). Curiously, the photograph of the two bio-energy plants recalls the traditional piles of charcoal that used to be so characteristic of Mediterranean landscapes. However, in this case these are two new plants set in cornfields in Bergisches Land, in the North Rhine-Westphalia region of Germany (Figure 5.3.3.2).

The set of photographs in figure 5.3.3.2 shows the contradictory relationship between agricultural landscapes and energy landscapes, the old vs. the new. The photograph of the Island of Milos (Figure 5.3.1) shows the island's traditional energy landscape with a set of old windmills in the hills above the village; this energy landscape has been given a makeover, with modern solar farms now supplying energy to the rural villages. The agro-energy landscapes in figure 5.3.3.2 are on a different scale. This technology for the production of solar thermal energy bears an obvious comparison to onshore wind farms, both in scale and also with respect to the end use that the energy production is put to. This is an outstanding example in the countryside near Sanlúcar la Mayor (Seville province, Spain) (Figure 5.3.3.2c). The height of the thermo-solar towers magnifies the effect of the photovoltaic plants on the landscape. The capacity to generate electricity is on a similar super-sized scale for the needs of these rural areas, all of which reveals a rather contradictory relationship: energy agro-landscapes as dependent spaces that are both a new economic activity and give the appearance of being green or sustainable.

The last series of photographs shows renewable energy installations in natural landscapes (Figure 5.3.3.3). This selection presents interesting examples in which water plays a role: Alqueva Dam in Portugal (Figure 5.3.3.3a). The size of these facilities makes it hard to ignore either the image of new landscapes created by hydroelectric energy production or of the public response. The use made of fast flowing water, whether natural or artificial, reveals both sides of the story: the beautiful view of the reservoirs upstream compared to downstream public reaction to the dams.

To conclude this collection of photographs, the last views of renewable energy landscapes take us to the Krafla geo-thermal power station: an extraordinary example that demonstrates how clearly geo-thermal power projects stand out in the Icelandic winter/ summer landscapes (Figure 5.3.3.3b and Figure 5.3.3.3c). The energy resources of the subsoil used for energy production are not as visible as the surface systems that exploit the resources.

Social acceptance of energy projects in the landscape depends not only on the aesthetics of the project, but also on the use made of the energy produced by the project (as heat or electricity). There is a much greater acceptance when the energy has a direct, positive impact on the local community.

The gap with public support and social acceptance of renewable energies requires new initiatives to improve recognition of responsive landscapes. This controversy is even more intense in the appreciation of rural landscapes, as their meanings and perceptions can be romanticised or fictionalised. As perception is influenced by social and cultural backgrounds, investigations into how landscapes are really perceived are invaluable for finding common ground on which further discussions, assessment methods, and landscape policies can be based.

Acknowledgements

This photo database would not have been possible without the collaboration of the COST Action RELY participants and their colleagues. The authors would like to express their gratitude for the photographs and metadata submitted.

SPEAKING A COMMON LANGUAGE: THE COST RELY GLOSSARY AS A BASIS FOR TRANSDISCIPLINARY AND INTERNATIONAL COLLABORATION

Naja Marot & Alexandra Kruse[1]

The dictionary is like a time capsule of all of human thinking ever since words began to be written down. And exploring where words have come from can increase your understanding of the words themselves and expand your understanding of how to use the words, and all of this change happens in your thinking when you read the words.

Andrew Clements

There is something like an explosion in the meaning of certain words: they have a greater value than their meaning in the dictionary.

Marcel Duchamp

5.4.1
Introduction

According to the summary of different definitions, a glossary is a collection of terms on a particular topic providing definitions of those terms, sorted in alphabetical order. In the special case where translations are provided, we talk about a bilingual or multilingual glossary, depending on the number of translations procured. Mostly, a glossary is prepared to provide a common ground for discussion and/or work in the frame of a project, a company or for research. In the frame of the COST Action TU1401 RELY (Renewable Energy and Landscape Quality), it was decided to prepare a multilingual glossary and to explore the terminology used when describing the relation between renewable energy (production) and its impacts on the landscape. The project focuses on a subject that will—with the constant increase of the renewable energy production (two-thirds of the newly installed power added to the world's grids in 2016, data by International Energy Agency, cited by Vaughan 2017)—continuously engage energy companies, researchers, and academics. Especially since the new production installations are mostly of solar and wind power, the two types of installations that cause a larger visual impact on the landscape, the question of societal perception of impacts has gained in importance (examples Mérida et al. 2012, Ruggiero and Scaletta 2014, Roth and Gruehn 2014). According to predictions, solar and wind could together supply up to 39 % of electricity by the year 2060 (Davies 2016). Therefore, the demand for a more rigorous regulation of this form of production is getting bigger and, consistently more and more people from different educational and professional backgrounds have to cooperate respectively and are confronted with each other in planning process, public hearings in particular.

Keyword-based on-line searches of English-language media (e.g. *The Guardian*, *New York Times*) have shown a good coverage of the renewable energy topic in general, but also revealed that landscape or more specifically, the term energy landscape is not mentioned very often. Entering 'energy landscape' as a search word in a search engine reveals that the first and most widely used meaning of energy landscape refers to biology, physics, chemistry, and biochemistry where 'an energy landscape is a mapping of all possible conformations of a molecular entity, or the spatial positions of interacting molecules in a system, or parameters and their corresponding energy levels, typically Gibbs free energy' (Wikipedia 2017). In another example from Vaughan and Hopkins (2017), the term 'energy landscape' indicates the energy field in general: 'While acknowledging that renewables are remaking the energy landscape, the professor of energy has criticised the cost of today's windfarms and solar technolog.' The reader gets more focalised hits if they enter the search word renewable energy landscape where most of the results are on the topic within the COST RELY context. In the scientific literature (search performed via Scopus) one can find a few renewable-energy-related terminology overview articles and chapters such as the one by Nadai and van der Horst (2010) or Nadai and Prados (2015) who introduce the terminology to describe the relation between renewable energy and landscape. Nadai and Prados furthermore mention the need for a wider recognition of the new phenomenon, both in research and practice.

These findings confirmed the need for the preparation of the glossary. The starting point of the work carried out within the COST RELY glossary (Kruse and Marot 2018) first involved browsing the existing on-line glossaries, both in the field of energy

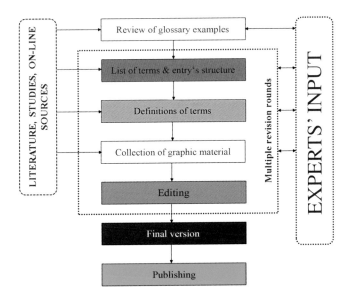

Figure 5.4.2.1
Glossary preparation process

and in the field of landscape/spatial planning. The overview shows there are several single topic glossaries, oriented either towards renewable energy or on landscape. However, the cross-over glossary remains missing and is badly needed. On one side, analysis of renewable energy or only energy-related glossaries (Eurostat 2017, IEA 2017, Integrated Energy Consultants 2017, USGBC 2017) shows that these glossaries mainly focus on the technical terms or terms related to statistical data for monitoring the production and use of energy, e.g. gross electricity production, KWh, net electricity production, biofuels, etc. Hence, they have very little connection to the territorial dimension of energy production. One of the rarest noticed exceptions was the integration of the term 'cultural landscape' into the U.S. Green Building Council glossary (2017) and mentioning of the 'landscaping' (defined as 'features and vegetation on the outside of or surrounding a building for aesthetics and energy conservation') in the U.S. Department of Energy glossary (2017). On the other side, the landscape and planning-related glossaries (COMMIN 2017, Scottish National Heritage 2017, Ministry of the

Environment and Spatial Planning 2017) introduce words to describe characteristics of the landscape, sometimes also elements or techniques of the planning process but surprisingly no energy-related terms. Technically, most of the glossaries are simple, alphabetically sorted lists of terms and definition, and contrary to our intentions do not include graphical material or other related sources. Therefore, we can conclude that COST RELY Glossary on renewable energies and landscape quality could present a welcome innovation among existing glossaries, both for the practitioners and researchers.

5.4.2
Methodological Approach to Glossary Preparation

The Glossary was prepared in the period from 2015 to 2017 (see Figure 5.4.2.1). Altogether, a core group of seven people closely worked on the preparation with 31 people contributing in total to 48 definitions. At the beginning, the team evaluated

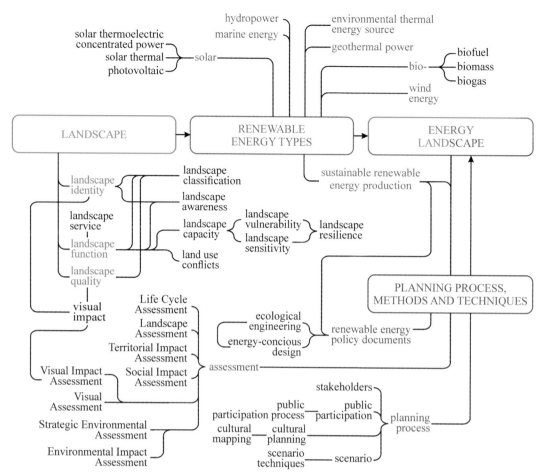

Figure 5.4.2.2
Semantic map of the glossary terms (Graphic: Blanca Del Espino)

existing glossaries on the coverage of the renewable energy landscape-related topics, their structure and elements, in order to define the structure of the glossary and the list of the terms to include. In particular, the glossary completed previously as part of the EUCALAND projects (Kruse et al. 2010, Kruse and Kruckenberg 2010) was chosen as the structural model.

The structure of each term consists of six elements: 1. the English name of the term, 2. definition, 3. related terms, 4. keywords, 5. illustration(s), and 6. sources. Definitions were formulated by using either COST RELY participants' knowledge or/ and various available sources (scientific articles, monographs, study reports, etc.). The aim was to provide short, comprehensive definitions that can

be used by different publics. Definitions can also consist of descriptions of sub-terms, as is the case of 'public' and 'the public concerned' described in the frame of in the term 'public participation'. The category 'related terms' was added to each entry to refer to thematically related terms in the COST RELY glossary itself, while the category 'keywords' implies related terms or words that can be used for search engines.

The list of 48 terms (see Figure 5.4.2.2) for the glossary was developed by consulting the whole group of COST RELY participants in various discussion rounds in order to get the list as complete and relevant as possible. Then, the terms were divided among the Action participants on the basis of their willingness and professional expertise

TOPIC	TERMS
LANDSCAPE	Energy landscapes, landscape, landscape awareness, landscape capacity, landscape character, landscape classification, landscape function, landscape identity, landscape quality, landscape resilience, landscape sensitivity, landscape service, landscape vulnerability, land use conflicts, visual impact
RENEWABLE ENERGY TYPES	Biofuel , biogas , biomass, environmental thermal energy source, geothermal power, hydropower, marine energy, photovoltaic, solar thermal, solar thermoelectric, sustainable renewable energy production, wind energy
PLANNING PROCESS, METHODS AND TECHNIQUES	Best Practice, cultural mapping, cultural planning, ecological engineering, environmental impact assessment, energy-conscious design, landscape assessment, landscape governance, life Cycle Assessment, planning process, public participation, public participation process, renewable energy policy documents, scenario, scenario techniques, social impact assessment, stakeholder, strategic environmental assessment, territorial impact assessment, visual assessment, visual impact assessment

Table. 5.4.2.1
Glossary terms, divided into three groups

to contribute. Each of these authors filled in the pre-designed word form. Simultaneously, representatives of each country were asked to provide translations for the whole list of terms into their native language. Altogether, translations for 28 languages including Esperanto were provided. The revision of terms was done in multiple rounds of discussion to endorse and secure a homogenous structure and content quality of the descriptions, as well as validate the references and illustrations (especially online links), and enable the consultation of different experts on the same term. The editing process finished in September 2017.

The cluster of terms (see Table 5.4.2.1) resembles the three dimensions which contribute to the project topic:

- Landscape,
- Renewable energy types, and
- Planning process, methods, and techniques.

Fifteen terms in the first group describe the characteristics of landscape, including landscape quality, landscape function, landscape identity, etc. In the second group of 12 terms, different types of renewable energies are described. In the last group with 21 terms, the user finds expressions concerning the planning process and as well entries on individual techniques used for enabling participative processes. These especially include assessment techniques used to assess social, environmental, economic, and administrative impacts of the renewable energy investments on landscape, such as environmental impact assessment, social impact assessment, and visual assessment.

With regards to the task of translation, participants encountered several problems:

- Particular term(s) do not exist in some of the languages,
- A term can exist but under a different name,
- A phenomenon does not exist but a descriptive expression is used instead of a term, and
- There is no translation but the English term is used instead.

Additionally, some of the methods are not used as investigation approaches in some of the countries, and therefore no term actually exists. Thus, for the languages covered in the translation part of the glossary, some translations are missing.

5.4.3
Overview of Glossary Terms with Examples

In the first part of the glossary, we find terms describing landscape characteristics. One of the terms is an 'energy landscape' that combines the energy production and its materialisation in the landscape. The definition: an 'energy landscape is characterised by one or more elements of the energy chain (e.g. energy extraction, assimilation, conversion, storage, transport or transmission of energy). The outcome can be a multi-layer energy landscape comprising combinations of technical and natural sources of energy within a landscape. In COST RELY, energy landscape is focused on renewable energy and the impact on landscape quality' (Kruse and Marot 2018). As visible in Figure 5.4.3.1, the term is presented with three related terms, six keywords, and three figures, and one source was added as a reference.

Most of the other terms under the umbrella of landscape give an overview, useful to a non-landscape researcher/practitioner, about how one can describe and valorise the landscape quality and conceptualise the impact of the renewable energy production on the landscape. For illustration, definitions of three selected terms:

- Landscape capacity: a degree to which a particular landscape character type or area is able to accommodate change without significant effects on its character, or overall change of landscape character type;
- Landscape character: a recognisable pattern of elements that occurs consistently in a particular type of landscape;
- Landscape quality: the perception of the holistic environmental, cultural, sensory, and psychological characteristics of a landscape with respect to their benefits or significance to people.

The second part of the glossary is in particular valuable for landscape researchers/practitioners since it introduces different types of renewable energy production they may not necessarily be familiar with. While the most common renewable energy production techniques such as wind turbines are well known, the COST RELY glossary also introduces less known types, such as marine energy, and diversifies between different types of solar energy use for the electricity production: photovoltaic, solar thermal, and solar thermoelectric (concentrated) power. As an added value, some of the definitions encompass technical details with which territorial-oriented researchers usually do not bother, though these details might have some territorial implications.

In the last part, landscape and spatial planning related terms are covered describing processes and methods to be used to assess, evaluate, and mitigate different types of impacts renewable energy production causes in the landscape. We get accustomed with the public participation process and different techniques, policy related terms, e.g. renewable energy policy document, and various types of assessments to be used for the identifi-

ENERGY LANDSCAPE

Definition

An energy landscape is characterised by one or more elements of the energy chain (e.g. energy extraction, assimilation, conversion, storage, transport, or transmission of energy). The outcome can be a multi-layer energy landscape comprising combinations of technical and natural sources of energy within a landscape. In COST RELY, energy landscape is focused on renewable energy and the impact on landscape quality.

Related terms

Landscape resilience, landscape sensitivity, landscape vulnerability

Keywords

Energyscapes, landscapes of carbon neutrality, multi-layer, multi-functional, renewable energy sources, sustainable energy landscape

Figure 3a Three layers of energy production in the area of Garzweiler II, Germany. Foreground: agriculture with oil pumping. Middle: open brown coal mining with a coal-fired power plant. Background: wind turbines. (Photo: Alexandra Kruse 2016)

Figure 3c This energy landscape in Carinthian Mölltal in Austria shows different layers of energy production and impact on the landscape. Foreground: electric train line and electricity high voltage cables. Middle: agriculture including modern hay balls; background, forestry. These very intensive and close layers dominate the Alpine valleys in Austria. (Photo: Alexandra Kruse 2016)

Figure 3b Wind energy landscape, Ore Mountains, Czech Republic. (Photo: Bohumil Frantal 2012)

Source

Definition developed by COST RELY Action.

Translations

Bosnia and Herzegovina Energetski pejzaži/Energetski krajolik	Hungarian Energiatáj *(the term is not in use)*
Bulgarian Ландшафт за производство на енергия	Italian Paesaggio dell'energia
Croatian Energetski krajolik	Islandic Orkulandslag
Czech Energetická krajina	Latvian Enerģijas ainava
Danish Energilandskab	Lithuanian Energijos gavybos kraštovaizdžiai
Dutch Energie landschap	Montenegrin Energetski pejzaž
Esperanto Energia pejzaĝo	Polish Krajobrazy energetyczne
Estonian Energiamaastik	Portuguese Paisagem d'energia
Finish Energiamaisemat *(the term is not in use)*	Romanian Peisaj energetic
French Paysage énergétique	Russian Энергетический пейзаж
German Energielandschaft	Slovenian Energetska krajina
Greek Ενεργειακό Τοπίο	Serbian Енергетски пејзажи *(земљишта)*
Hebrew נופי אנרגיה	Spanish Paisajes de las energías/ Paisajes energéticos
	Swedish Energi landskap

Figure 5.4.3.1
Example of a term entry into the glossary.

253

cation and analysis of the impacts which power plants can have on territory, environment, and society. These assessments have different basis and background—which can be embedded into EU legislation, e.g. Strategic Environmental Assessment in Directive 2001/42/EC on the assessment of the effects of certain plans and programmes on the environment (SEA Directive) (The European Parliament 2001), or they can be the result of a practice or a research pursuit, e.g. territorial impact assessment (Golobic and Marot 2010). An important emphasis is put on the visual aspect of the relation between renewable energy production and the landscape as the most obvious impact of the installations. Therefore, definitions of terms like visual impact, visual assessment, and visual impact assessment are available to the reader.

5.4.4
Discussion

Deriving definitions from a big group of experts with different backgrounds (landscape planners, landscape architects, technical engineers, geographers, sociologist, etc.) means viewing the same phenomenon with different goggles. Epistemic communities differ from one discipline and even nation to another. The main effort of the revision process and the group discussion was to unify all definitions. However, the translation process revealed differences in understanding and daily use of the selected terms. Such an example is the term 'landscape classification' which could be replaced by 'landscape typology', although only the first one was integrated into the glossary. Such differences needed to be understood and respected. Furthermore, in the case of the assessments we can point out the opposite case: the definitions might be the same across the European countries; however, the

outcomes of the assessments are not always perceived as the same. While in some countries strategic environmental assessment results must be taken into account, e.g. in Slovenia, in some countries, e.g. the UK they are not (Fischer et al. 2015). Formulating the definitions also revealed that while landscape and planning terms were more or less common knowledge across the regions, the most foreign concepts to the action participants were the technological terms related to renewable energy types. To a certain extent this can be explained by the professional background of the Action participants, but it might also be true that landscape terms have been (mis)used in many public discussions and the media. Therefore, they belong to the common lexicon, while technical terms are still reserved for professionals working on these topics.

Diversity among the countries additionally displayed itself especially through translation work. Feedback showed multiple situations which can occur:

1. A translation for an individual term exists, mostly it comes from a direct translation, but the term is not recognised by the epistemic community or the English term is used instead. For instance, nearly all languages can translate stakeholder(s), but many use the English term instead of the translation.

2. The phenomenon (methods, concept, approach) exists but under a different name. to give one example from Germany: The literal translation of 'assessment' would be 'Bewertung', but in this context, assessment has to be translated as 'Prüfung' which means examination. For example, Social Impact Assessment is translated as 'Sozialverträglichkeitsprüfung'.

Sometimes, even terms which seemed trivial at the beginning turned out to be far more complex, e.g. landscape identity, where two interpretations are possible:

- The identity/image of the landscape itself (in the minds of those who experience it). This seems to be the meaning implied by e.g. the Czech, French, Italian, and Portuguese translations.

- The landscape component of the (self-)identity of a social group, e.g. locals living in a particular landscape (Loupa Ramos et al. 2016).

3. The term and definition do not exist in the assessment tool 'Sustainability Appraisal (SA)' for the Netherlands and Serbia. Therefore, the glossary might give a chance to introduce terms in the national praxis and also to introduce new tools into the national planning processes.

4. Some of the given terms are well-known in the English language but fall more into the category of expressions or descriptions (and are not considered a real term) and therefore appear difficult to translate effectively. To give one example: if one were to translate 'landscape compatible renewable energy production systems' into Icelandic it could be only done in a form of a long sentence and not as a single clearly defined term.

Since new terminology mostly originates from the English language, the non-English languages also face the challenge of either directly translating the terms or inventing a completely new expression. The most known examples from the last decades are the word sustainability and its later 'cousin'—resilience. Epistemic communities from different territory-focused professions, e.g. geography, landscape architectures, architecture, spatial planning, sociology, etc. led years-long discussions on how to translate or name a certain concept. Overall, the most problematic glossary terms for translation were ecological engineering, cultural mapping, and cultural planning which are among the new planning approaches. In the end, it is not only a translation but it is also a question of conceptualisation of the term joined with the reflection on its (potential) use.

5.4.5
Conclusion: Future Use

In the end, the glossary was a good exercise of bringing to the same table interdisciplinary terminology used by different countries' to describe the relation between the renewable energy sector and landscape. This exercise consisted of the review of existing literature and other valuable resources, as well as being aimed to reach consensus on the most common terminology of the COST RELY project among the participants who used it already during the Action's lifetime and will certainly use it for the dissemination activities. The glossary, available both in the printed and on-line version (see Kruse and Marot 2018), brings together landscape and the technical dimension of the energy production and it is the intention of its authors, that it present an important and comprehensive source for the practitioners, researchers, academics, and policy-makers in the respected fields. Set-up to enable the heterogeneous group of Action participants to work and communicate together, the COST RELY Glossary on Renewable Energy and Landscape Quality is meant to enable politicians, laypeople, scientists, and practitioners to work together without losing time and energy due to the potential misunderstanding and misinterpretation. The glossary will be added to the Action website and in that way made accessible for a larger audience with the aim of fostering exchange and discussion among different disciplines as well as between stakeholders from different backgrounds.

1   We would like to add acknowledgements to all those who contributed to the COST RELY glossary in manifold ways, 31 Action participants altogether.

5.5

OUTREACH OF THE COST ACTION RELY: SUMMARY OF DISSEMINATION ACTIVITIES

Bénédicte Gaillard, Sina Röhner & Alexandra Kruse

The dissemination plan of the Action addresses the following six main target groups: policy-makers, decision-makers, planners, researchers, non-governmental organisations (NGOs), and the general public. In order to reach these different groups, it was necessary to develop various materials and to use different means of dissemination, which are described in this chapter. While it is quite easy to reach scientists (through scientific articles and conferences), it is much more difficult to reach policy-makers at EU and national levels, drawn from relevant EU DGs (e.g. REGIO, ENERGY, ENV, AGRI) as well as European bodies (e.g. European Environment Agency). It is preferable to address national bodies as well as the public in their respective national language. The aim is that all of these constituencies will receive a profound scientific and empirical (practice-based) knowledge base for embedding procedures of public participation and landscape-sensitive planning for renewable energy systems in legislation, subsidy systems and strategic renewable energy plans. The Action therefore has developed an adjusted dissemination strategy, which is presented in the following chapter. In addition to the mentioned target groups, scientists from participating and other European countries will be encouraged to further build on the findings of the Action, and academics from other continents can follow the example in terms of landscape-compatible participatory planning for renewable energy production.

Action Website

Right at the Action's initiation, a website[1] was set-up and developed. It is a public site for information and at the same time it provides content and records the results of the work that has been done. It also functions as a forum for a constant moderated dialogue between the Action and the general public, alongside external NGOs. In this respect, it goes far beyond the mere dissemination and publication of the Action's results, but provides a valuable input to further research by preparing and providing communication channels for synchronous and asynchronous dialogue between various partners within this Action and across external partners. The website furthermore provides information about the

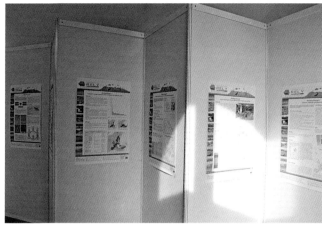

Figure 5.5.1
The COST RELY travelling exhibition rises attention and informs at fairs, universities or in town halls.
a) Banja Luka / Bosnia & Herzegovina (Photo: Alexandra Kruse)
b) Sevilla / Spain (Photo: Juan José Gonzalez)
c) Bucharest / Romania (Photo: Maria Bostenaru Dan)

Action members and interacts with the other three WGs to establish and constantly maintain the multilingual glossary (see chapter 5.4) for scientific collaboration and trans-border public participation. The website will be maintained for at least three years after the Action's final delivery.

Travelling Exhibition and Leaflets
Posters and leaflets, translated into the Action member's languages, are a useful and efficient means of raising awareness on the subject and in helping to introduce the Action to a larger, non-scientific audience. The aim is to inform the different target group of results as efficiently as possible:
Decision-makers in public agencies (e.g. natural heritage protection and regulation), in relevant regional and local governments, and representative business bodies (e.g. European Wind Energy Association) will have typologies of best practices, guidelines for assessment, frameworks, and tool-boxes to improve actual project planning in terms of public participation and consideration of landscape quality objectives.

Planners active in the fields of energy, landscape and agriculture as well as urban and regional planners will also benefit from the consolidated and extended knowledge and applicable toolkits for the realisation of the transition towards renewable energy systems.

The general public, which acts both as a consumer and also as a producer of energy (e.g. photovoltaic installation on rooftops, cooperative wind turbines, etc.), will profit from documented best practices and toolkits for comprehensive public participation in the planning of renewable energy systems.

NGOs with responsibilities for sectoral interests, or communities of interest or for specific places, will be given knowledge-based tool such as a multilingual glossary to extend their role in formal and informal planning processes. The Action leaflet was first produced in English and then translated into a number of other languages (BiH, DE, ES, FR, GR, HU, IT, NL, PT) by Action participants and distributed at numerous events and conferences across a diverse audience. They can be downloaded from the Ac-

Figure 5.5.2
The Action flyer is travelling through the World:
a) Forbidden City; Bejing. Photo: David Miller.
b) New Orleans, US. Photo: Bohumil Frantál.
c) Aberdeen, UK. Photo: David Miller.
d) Swabian Alp, Germany. Photo: Sina Röhner

tion website.[2] The travelling exhibition, consisting of several posters on the general background of the Action, its members, and participating countries, activities, work, and results of the WGs was initiated at the beginning of the Action. Further panels were added as the work of the Action proceeded. To inform different user groups and get in contact with laypeople as well as a scientific audience, the posters were exhibited across different events; for example in Austria, Bosnia-Herzegovina, France, Malta, Romania, and Spain.

Photo Database
The photo database contains the photos taken during the various project activities carried out from Portugal to Hungary, and from Greece to Iceland. Some hundred photos have been chosen in all, representing a new vision of European energy landscapes, considering also the varying aspects of landscape quality. Opening windows on different renewable energy sources and systems will enable their broad regional diversity to be documented. The photo database was used for Action use in particular (such as illustrating the glossary and travelling exhibition) and will be available for public use via the Action homepage.

Photo Competition
Several public photo competitions have been organised and publicised via different channels in order to collect poems or pictures of renewable energy installations and landscape quality accompanied by a text. We also asked for pictures of renewable energy installations found in unexpected places.

Figure 5.5.3
Photos from the photo competition on renewable energy and landscape quality and RE in unexpected places.
a) Palm Spring wind fields in the sunset. Photo: Naja Marot.
b) Wind turbines in Romania. Photo: Marina Mihaila.
c) Biofuels. Photo: Naja Marot.
d) PV in unexpected place. Photo: Csaba Centeri.

All members of COST RELY were entitled to vote among the contributions. The three winners are Naja Marot, Stefanie Müller, and Alexandra Kruse. The winners will see their pictures used as illustrations in this Action book.

Policy Papers
In the last year of the Action's lifetime, policy papers addressed to politicians and policy-makers were produced and can be downloaded from the Action homepage. According to the COST RELY glossary, they address target groups on different planning levels: the first level is the national level, where countries can adopt national renewable energy strategies, presenting their objectives and measures to be implemented to help meeting their renewable energy targets. These strategies include a description of the mix of renewable energy technologies to be employed and the aims for the electricity, heating, and transport sectors. The second level is the regional level with regional renewable energy plans, and finally on the third and local level, communities can adopt local energy concepts. These local energy concepts (LEC) assess opportunities and propose solutions for the energy supply of the respective community, taking into account its long-term development and existing energy capacity (Kruse and Marot 2018).

1   Cost-rely.eu.
2   http://cost-rely.eu/2-uncategorised/153-information-material-downloads.

5.6

IMPACT THROUGH EDUCATION

Isidora Karan

5.6.1
Awareness-raising

Renewable energy is widely considered a desirable way of production energy in the context of sustainable and environmentally responsible development, even though there is community resistance towards renewable energy systems construction. This is mostly related to transformation of recognisable landscape characteristics, but also to negative perceptions of energy landscapes in general (Kontogianni et al. 2014, Silva, and Delicado 2017, see 4.1.). Energy landscape potentials and qualities are not recognised as such and a better understanding of how renewable energy deployment can be reconciled to contribute to the sustainable transformation of energy landscapes. Recent researcher has indicated that there is a yawning gap which needs to be addressed in the area of energy education and awareness on different levels. There is also a lack

of expertise when it comes to comprising combinations of technical and natural sources of energy within a landscape (Sodha 2014). In order to further improve renewable energy development, the awareness of the two-way interaction between renewable energy systems and landscape quality should be increased. That implies an extra effort that should be added to the education of both experts and general public. It is important to recognised that awareness-raising cannot be a purely top-down process but needs to be seen as a 'multi-directional transfer of knowledge' and 'co-creation of meaning' (Council of Europe 2002).

5.6.2 RELY Education

In the framework of the RELY (Renewable Energy and Landscape Quality) COST Action, various educational activities have been undertaken in or-

Figure 5.6.1
Participants of the COST RELY training school in Dublin, Ireland, 2016. Photo: Michael & Sandra Roth.

der to increase awareness of positive relationships between renewable energy and landscape quality. Those activities were orientated towards the education of the general public in different European countries (e.g. traveling exhibition, flyers, etc.; see 5.5), but the main focus was placed on the education of young scientists and future experts from all around Europe. Different types of educational activities have been realised, such as e-lectures or series of online sessions for all RELY Action participants and other interested parties, short-term scientific missions that allow knowledge transfer between individual researchers and scientific institutions, and training schools for students and early stage researchers coming from different European countries.

STSM

Short-term scientific missions (STSM) are aimed at supporting individual mobility and professional growth and at strengthening existing networks and fostering collaborations, allowing scientists to visit an institution in another participating COST RELY country. In the period 2015–2018, 17 STSMs were realised within the COST Action RELY (five in 2015, five in 2016, four in 2017, and three in 2018). The early stage researchers came from nine European countries and were hosted at universities and institutes in ten European countries; 50 % were female; 60 % came from inclusiveness target countries. The exchange visits lasted from two weeks up to three months. Young researchers tutored by experienced researchers from host institutions were working on important issues related to renewable energy and landscape quality (e.g. methods for assessing the suitability of landscapes for renewable energy systems, smart practices to smart visibility of renewable energy, survey on participatory RE planning in the European countries and subjective aspects of participation, etc.).

Early stage researchers, such as Georgia Sismani from Greece, were particularly invited to participate.

The greatest benefit of attending an STSM is the chance to work in person with other experts of your field and thus gain valuable knowledge and experience. … It offers an opportunity to see how the host institution works, to come in contact with new methodologies and software and to exchange knowledge between the two institutions. This may lead to further research ideas and publications. … The most important points I kept from this experience are the contacts I made with the host institution and the valuable experience I gained for future research. Overall, an STSM establishes the opportunity to foster the collaboration between the two institutions. Georgia Sismani, Aristotle University of Thessaloniki

Outcomes of the STSMs were published in international scientific papers and/or presented at academic conferences, some were also disseminated to a wider audience (e.g. articles translated in national languages and published in national magazines). This in turn contributed to further dissemination of the RELY results and to raising awareness among societies.

Training Schools

Training schools provide intensive training in emerging research topics, but in the same time also cover appropriate retraining as part of life-long learning. Furthermore they serve to create networks for future cooperation. Training schools were addressed mainly at early stage researchers, but also at PhD students and MSc students. The two training schools organised within the COST Action RELY gathered more than 40 participants: young researchers and practitioners from diverse disciplines and from many parts of Europe. The first training school on Renewable Energy and Landscape quality: Techniques, Communities and Planning, was held at the Dublin Institute of Technology and Dublin City University in August 2016 and organised by Ken Boyle and Pat Brereton. The second training school was organised by Karl Benediktsson and Edda Waage from the University of Iceland and took place in Reykjavík in May 2017, where participants dealt with the Questions of Power and Participation: Renewable Energy and Landscape in Policy and Planning. Both training schools included lectures, a field trip, and a series of workshops related to the specific site. Participants who had completed the training schools gained a deeper understanding of the strategic and spatial planning for renewable energy and landscape values and were able to critically analyse particular cases of renewable energy planning proposals at different levels (K. Benediktsson, 2017, K. Boyle & P. Brereton 2016). Besides professional and knowledge growth, participants stressed the benefits of collaboration with young researchers coming from different regions and with different background.

Participating in COST Rely training schools really broadened my understanding of energy transition towards renewables. It is all too easy to get caught up only in your own ideas when doing desk research and it's therefore encouraging to meet other people dealing with the similar topic but from another perspective. The almost flawless organization facilitated exchange of knowledge, networking and research at and after the training schools. I met some great people there, resulting in fruitful collaboration that will probably continue in the years to come. The various new perspectives I got from other participants and trainers had a great impact on my own research. Tadej Bevk, PhD student, Department of Landscape Architecture, University of Ljubljana, Slovenia.

Some participants continued to cooperate after the training schools ended and published articles on the topics related to renewable energy and landscape quality (e.g. http://gispoint.de/fileadmin/user_up-load/paper_gis_open/DLA_2017/537629003.pdf and http://gispoint.de/fileadmin/user_upload/paper_gis_open/DLA_2017/537629024.pdf). Some became active members of the RELY Action. It should be stressed that the training schools were a great success and highly appreciated not only among participants and trainers, but also recognised among local residents and media (e.g. https://www.hi.is/frettir/stodu_fyrir_namskei-di_um_landslag_og_orkuvinnslu). That way, the awareness of landscape quality in relation to renewable energy was raised among general public.

5.6.3
Final Remarks

RELY educational activities were primarily oriented towards the education of young scientists and future experts from diverse disciplines in order to contribute to awareness of the two-way interaction between renewable energy and landscape quality and to establish new collaborative links. However, outcomes of these activities have broader impact for academia in general and awareness-raising through dissemination in different types of media (scientific journals, internet, and local newspapers). Even more important than education are the connections established among participants of the RELY educational activities which are still remaining and that the new meaning of the relationship of renewable energy landscape quality is created.

Acknowledgement
Acknowledgements to all those who participated in Short Term Scientific Missions and Training Schools and contributed with their remarks and experiences.

References

Mata Olmo, R. & Sanz Herraiz, R. (eds.). (2004). *Atlas de los Paisajes de España*. Madrid: Ministerio de Medio Ambiente, Medio Rural y Marino.

Aasetre, J. (2013). 'Chapter 16: Nature Conservation in Norway'. In Rusten, G., Potthoff, K. & Sangolt, L. (eds.). *Norway: Nature, Industry and Society*. Bergen: Fagbokforlaget.

Abbasi, T. & Abbasi, S. A. (2011). 'Small Hydro and the Environmental Implications of Its Extensive Utilization'. *Renewable and Sustainable Energy Reviews* 15(4): 2134–2143.

Abdmouleh, Z., Alammari, R. A. M. & Gastli, A. (2015). 'Review of Policies Encouraging Renewable Energy Integration and Best Practices'. *Renewable and Sustainable Energy Reviews* 45C: 249–262.

Adelaja, S., Shaw, J., Beyea, W. & Mckeown, J. D. Ch. (2010). 'Renewable Energy Potential on Brownfields Sites: A Case Study of Michigan'. *Energy Policy* 38: 7021–7030.

Aitken, M., McDonald, S. & Strachan, P. (2008). 'Locating "Power" in Wind Power Planning Processes: The (Not So) Influential Role of Local Objectors'. *Journal of Environmental Planning and Management* 51(6): 777–99.

Aliaj, B. & Lulo, K. (2003). 'Qyteti i Tiranës: Histori mbi Urbanistiken dhe Arkitekturen'. In Aliaj, N., Lulo, K. & Myftiu, G. *The Challenge of Urban Development*. Tirana: Seda and Co-Plan.

Allen, B., Maréchal, A., Nanni, S., Pražan, J., Baldock, D. & Hart, K. (2015). *Data Sources to Support Land Suitability Assessments for Bioenergy Feedstocks in the EU: A Review*. London: Institute for European Environmental Policy (IEEP).

Amaducci, S., Yin, X. & Colauzzi, M. (2018). 'Agrivoltaic Systems to Optimise Land Use for Electric Energy Production'. *Applied Energy* 220: 545–561.

Amankwah-Amoah, J. (2015). 'Solar Energy in Sub-Saharan Africa: The Challenges and Opportunities of Technological Leapfrogging'. *Thunderbird International Business Review* 57(1): 15–31.

Andrés-Ruiz (de), C., Iranzo-García, E. & Espejo-Marín, C. (2015). 'Solar Thermoelectric Power Landscapes in Spain'. In Frolova, M., Prados, M-J., Nadaï, A. (eds.). *Renewable Energies and European Landscapes: Lessons from Southern European Cases*, 237–254. New York/London: Springer.

Andritsos, N., Dalambakis, P., Arvanitis, A., Papachristou, M. & Fytikas, M. (2015). 'Geothermal Developments in Greece: Country Update 2010–2014'. *Proceedings World Geothermal Congress* 2015: 19–24.

Angelopoulos, D., Doukas, H., Psarras, J. & Stamtsis, G. (2017). 'Risk-based Analysis and Policy Implications for Renewable Energy Investments in Greece'. *Energy Policy* 105: 512–523.

Aničić B., Pereković, P. & Tomić D. (2013). 'Criteria for the Integration of Wind Farms into Landscape'. *Prostor* 1(45): 116–127.

Anonymous (2009). 'Turkey's First Geothermal Power Plant Opened', 17 March 2009 (online). Accessed 9 January 2018. Available at <http://yesilekonomi.com/turkiyenin-ilk-jeotermal- elektrik-santrali-acildi>.

Anonymous (2014). '38 Solar Power Plants Produce Electricity in Turkey', 7 June 2014 (online). Accessed 8 January 2018. Available at <http://enerjienstitusu.com/2014/06/07/turkiyede-38- gunes-enerjisi-santrali-elektrik-uretiyor/>.

Anonymous (2015). 'Report on the Electricity Production Sector: Office of the Head of Department, Electricity Production Companies Research, Planning and Coordination' (in Turkish) (online). Accessed 8 January 2018. Available at <http://www.enerji.gov.tr/File/?-path= ROOT% 2F1%2FDocuments%2F-Sekt%C3%B6r%20Raporu%2FEUAS-Sektor_ Raporu2014.pdf>.

Anonymous (2017). 'Guidelines for the Preparation and Evaluation of Urban Design Projects for Approval by the Ministry of the Environment and Urban Planning' (in Turkish) (online). Accessed 7 January 2018. Available at <http://www.csb.gov.tr/db/ mpgm/webmenu/webmenu 15609.pdf>.

Antrop, M. (2000). 'Background Concepts for Integrated Landscape Analysis'. *Agriculture, Ecosystems & Environments* 77(1–2): 17–28.

Antrop, M. (2005). 'Why Landscapes of the Past Are Important for the Future'. *Landscape and Urban Planning* 70(1–2): 21–34.

Antrop, M. & Van Eetvelde, V. (2017). *Landscape Perspectives: The Holistic Nature of Landscape*. Dordrecht: Springer.

Antrop, P. (2004). 'Landscape Change and the Urbanization Process in Europe'. *Landscape and Urban Planning* 67: 9–26.

APERe. (2017). 'Observatoire belge des énergies renouvelables' (online). Accessed 23 December 2017. Available at <https://www.apere. orrg/>.

Apostol, D., Palmer, J., Pasqualetti, M., Smardon, R. & Sullivan, R. (2017). 'Introduction to the Changing Landscapes of Renewable Energy'. In Apostol, D. et al. (eds.). *The Renewable Energy Landscape: Preserving Scenic Values in Our Sustainable Future*, 1–13. Abingdon: Routledge.

Arnórsson, S. (2011). *Jarðhiti á Íslandi: Eðli auðlindar og ending—Verklag við undirbúning að vinnslu—Umhverfisáhrif af nýtingu*. Reykjavík: Rammaáætlun.

Arnstein, S. R. (1969). 'A Ladder of Citizen Participation'. *Journal of the American Planning Association* 35(4): 216–224.

B&H CPNM (2017). 'Hydro Power Plants Mala HE "Bihac"' (online). Accessed 23 December 2017. B&H Commission to Preserve National Monuments Website. Available at <http://old.kons.gov.ba/main.php?id_struct=6&lang=1&action=view&id=3332>.

Bakken, T. H., Aase, A. G., Hagen, D., Sundt, H., Barton, D. N. & Lujala, P. (2014). 'Demonstrating a New Framework for the

Comparison of Environmental Impacts from Small- and Large-Scale Hydropower and Wind Power Projects'. *Journal of Environmental Management* 140: 93–101.

Baraja-Rodríguez, E., Herrero-Luque, D. & Pérez-Pérez, B. (2015). 'A Country of Windmills: Wind Energy Development and Landscape in Spain'. In Frolova, M., Prados, M. & Nadaï, A. (eds.). *Renewable Energies and European Landscapes: Lessons from Southern European Cases*. New York/London: Springer.

Barbier, E. (2002). 'Geothermal Energy Technology and Current Status: An Overview'. *Renewable and Sustainable Energy Reviews* 6(1): 3–65.

Bardach, E. (2000). *A Practical Guide for Policy Analysis: The Eightfold Path to More Effective Problem Solving*. New York: Chatham House Publishers.

Battaglini, A., Komendantova, N., Brtnik, P. & Patt, A. (2012). 'Perception of Barriers in Expansion of Electricity Grids in the European Union'. Energy Policy 47: 254–259.

Baudou, Evert (1994). 'The Luleälv Project'. In Helmfrid, Staffan (1994). *Landscape and Settlements: National Atlas of Sweden*, 22ff.

Bender, O. & Schumacher, K. (2008). 'From Protection to Management: Policies to Sustain (Pre-Industrial) Landscapes in Austria and Germany'. In Fairclough, G., Grau Møller, P. (eds.). *Landscape as Heritage: The Management and Protection of Landscape in Europe; A Summary by the COST A27 Project 'Landmarks'*, 77–127. Bern: COST.

Bender, B. (1998). *Stonehenge: Making Space*. Oxford, Berg.

Benediktsson, K. (2017). 'COST-RELY Training School about Policy and Planning for Renewable Energy'. *Tájökológiai Lapok* (Hungarian journal of landscape ecology) 15(2): 161–162.

Bergens Tidende (2018). 'Vinden blåser ut av Norge'. https://www.bt.no/btmagasinet/i/1kXvdG/Vinden-blaser-ut-av-landet. Accessed 10 March 2018.

Bergström, L., Kautsky, L., Malm, T., Rosenberg, R., Wahlberg, M., Capetillo, N. A. & Wilhelmsson, D. (2014). 'Effects of Offshore Wind Farms on Marine Wildlife: A Generalized Impact Assessment'. *Environmental Research Letters* 9(3): 034012.

Berka, A. L. & Creamer, E. (2017). 'Taking Stock of the Local Impacts of Community Owned Renewable Energy: A Review and Research Agenda'. *Renewable and Sustainable Energy Reviews* 82(3): 3400-3419.

Bezrukovs, V., Bezrukovs, Vl., Zacepins, A. & Komashilovs, V. (2015). 'Assessment of Wind Shear and Wind Energy Potential in the Baltic Sea Region of Latvia'. *Latvian Journal of Physics and Technical Sciences* 52(2): 26–39.

BfN (n.d.). *Landschaftsschutzgebiete* (online). Accessed 21 December 2017. Available at <https://www.bfn.de/themen/gebietsschutz-grossschutzgebiete/landschaftsschutzgebiete.html>.

Bibby, J. S., Douglas, H. A., Thomasson, A. J. & Robertson, J. S. (1991). *Land Capability Classification for Agriculture*. Macaulay Land Use Research Institute.

Bickerstaff, K. (2012). 'Because We've Got History Here: Nuclear Waste, Cooperative Siting, and the Relational Geography of a Complex Issue'. *Environment and Planning* A 44: 2611–2628.

Bidwell, D. (2013). 'The Role of Values in Public Beliefs and Attitudes Towards Commercial Wind Energy'. *Energy Policy* 58: 189–199. doi: 10.1016/j.enpol.2013.03.010

Bilgili, M., Yasar, A. & Simsek, E. (2011). 'Offshore Wind Power Development in Europe and Its Comparison with Onshore Counterpart'. *Renewable and Sustainable Energy Reviews* 15: 905–915.

Bishop, I. D. & Miller, D. R. (2006). 'Visual Assessment of Off-Shore Wind Turbines: The Influence of Distance, Contrast, Movement and Social Variables'. *Renewable Energy* 32: 814–831.

Bluemling, B., Mol, A. P. J. & Tu, Q. (2013). 'The Social Organization of Agricultural Biogas Production and Use'. *Energy Policy* 63: 10–17.

BMWI (2010). 'Energy Concept for an Environmentally Sound, Reliable and Affordable Energy Supply'. Available at <http://www.osce.org/eea/101047?download=true>.

BMWI (2015). 'Erneuerbare Energien in Zahlen: Nationale und internationale Entwicklung im Jahr 2014'. Available at <http://docplayer.org/11918478-Erneuerbare-energien-in-zahlen- nationale-und-internationale-entwicklung-im-jahr-2014.html>.

Bomberg, E. & McEwen, N. (2012). 'Mobilizing Community Energy'. *Energy Policy* 51: 435–444.

Bosch & van Rijn (2018). *Windstats Satistieken* (online). Available at <https://windstats.nl/statistieken/>.

Botinčan, B., Kušan, V., Koren, Ž., Mesić, Z., Šteko, V., Birov, T., Rapić, S., Haramina, T., Antunović, B., Grgurić, Z., Borić, B., Grgurić, S., Kurevija, T. & Antonić, O. (2015). 'The Plan of Use of Renewable Energy Sources in Dubrovačko–Neretvanska County' (online). Draft Plan Proposal, OIKON Ltd., Institute of Applied Ecology, Zagreb. Accessed 20 November 2017. Available at <http://www.edubrovnik.org/wp-content/uploads/2016/03/Nacrt-Plana-kori%C5%A1tenja-OIE-DN%C5%BD-26-1-2015.pdf>.

Bottero M. (2013). 'Sustainability Assessment of Large Dams: The Case of a Hydropower Plant in Bulgaria'. *Manag Environ Qual An Int J* 24: 178–98.

Boyle, K. & Brereton, P. (2016). *Training School Preparatory Meeting 2015*, Dublin.

Bretschneider, S., Marc-Aurele, F. J. & Wu, J. (2005). '"Best practices" Research: A Methodological Guide for the Perplexed'. *Journal of Public Administration Research and Theory* 15(2): 307–323.

Breukers, S. & Wolsink, M. (2007a). 'Wind Power Implementation in Changing Institutional Landscapes: An International Comparison'. *Energy Policy* 35(5): 2737–50.

Breukers, S. & Wolsink, M. (2007b). 'Wind Energy Policies in the Netherlands: Institutional Capacity-Building for Ecological Modernisation'. *Environmental Politics* 16(1): 92–112.

Bridge, G., Bouzarovski, S., Bradshaw, M. & Eyre, N. (2013). 'Geographies of Energy Transition: Space, Place and the Low-Carbon Economy'. *Energy Policy* 53: 331–340. doi: 10.1016/j.enpol.2012.10.066

Bright, R. M., Anton-Fernandez, C., Astrup, R., Cherubini, F., Kvalevag, M. & Stromman, A. H. (2014). 'Climate Change Implications of Shifting Forest Management Strategy in a Boreal Forest Ecosystem of Norway'. *Global Change Biology* 20(2): 607–621.

Bürgi, M., Ali, P., Chowdhury, A., Heinimann, A., Hett, C., Kienast, F., Mondal, K. M., Upreti, B. R. & Verburg, P. H. (2017). 'Integrated Landscape Approach: Closing the Gap between Theory and Application'. *Sustainability* 9(8): 1371.

Buchecker, M., Menzel, S. & Home, R. (2013). 'How Much Does Participatory Flood Management Contribute to Stakeholders' Social Capacity Building? Empirical Findings Based on a Triangulation of Three Evaluation Approaches'. *Natural Hazards and Earth System Sciences* 13(6): 1427–44.

Bullen, J. (2017). 'LandMAP Methodology Overview'. Natural Resources Wales. Accessed 1 April 2018. Availalable at <https://naturalresources.wales/media/681752/landmap-methodology-overview-2017-eng.pdf>.

Bundi, S. (2016). 'Landschaft bewahren: Natur- und Heimatschutz'. In Jon, M., Backhaus, N., Hürlimann, K. & Bürgi, M. (eds.). *Geschichte der Landschaft in der Schweiz: Von der Eiszeit bis zur Gegenwart*, 206–218. Zurich: Orell Füssli.

Burkhard, B., Kroll, F., Müller, F. & Windhorst, W. (2009). 'Landscapes' Capacities to Provide Ecosystem Services: A Concept for Land-Cover Based Assessments'. *Landscape Online* 15: 1–22. doi: 10.3097/LO.200915

Burkhard, B., Kroll, F., Nedkov, S. & Müller, F. (2012). 'Mapping Ecosystem Service Supply, Demand and Budgets'. *Ecological Indicators* 21: 17–29.

Burningham, K., Barnett, J. & Thrush, D. (2006). *The Limitations of The NIMBY Concept for Understanding Public Engagement with Renewable Energy Technologies: A Literature Review*, 1–20. Working Paper. Manchester: School of Environment and Development, University of Manchester.

Burningham, K., Barnett, J. & Walker, G. (2015). 'An Array of Deficits: Unpacking NIMBY Discourses in Wind Energy Developers Conceptualizations of Their Local Opponents'. *Society & Natural Resources* 28(3): 246–260.

Böck, H. (2012). 'Wie aus Wasserkraft in Deutschland Strom wird' (online). Accessed 10 April 2018. Available at <http://www.dw.com/de/wie-aus-wasserkraft-in-deutschland-strom-wird/a-15951584>.

Cabraal, A., Cosgrove-Davies, M. & Schaeffer, L. (1996). 'Best Practices for Photovoltaic Household Electrification Programs: Lessons from Experiences in Selected Countries'. *Technical Paper Number 324*. Washington, D.C.: World Bank.

Cairngorms National Park Authority (2015). 'Spccial Landscape Qualities of the Cairngorms National Park'. Accessed 1 April 2018. Available at <http://cairngorms.co.uk/caring-future/cairngorms-landscapes/special-landscape-qualities/>.

Calvert, K., Birch, K. & Mabee, W. (2017). 'New Perspectives on an Ancient Energy Resource'. In Bouzarovski, S., Pasqualetti, M. J., & Castán Broto, V. (2017). *The Routledge Research Companion to Energy Geographies*, 47–60. London/New York: Routledge Publishing.

Calvert, K. & Mabee, W. (2015). 'More Solar Farms or More Bioenergy Crops? Mapping and Assessing Potential Land-Use Conflicts among Renewable Energy Technologies in Eastern Ontario, Canada'. *Applied Geography* 56: 209–221.

Câmpeanu, V. & Pencea, S. (2014). 'Renewable Energy Sources in Romania: From a "Paradise" of Investors to a Possible Abandon or to Another Boom? The Impact of a New Paradigm in Romanian Renewable Sources Policy'. *Procedia Economics and Finance* 8: 129–137. doi: 10.1016/S2212-5671(14)00072-0

Carfora, A., Pansini, R. V., Romano, A. A. & Scandurra, G. (2018). 'Renewable Energy Development and Green Public Policies Complementarities: The Case of Developed and Developing Countries'. *Renewable Energy* 115: 741–749.

Carlisle, J. E., Kane, S. L., Solan, D. & Joe, J. C. (2014). 'Support for Solar Energy: Examining Sense of Place and Utility-Scale Development in California'. *Energy Research & Social Science* 3: 124–130. doi: 10.1016/j.erss.2014.07.006

Carrosio, G. (2013). 'Energy Production from Biogas in the Italian Countryside: Policies and Organization Models'. *Energy Policy* 63: 3–9.

Castan Broto, V. (2017). 'Energy Landscapes and Urban Trajectories Towards Sustainability'. *Energy Policy* 108: 755–764.

Central Statistics Office (CSO) (2018). Available at <https://www.cbs.nl/nl-nl/nieuws/2017/22/aandeel- hernieuwbare-energie-5-9-procent-in-2016>.

CETMA (2007a). Italian Ministry for the Environment, Land and Sea, CETMA. *Renewable Energy Resource Assessment: Wind, Solar, and Biomass Energy Assessment, Republic of Montenegro*, 17–18 and 130–133. Accessed 11 September 2017.

CETMA (2007b). Italian Ministry for the Environment, Land and Sea, CETMA, Renewable

Energy Resource Assessment, Wind, Solar, and Biomass Energy Assessment, Republic of Montenegro. Accessed 21 September 2017.

Chomjak, P. & Tomič, P. (2008). 'Environmental Presumption of the Wind Power Turbines Building in Slovakia' (Environmentálne predpoklady výstavby veterných elektrární na Slovensku). *Životné prostredie* 42(6): 297–301.

Christensen, E. D., Johnson, M., Sørensen, O. R., Hasager, C. B., Badger, M. & Larsen, S. E. (2013). 'Transmission of Wave Energy through an Offshore Wind Turbine Farm'. *Coastal Engineering* 82: 25–46.

Christensen-Dalsgaard, S., Lorentsen, S.-H., Dahl, E. L., Follestad, A., Hanssen, F. & Systad, G. H. (2010). 'Marine Wind Farms: Seabirds, White-Tailed Eagles, Eurasian Eagle-Owl and Waders; A Screening of Potential Conflict Areas'. *Norwegian Institute for Nature Research (NINA) Report 557*. Norwegian Institute for Nature Research, Trondheim.

Ciervo, M. & Schmitz, S. (2017). 'Sustainable Biofuel: A Question of Scale and Aims'. *Moravian Geographical Reports* 25(4): 220–233.

Clare County Council (2017). *Clare County Development Plan 2017–2023*, vol. 6, *Clare Renewable Energy Strategy 2017–2023*.

CleanTechnica (2017). 'Dutch Solar Bike Path SolaRoad Successful & Explanding' (online). Available at <https://cleantechnica.com/2017/03/12/dutch-solar-bike-path-solaroad-successful-expanding/>.

Cohen, J. J., Reichl, J. & Schmidthaler, M. (2014). 'Re-Focussing Research Efforts on the Public Acceptance of Energy Infrastructure: A Critical Review'. *Energy* 76: 4–9. doi: 10.1016/j.energy.2013.12.056

Colman, J. E., Eftestøl, S., Finne, M. H., Huseby, K. & Nybakk, K. (2008). Fagrapport reindrift: Konsekvenser av vindkraft- og kraftledningsprosjekter på Fosen. ASK Rådgivning AS and SWECO Norge AS.

COM (2011). 112. 'Roadmap for Moving to a Competitive Low Carbon Economy in 2050'.

Accessed 20 February 2018. Available at <http://eur-lex.europa.eu/legal-content/EN/TXT/?uri=CELEX:52011DC0112>.

COMMIN (2017). 'BSR Glossaries'. Accessed 20 September 2017. Available at <http://commin.org/en/bsr-glossaries/national-glossaries/index.html>.

Committee on Climate Change (CCC) (2014). 'Meeting Carbon Budgets: 2014 Progress Report to Parliament' (online). Accessed 1 December 2017. Available at <www.theccc.org.uk/wp-content/uploads/2014/07/CCC-Progress-Report-2014_web_2.pdf>.

Comodi, G., Bevilacqua, M., Caresana, F., Pelagalli, L., Venella, P. & Paciarotti, C. (2014). 'LCA Analysis of Renewable Domestic Hot Water Systems with Unglazed and Glazed Solar Thermal Panels'. Energy Procedia 61: 234–237.

Cornwall County Council (2014). 'Judging Landscape Capacity: A Development Management Toolkit'. Accessed 1 April 2018. Available at <www.cornwall.gov.uk/media/23827847/assessing-landscape-character-a-dmt-endorsed-by-pac-16-october-2014-2015-publication-version.pdf>.

Council of Europe (CoE) (2000). 'The European Landscape Convention, ETS No. 176' (online). Accessed 12.12.2017. Available at <http://www.coe.int/t/dg4/cultureheritage/heritage/Landscape>.

Council of Europe (CoE) (2002). *European Landscape Convention*. Florence.

Cowell, R., Bristow, G. & Munday, M. (2011). 'Acceptance, Acceptability and Environmental Justice: The Role of Community Benefits in Wind Energy Development'. *Journal of Environmental Planning and Management* 54: 539–557. doi 10.1080/09640568.2010.521047

Croenergo (2015). 'Croatian Energy in 2014: Import of Electrical Energy 13%, Gas 41.7%; The Year Marked by Continued Drop of Consumption' (online). Accessed 20 November 2017. Available at <http://www.croenergo.eu/Hrvatska-energetika-u-2014-Uvoz-elek-

tricne-energije-13-plina-417;-godinu-obiljezio-daljnji-pad-potrosnje-27450.aspx>.

Csemez, A. (1996). *Tájtervezés—tájrendezés*. Budapest: Mezőgazda.

CSP 2016. 'Latvijas energobilance 2015. gadā' (Energy audit in Latvia in 2015) (online). Central Statistical Bureau of Latvia Website. Accessed 16 December 2017. Available at <http://www.csb.gov.lv/sites/default/files/nr_33_latvijas_energobilance_2015_1 6_00_lv.pdf>.

Cultural Heritage Agency (2018). Available at <https://landschapinnederland.nl/panorama-landschap, https://landschapinnederland.nl/bronnen-en-kaarten>.

Czech Biomass Association (CZ Biom) (2016). 'Map of Biogas Stations' (online). Available at <http://www.czba.cz/mapa-bioplynovych-stanic/; 2015>.

Czech Statistical Office (CZSO) (2015). 'Agriculture: Time Series' (online). Available at <https://www.czso.cz/csu/czso/zem_cr; 2015>.

Czech Wind Energy Association (CWEA) (2017). 'Wind Power Plants in the Czech Republic: Current Installations' (online). Available at <http://www.csve.cz/clanky/aktualni-instalace-vte-cr/120>.

Dai, K., Bergot, A., Liang, C., Xiang, W.-N. & Huang, Z. (2015). 'Environmental Issues Associated with Wind Energy: A Review'. *Renewable Energy* 75: 911–921. doi: 10.1016/j.renene.2014.10.074

Dale, V. H., Kline, K. L., Buford, M. A., Volk, T. A., Smith, C. T. & Stupak, I. (2016). 'Incorporating Bioenergy into Sustainable Landscape Designs: Renewable and Sustainable Energy Reviews.' *Pergamon* 56: 1158–1171.

Danish Energy Agency (2009). 'Wind Turbines in Denmark'. Danish Energy Agency and Kommunikationsbureauet Rubrik. Available at <www.ens.dk>.

Davies, R. (2016). 'Global Demand for Energy Will Peak in 2030, Says World Energy Council'. Accessed 5 October 2017. Available at

269

<https://www.theguardian.com/business/2016/oct/10/global-demand-for-energy-will-peak-in-2030-says-world-energy-council>.

Davodeau, H. (2001). 'Politiques publiques et paysages: Mise en oeuvre d'un travail de recherche ESTHUA'. *ESO* (ESTHUA Université d'Angers ESO—UMR 6590) 15: 105⬚108.

De Marco, A., Petrosillo, I., Semeraro, T., Pasimeni, M. R., Aretano, R., & Zurlini, G. (2014). 'The Contribution of Utility-Scale Solar Energy to the Global Climate Regulation and Its Effects on Local Ecosystem Services'. *Global Ecology and Conservation* 2: 324–337.

Deemer, B., Harrison, J., Siyue, L., Beaulieu, J., DelSontro, T., Barros, N., Bezerra-Neto, J., Powers, S., dos Santos, M. & Vonk, J. (2016). 'Greenhouse Gas Emissions from Reservoir Water Surfaces: A New Global Synthesis'. *BioScience* 66(11): 949–964.

DEHLG, Department of the Environment, Heritage and Local Government (2006). 'Wind Energy Planning Guidelines, Environment, Heritage and Local Government, Irlanda'. Available at <http://www.environ.ie/en/>.

Del Río González, P. (2008). 'Ten Years of Renewable Electricity Policies in Spain: An Analysis of Succesive Feed-In Tariff Reforms'. *Energy Policy* 36(8): 2917–2929.

Del Rio, P. & Burguillo, M. (2009). 'An Empirical Analysis of the Impact of Renewable Energy Deployment on Local Sustainability'. *Renewable and Sustainable Energy Reviews* 13(6–7): 1314–1325.

Delfi GRYNAS (2016). 'Apklausa: dauguma lietuvių nori, kad elektra būtų gaminama iš atsinaujinančių išteklių' (Survey: most Lithuanians want electricity to be produced from renewable resources) (online). Accessed 30 December 2017. Available at <https://www.delfi.lt/grynas/gyvenimas/apklausa-dauguma-lietuviu-nori-kad-elektra-butu-gamina-is-atsinaujinanciu-istekliu.d?id=70355692#ixzz3zlqQKCNg>.

Delicado, A., Junqueira, L., Fonseca, S., Truninger, M., Silva, L., Horta, A. & Figueiredo, E. (2014). 'Not in Anyone's Backyard? Civil Society Attitudes towards Wind Power at the National and Local Levels in Portugal'. *Science & Technology Studies* 27 (2): 49-71.

Delicado, A., Figueiredo, E., Silva, L. (2016). 'Community Perceptions of Renewable Energies in Portugal: Impacts on Environment, Landscape and Local Development'. *Energy Research & Social Science* 13: 84–93.

Department for Business, Energy and Industrial Strategy (2017a). 'Digest of UK Energy Statistics (DUKES): Renewable sources of Energy, Chapter 6'. Available at <www.gov.uk/government/statistics/renewable-sources-of-energy-chapter-6-digest-of-united-kingdom-energy-statistics-dukes>.

Department for Business, Energy and Industrial Strategy (2017b). 'UK Energy Statistics, Q2 2017'. Accessed 28 September 2017.

Department of Enterprise, Trade and Investment (DETI) (2010). 'Energy: A Strategic Framework for Northern Ireland'. Available at <www.detini.gov.uk/strategic_energy_framework__sef_2010_-3.pdf>.

Deputy State Secretariat for Environment Protection (2018). 'Országos jelentőségű, egyedi jogszabállyal védett természeti területek'. Accessed 26 March 2018. Available at <http://www.termeszetvedelem.hu/orszagos-jelentosegu-egyedi-jogszaballyal-vedett-termeszeti-teruletek>.

Devine-Wright, P. (2005). 'Beyond NIMBYism: Towards an Integrated Framework for Understanding Public Perceptions of Wind Energy'. *Wind Energy* 8(2): 125–139.

Devine-Wright, P. (2005). 'Local Aspects of UK Renewable Energy Development: Exploring Public Beliefs and Policy Implications'. *Local Environment* 10(1): 57–69.

Devine-Wright, P. (2011). 'Place Attachment and Public Acceptance of Renewable Energy: A Tidal Energy Case Study'. *Journal of Environmental Psychology* 31(4): 336–343.

Devine-Wright, P. (2011). 'Public Engagement with Large-Scale Renewable Energy Technologies: Breaking the Cycle of Nimbyism'. *Climate Change* 2(1):19–26.

Devine-Wright, P. & Batel, S. (2017). 'My Neighbourhood, My Country or My Planet? The Influence of Multiple Place Attachments and Climate Change Concern on Social Acceptance of Energy Infrastructure'. *Global Environmental Change* 47: 110–120.

Devine-Wright, P. & Howes, Y. (2010). 'Disruption to Place Attachment and the Protection of Restorative Environments: A Wind Energy Case Study'. *Journal of Environmental Psychology* 30: 271–280.

Dijkman, H. (2015). 'Investigating the Spatial Impacts of the Energy Transition at the Local Scale'. Available at <http://edepot.wur.nl/390140).stir>.

Dijkman, T. J., & Benders, R. M. J. (2010). 'Comparison of Renewable Fuels Based on Their Land Use Using Energy Densities'. *Renewable and Sustainable Energy Reviews* 14(9): 3148–3155.

Dinesh, H., & Pearce, J. M. (2016). 'The Potential of Agrivoltaic Systems'. *Renewable and Sustainable Energy Reviews* 54: 299–308.

DiPippo, R. (2015). 'Geothermal Power Plants: Evolution and Performance Assessments'. *Geothermics* 53: 291–307.

Dolman, S. & Simmonds, M. (2010). 'Towards Best Environmental Practice for Cetacean Conservation in Developing Scotland's Marine Renewable Energy'. *Marine Policy* 34(5): 1021–1027.

Domac, J., Risović, S., Segon, V., Pentek, T., Safran, B. & Papa, I. (2015). 'Can Biomass Trigger an Energy-wise Transition in Croatia and Rest of South-Eastern Europe?' *Journal of Forestry* 11(22): 561–569.

Dragomir, G., Șerban, A., Năstase, G. & Brezeanu, A. I. (2016). 'Wind Energy in Romania: A Review from 2009 to 2016'. *Renewable and Sustainable Energy Reviews* 64: 129–143. doi: 10.1016/J.RSER.2016.05.080

Dupraz, C., Marrou, H., Talbot, G., Dufour, L., Nogier, A. & Ferard, Y. (2011). 'Combining Solar Photovoltaic Panels and Food Crops for Optimising Land Use: Towards New Agrivoltaic Schemes'. *Renewable Energy* 36(10): 2725–2732.

Dutch National Spatial Data Service (2018). Available at <https://www.pdok.nl/>.

Dvořák, P., Martinát, S., Van der Horst, D., Frantál, B. & Turečková, K. (2017). 'Renewable Energy Investment and Job Creation: A Cross-sectoral Assessment for the Czech Republic with Reference to EU Benchmarks'. *Renewable and Sustainable Energy Reviews* 69: 360–368.

Dövényi, Z. (ed.) (2010). *Magyarország kistájainak katasztere*. Budapest: MTA Földrajztudományi Kutatóintézet.

Ebinger, J. (2010). *Albania's Energy Sector: Vulnerable to Climate Change*. Europe and Central Asia Knowledge Brief, vol. 29. Washington, D.C.: The World Bank.

Eiter, S. & Vik, M. L. (2015). 'Public Participation in Landscape Planning: Effective Methods for Implementing the European Landscape Convention in Norway'. *Land Use Policy* 44: 44–53.

Electrica Furnizare SA Website, Accessed 11 March 2018. Available at <http://www.electricafurnizare.ro/informatii–utile/etichetarea-energiei/>. 'Romania's Energy Strategy for 2007–2020' (in Romanian: Strategia Energetică a Româ niei pentru perioada 2007–2020). Accessed 4 December 2017. Available at <http://www.minind.ro/energie/STRATEGIA_energetica_actualizata.pdf>.

ELEM, Macedonian Power Plant's Website. Accessed December 2017. Available at <http://www.elem.com.mk/?lang=en>.

Ellis G., Barry, J. & Clive. R. (2007). 'Many Ways to Say "No", Different Ways to Say "Yes": Applying Q-Methodology to Understand Public Acceptance of Wind Farms Proposals'. *Journal of Environmental Planning and Management* 50(4): 517–551.

Ellis, G. & Ferraro, G. (2016). *The Social Acceptance of Wind Energy: Where We Stand and the Path Ahead*. Luxembourg: Publications Office.

Emmann, C. H., Arens, L. & Theuvsen, L. (2013). 'Individual Acceptance of the Biogas Innovation: A Structural Equation Model'. *Energy Policy* 62: 372–378. doi: 10.1016/j.enpol.2013.07.083

Energostat (2017). 'The Share of Sources on Electricity Production' (online). Available at <http://oenergetice.cz/energostat/>.

Energy Agency of the Republic of Macedonia Website. Accessed December 2017. Available at <2http://www.ea.gov.mk/>.

Energy and Water Agency. 'National Renewable Energy Action Plan II'. Accessed 1 December 2017. Available at <https://energywateragency.gov.mt/en/Documents/National%20Renewable%20Energy%20 Action%20Plan%20(2017).pdf>.

Energy Regulatory Office (ERO) (2016). 'Roční zpráva o provozu ES ČR 2015' (online). Available at <http://www.eru.cz/documents/10540/462820/Rocni_zprava_provoz_ES_2015.pdf>.

ENTSO-E (2016). 'European Network of Transmission System Operators for Electricity'. Database. Available at <https://www.entsoe.eu/data/statistics/Pages/default.aspx>.

Environmental Protection Agency (EPA) (2010). 'Energy Department Announces National Initiative to Redevelop Brownfields with Renewable Energy' (online). Accessed 20 March 2018. Available at <http://epa.gov/brownfields/partners/brightfd.htm>.

Enviroportal (2010). 'Standards and Limits for the Installation of Wind Power Plant and Wind Farms in the Slovak Republic' (Štandardy a limity pre umiestňovanie veterných elektrární a veterných parkov v SR) (online). Accessed 12 December 2017. Available at <https://www.enviroportal.sk/standardy-a-limity-pre-umiestnovanie-veternych-elektrarni- a-veternych-parkov-v-sr>.

Enviroportal Website (online). Accessed 14 December 2017. Available at <https://www.enviroportal.sk/indicator/detail?id=2601&pdf=true>.

ETEK (2014). *Estonian Renewable Energy Sector Yearbook*. Estonian Renewable Energy Association.

European Commission (2012). 'Directive of the European Parliament and of the Council, Amending Directive 98/70/EC Relating to the Quality of Petrol and Diesel Fuels and Amending Directive 2009/28/EC on the Promotion of the Use of Energy from Renewable Sources'.

European Commission (2013a). 'Environment Action Programme to 2020, Decision No 1386/2013/EU of the European Parliament and of the Council'. *Official Journal of the European Union*.

European Commission (2013b). 'A New EU Forest Strategy: For Forests and the Forest-Based Sector'. *Communication from the Commission to the European Parliament, the Council, the European Economic and Social Committee and the Committee of the Regions*.

European Commission (2014). 'A Policy Framework for Climate and Energy in the Period from 2020 to 2030'. *Communication from the Commission to the European Parliament, the Council, the European Economic and Social Committee and the Committee of the Regions*.

European Commission (2016). 'Proposal for a Directive of the European Parliament and of the Council amending Directive 2012/27/EU on Energy Efficiency'. *COM/2016/761 Final*.

European Commission (2017a). *EU Energy in Figures: Statistical Pocketbook*. Brussels: European Commission.

European Commission (2017b). 'Report from the Commission to the European Parliament, the Council, the European Economic and

Social Committee and the Committee of the Regions: Renewable Energy Progress Report' (online). Accessed 20 November 2017. Available at <https://ec.europa.eu/transparency/regdoc/rep/1/2017/EN/COM-2017-57-F1-EN-MAIN-PART-1.PDF>.

European Cooperation in the Field of Scientific and Technical Research—COST (2014). *Memorandum of Understanding for the Implementation of a European Concerted Research Action Designated as COST Action TU1401: Renewable Energy and Landscape Quality (RELY)*. Brussels: COST.

European Council (2014). 'Conclusions on 2030 Climate and Energy Policy Framework. 23–24.10.2014' (online). Accessed 24 November 2017.

European Parliament, Council of the European Union (2001). *Directive 2001/42/EC on the Assessment of the Effects of Certain Plans and Programmes on the Environment (SEA Directive)*. Brussels: European Union.

European Union (2017). *Global Stocktake: Submission by the Republic of Malta and The European Commission on behalf of the European Union and its Member States, to the United Nations Framework Convention on Climate Change*. Brussels: European Union.

European Union Court of Auditors (2017). 'Landscape Review: EU Action on Energy and Climate Change'. Available at <www.eca.europa.eu/en/Pages/DocItem.aspx?-did=41824>.

European Union, 2011. 'Environment Impact Assessment Directive' (online). Brussels: European Union. Accessed December 2016. Available at file:///C:/Users/Gisele/Downloads/Directives%20EIA.pdf.

Eurostat (2016a). *Agriculture, Forestry and Fishery Statistics 2015 Edition*.

Eurostat (2016b). 'Energy from Renewable Sources'. Accessed 1 July 2018. Available at <http://ec.europa.eu/eurostat/statistics-explained/index.php/Energy_from_renewable_sources>.

Eurostat (2017). 'Renewable Energy Statistics'. Accessed 9 January 2018. Available at <http://ec.europa.eu/eurostat/statistics-explained/index.php/Renewable_energy_statistics/de>.

Eurostat (2017). 'Thematic Glossaries, Energy'. Accessed 20 September 2017. Available at <http://ec.europa.eu/eurostat/statistics-explained/index.php/Thematic_glossaries>.

Eurostat (2017a). 'Energy Balance Flow for European Union (28 Countries) 2015'. Accessed 23 December 2017. Available at <http://ec.europa.eu/eurostat/cache/sankey/sankey.html?geo=EU28&year=2015&unit=KTOE&fuels=0000&highlight=&nodeDisagg=0101000000&flowDisagg=false>.

Eurostat (2017b). 'Share of Energy from Renewable Sources'. Accessed 23 December 2017. Available at <http://appsso.eurostat.ec.europa.eu/nui/show.do?dataset=nrg_ind_335a&lang=en>.

Eurostat (2018). 'Energy from Renewable Sources'. Available at <http://ec.europa.eu/eurostat/statistics-explained/index.php/Renewable_energy_statistics>.

Eurostat (2018). 'Renewable Energy Statistics'. Available at <http://ec.europa.eu/eurostat/statistics-explained/index.php/Renewable_energy_statistics>.

Evans, A., Strezov, V. & Evans, T. J. (2009). 'Assessment of Sustainability Indicators for Renewable Energy Technologies'. *Renewable and Sustainable Energy Reviews* 13(5): 1082–1088.

EWEA, European Wind Energy Association (2016). 'The European Offshore Wind Industry: Key Trends and Statistics 2015'. Accessed 1 July 2018. Available at <https://www.ewea.org/fileadmin/files/library/publications/statistics/EWEA-European-Offshore-Statistics-2015.pdf>.

Facchini, F. (1995). 'L'évaluation du paysage: une revue critique de littérature'. *Revue d'économie régionale et urbaine* 3 <hal-01350621>.

Faugli, P. E. (1994). 'The Aurland Catchment Area: The Watercourse and Hydropower Development'. *Norsk Geografisk Tidsskrift—Norwegian Journal of Geography* 48: 3–7.

Fazey, I., Gamarra, J. G. P., Fischer, J., Reed, M. S., Stringer, L. C. & Christie, M. (2009). 'Adaptation Strategies for Reducing Vulnerability to Future Environmental Change'. *Frontiers in Ecology and the Environment* 8(8): 414–422.

Fechner, H., Mayr, C., Schneider, A., Rennhofer, M. & Peharz, G. (2016). *Technologie Roadmap für Photvoltaik in Österreich*. Vienna: BVIT.

Federal Department of Foreign Affairs FDFA (2017). 'Geografie: Fakten und Zahlen (online)'. Accessed 24 December 2017. Available at <https://www.FDFA.admin.ch/aboutswitzerland/de/home/umwelt/geografie/geografie--- fakten-und-zahlen.html>.

Federal Office of Spatial Development ARE (2017). *Konzept Windenergie: Basis zur Berücksichtigung der Bundesinteressen bei der Planung von Windenergieanlagen*. Bern.

Ferrario, V. & Castiglioni, B. (2017). 'Visibility/Invisibility in the "Making" of Energy Landscape: Strategies and Policies in the Hydropower Development of the Piaveriver (Italian Eastern Alps)'. Energy Policy 108: 829–835.

Ferrario, V. & Reho, M. (2015). 'Looking beneath the Landscape of Carbon Neutrality: Contested Agroenergy Landscapes in the Dispersed City'. In Frolova, M. et al. (eds.). Renewable Energies and European Landscapes, 95–113. Dordrecht: Springer.

Filipovski, G., Mitrikeski, J., Mitkova, T., Markoski, M., Petkovski, D., Mukaetov, D., Andreevski, M. & Vasilevski, K. (2015). *Pedological (Soil) Map of Macedonia and Its Interpreters*. Skopje: Macedonian Academy of Sciences and Arts.

Fiorino, D. J. (1990). 'Citizen Participation and a Survey of Institutional Mechanisms'. *Science, Technology, & Human Values* 15(2): 226–243.

Firestone, J., Bates, A., Knapp, L. A. (2015). 'See Me, Feel Me, Touch Me, Heal Me: Wind Turbines, Culture, Landscapes, and Sound Impressions'. *Land Use Policy* 46: 241–249. doi: 10.1016/j.landusepol.2015.02.015

Fischer, T. B., Sykes, O., Gore, T., Marot, N., Golobič, M., Pinho, P. & Perdicoulis, A. (2015). 'Territorial Impact Assessment of European Draft Directives: The Emergence of a New Policy Assessment Instrument'. *European Planning Studies* 23(3): 433–451. doi: 1 0.1080/09654313.2013.868292

Flacke, J. & de Boer, C. (2017). 'An Interactive Planning Support Tool for Addressing Social Acceptance of Renewable Energy Projects in the Netherlands'. *InIsprs International Journal of Geo-Information* 6(10).

Flash, B., Lieberz, S., Rondon, M., Williams, B. & Wilson, C. (2016). *EU Biofuel Annual 2016. GAIN Report Number: NL6021*. The Hague: USDA Foreign Agricultural Service.

Fournis, Y. & Fortin, M.-J. (2017). 'From Social "Acceptance" to Social "Acceptability" of Wind Energy Projects: Towards a Territorial Perspective'. *Journal of Environmental Planning and Management* 60: 1–21. doi: 10.108 0/09640568.2015.1133406

Frantál, B. (2015). 'Have Local Government and Public Expectations of Wind Energy Project Benefits Been Met? Implications for Repowering Schemes'. *Journal of Environmental Policy & Planning* 17(2): 217–236.

Frantál, B. & Kunc, J. (2010). 'Factors of the Uneven Regional Development of Wind Energy Projects (A Case of the Czech Republic)'. *Geografický Časopis* 62(3): 183–201.

Frantál, B., Kunc, J. (2011). 'Wind Turbines in Tourism Landscapes: Czech Experience'. *Annals of Tourism Research* 38(2): 499–519.

Frantál, B. & Prousek, A. (2016). 'It's Not Right, But We Do It: Exploring Why and How Czech Farmers Become Renewable Energy Producers'. *Biomass & Bioenergy* 87: 26–34.

Frantál, B. & Urbánková, R. (2017). 'Energy Tourism: An Emerging Field of Study'. *Current Issues in Tourism* 20(13): 1395–1412,

Frantál, B., Bevk, T., van Veelen, B., Harmanescu, M. & Benediktsson, K. (2017). 'The Importance of On-Site Evaluation for Placing Renewable Energy in the Landscape: A Case Study of the Burfell Wind Farm (Iceland)'. *Moravian Geographical Reports* 25(4): 234–247.

Frantál, B., Van der Horst, D., Martinát, S., Schmitz, S., Teschner, N., Silva, L., Golobic, M., Roth, M. (2018). *Spatial Targeting, Synergies and Scale: Exploring the Criteria of Smart Practices for Siting Renewable Energy Projects. Energy Policy*, 120: 85–93.

Fraunhofer Gesellschaft (2017). 'Sonne ernten auf zwei Etagen: Agrophotovoltaik steigert die Landnutzungseffizienz um über 60 Prozent' (online). Accessed 10 April 2018. Available at <https://www.ise.fraunhofer.de/de/presse-und-medien/presseinformationen/2017/sonne-ernten-auf-zwei-etagen-agrophotovoltaik-steigert-landnutzungseffizienz-um-ueber-60-prozent.html>.

Fraunhofer-Institute für Solare Energiesysteme ISE (2016). 'Die Machbarkeit von Agrophotovoltaik wird demonstriert' (online). Accessed 10 April 2018. Available at <http://www.agrophotovoltaik.de/machbarkeit/modellprojekt/>.

Fraunhofer-ISE (2016). 'Net Installed Electricity Generation Capacity in Germany in 2014'. Accessed 15 March 2018. Available at <https://www.energy-charts.de/power_inst.htm?year=2014&period=annual&type=power_inst>.

Frey, B. S. & Oberholzer-Gee, F. (1997). 'The Cost of Price Incentives: An Empirical Analysis of Motivation Crowding-Out'. *American Economic Review* 87(4):746–55.

Frolova, M. (2017). 'Construction of Hydropower Landscapes through Local Discourses: A Case Study from Andalusia (Southern Spain)'. In Bouzarovski, S., Pasqualetti, M. J. & Castán Broto, V., *Routledge Research Companion to Energy Geographies*. Abingdon: Routledge Publishing.

Frolova, M., Centeri, C., Benediktsson, K., Hunziker, M., Kabai, R., Sismani, G. & Martinopoulos, G. (2016). 'Renewable Energy Systems and Mountain Landscapes: An Overview of European Research'. In Bender, O., Baumgartner, J., Heinrich, K., Humer-Gruber, H., Scott, B., Töpfer & T. (eds.). *Mountains, Uplands, Lowlands: European Landscapes from an Altitudinal Perspective*, Abstract Book of the PECSRL Conference, 90–91. Innsbruck: Austrian Academy of Sciences Press.

Frolova, M., Jiménez-Olivencia, Y., Sánchez-del Árbol, M-Á., Requena-Galipienso, A., Pérez-Pérez, B. (2015a). 'Hydropower and Landscape in Spain: Emergence of the Energetic Space in Sierra Nevada (Southern Spain)'. In Frolova, M., Prados, M.-J. & Nadaï, A. (eds.). *Renewable Energies and European Landscapes: Lessons from Southern European Cases*, 117–134. New York/London: Springer.

Frolova, M., Prados M.-J. & Nadaï, A. (2015b). 'Emerging Renewable Energy Landscapes in Southern European Countries'. In Frolova, M., Prados, M.-J., Nadaï, A. (eds.). *Renewable Energies and European Landscapes: Lessons from Southern European Cases*, 3–24. New York/London: Springer.

Fürst, H. (2003). 'The Hungarian-Slovakian Conflict over the Gabcikovo-Nagymaros Dams: An Analysis'. Accessed 26 March 2018. Available at <http://www.columbia.edu/cu/ece/research/intermarium/vol6no2/furst3.pdf>.

García, C. (2011). 'Grid-Connected Renewable Energy in China: Policies and institutions under Gradualism, Developmentalism, and Socialism'. *Energy Policy* 39(12): 8046–8050.

García-Frapolli, E., Schilmann, A., Berrueta, V., Riojas-Rodríguez, H., Edwards, R., Johnson, M., Guevara-Sanginés, A., Armendariz, C. & Masera, O. (2010). 'Beyond Fuel Wood Savings: Valuing the Economic Benefits of Introducing Improved Biomass Cook Stoves in the Purépecha Region of Mexico'. *Ecol Econ* 69(12): 2598–2605.

Gastner, M. T. & Newman, M. E. J. (2004). 'Diffusion-Based Method for Producing Density Equalizing Maps.' *Proc. Natl. Acad. Sci. USA (PNAS)* 101(20): 7499–7504.

Geissler, G., Koppel, J. & Gunther, P. (2013). 'Wind Energy and Environmental Assessments—A Hard Look at Two Forerunners' Approaches: Germany and the United States'. *Renewable Energy* 51: 71–78.

Gibbons, S. (2015). 'Gone with the Wind: Valuing the Visual Impacts of Wind Turbines through House Prices'. *Journal of Environmental Economics and Management* 72: 177–196.

Gill, A. B. (2005). 'Offshore Renewable Energy: Ecological Implications of Generating Electricity in the Coastal Zone'. *Journal of Applied Ecology* 42: 605–615.

Golobic, M. & Marot, N. (2011). 'Territorial Impact Assessment: Integrating Territorial Aspects in Sectoral Policies'. *Evaluation and Program Planning* 34(3): 163–173.

Granhus, A., Lüpke, N. v., Eriksen, R., Søgaard, G., Tomter, S., Antón-Fernández, C. & Astrup, R. (2014). 'Tilgang på hogstmoden skog fram mot 2045'. *Ressursoversikt fra Skog og landskap* 3: IV. Norwegian Forest and Landscape Institute. Ås: Norway.

Gregory, A. J., Atkins, J. P., Burdon, D. & Elliott, M. (2013). 'A Problem Structuring Method for Ecosystem-Based Management: The DPSIR Modelling Process'. *European Journal of Operational Research* 227(3): 558–69.

Grin, J., Rotmans, J. & Schot, J. (2010). *Transitions to Sustainable Development: New Directions in the Study of Long Term Transformative Change*. New York/London: Routledge.

Gross, C. (2007). 'Community Perspectives of Wind Energy in Australia: The Application of a Justice and Community Fairness Framework to Increase Social Acceptance'. *Energy Policy* 35(5): 2727–36.

Gulati, R. & Smith, R. (2009). *Maintenance and Reliability Best Practices*. New York: Industrial Press Inc.

Gundersen, V., Clarke, N., Dramstad, W. & Fjellstad, W. (2016). 'Effects of Bioenergy Extraction on Visual Preferences in Boreal Forests: A Review of Surveys from Finland, Sweden and Norway'. *Scandinavian Journal of Forest Research* 31(3): 323–334.

Gvero, P. (2017). 'Energy System of Bosnia and Herzegovina'. Presentation for the RELY COST WG Meeting in Banjaluka (unpublished).

Haas, R., Eichhammer, W., Huber, C., Langniss, O., Lorenzoni, A., Madlener, R., Menanteau, P., Morthorst, P.-E., Martins, A., Oniszk, A., Schleich, J., Smith, A., Vass, Z. & Verbruggen, A. (2004). 'How to Promote Renewable Energy Systems Successfully and Effectively'. *Energy Policy* 32: 833–839. doi: 10.1016/ S0301-4215(02)00337-3

Habermas, J. (1981). *Theorie des kommunikativen Handelns*, 2 vols. Frankfurt am Main: Suhrkamp.

Hadwan, M. & Alkholidi, A. (2016). 'Solar Power Energy Solutions for Yemeni Rural Villages and Desert Communities'. *Renewable and Sustainable Energy Reviews* 57: 838–849.

Hagen, D. & Erikstad, L. (2013). 'Arealbrukens betydning for miljøprofil i småkraftbransjen, med vekt på vei og rørgate' (Land-use and the environmental profile in development of small scale hydropower plants). *Kart og plan* 73: 297–308.

Haggett, C. (2008). 'Over the Sea and Far Away? A Consideration of the Planning, Politics, and Public Perceptions of Offshore Wind Farms'. *Journal of Environmental Policy and Planning* 10: 289–306.

Hagstofa Íslands (Statistics Iceland) (2017). 'Orkumál' (Energy). Accessed 2 January 2018. Available at <https://hagstofa.is/talnaefni/ atvinnuvegir/orkumal/orkumal/>.

Hall, T., Ponzio, M. & Tonell, L. (1994). In Helmfrid, Staffan (1994). *Landscape and Settlements: National Atlas of Sweden*, 140ff.

Hansen, G. H. (2013). 'Chapter 9: New Renewable Energy and the Norwegian Policy Triangle'. In Rusten, G., Potthoff, K. & Sangolt, L. (eds.). *Norway: Nature, Industry and Society*. Bergen: Fagbokforlaget.

Haraldsson, I. G. & Ketilsson, J. (2010a). 'Jarðhitanotkun til raforkuvinnslu og beinna nota til ársins 2009' (Use of geothermal resources for electricity production and direct utilisation up to 2009). Reykjavík: Orkustofnun (report OS-2010/02).

Haraldsson, I. G. & Ketilsson, J. (2010b). 'Frumorkunotkun jarðvarmavirkjana og hitaveitna á Íslandi til ársins 2009' (Primary energy use of geothermal electricity plants and district heating utilities up to 2009). Reykjavík: Orkustofnun (report OS-2010/03).

Haska, H. (2010). 'The Status of European Beech (Fagus Sylvatica L.) in Albania and Its Genetic Resources'. *Communicationes Instituti Forestalis Bohemicae* 25: 10–24.

Hastik, R., Basso, S., Geitner, C., Haida, C., Poljanec, A., Portaccio, A., Vrščaj, B. & Walzer, C. (2015). 'Renewable Energies and Ecosystem Service Impacts'. *Renewable and Sustainable Energy Reviews* 48: 608–623.

He, Z. X., Xu, S. C., Shen, W. X., Zhang, H., Long, R. Y., Yang, H. & Chen, H. (2016). 'Review of Factors Affecting China's Offshore Wind Power Industry'. *Renewable and Sustainable Energy Reviews* 56: 1372–1386.

Heeb, J. & Hindenlang, K. (2008). 'Negotiating Landscape in the Swiss Alps: Experience with Implementation of a Systemic Landscape Development Approach'. *Mountain Research and Development* 28(2): 105–09.

Hellenic Operator of Electricity Market (2017). RES-CHP *Monthly Statistics*.

Helmfrid, S. (1994). 'Patterns in the Cultural Landscape'. In Helmfrid, S. (1994). *Land-*

scape and Settlements: National Atlas of Sweden, 10–11.

Helmfrid, S., Sporrong, U., Tollin, C. & Windgren, M. (1994). 'Sweden's Cultural Landscape: A Regional Description'. In Helmfrid, S. (1994). Landscape and Settlements: National Atlas of Sweden, 60ff.

Henderson, A. R., Morgan, C., Smith, B., Sørensen, H., Barthelmie, R. & Boesmans, B. (2003). 'Offshore Wind Energy in Europe: A Review of the State-of-the-Art'. Wind Energy 6: 35–52.

Hennig, B. D. (2013). Rediscovering the World: Map Transformations of Human and Physical Space. Heidelberg/New York/Dordrecht/London: Springer.

Heras-Saizarbitoria, I., Cilleruelo, E. & Zamanillo, I. (2011). 'Public Acceptance of Renewables and the Media: An Analysis of the Spanish PV Solar Experience'. Renewable and Sustainable Energy Reviews 15: 4685–4696. doi: 10.1016/j.rser.2011.07.083

Herranz-Surrallés, A. (2016). 'An Emerging EU Energy Diplomacy? Discursive Shifts, Enduring Practices'. Journal of European Public Policy 23(9): 1386–1405.

Hillestad, K. O. (1992). Landscape Design in Hydropower Planning. Hydropower Development Series No. 4. Trondheim, Norwegian Institute of Technology: Hydraulic Engineering.

HM Government (2011). 'The Carbon Plan: Delivering Our Low Carbon Future' (online). Accessed 1 December 2017. Available at <www.gov.uk/government/uploads/system/uploads/attachment_data/ file/47613/3702-the-carbon-plan-delivering-our-low-carbon-future.pdf>.

HM Government (2013). 'Energy Act 2013'. Accessed 1 April 2018. Available at <www.legislation.gov.uk/ukpga/2013/32/contents/enacted>.

Holland, R., Eigenbrod, F., Muggeridge, A., Brown, G., Clarke, D. & Taylor, G. (2015).

'A Synthesis of the Ecosystem Services Impact of Second Generation Bioenergy Crop Production'. Renewable and Sustainable Energy Reviews 46: 30–40.

Hostmann, M., Buchecker, M., Ejderyan, O., Geiser, U., Junker, B., Schweizer, S., Truffer, B. & Zaugg Stern, M. (2005). Wasserbauprojekte gemeinsam planen: Handbuch für die Partizipation und Entscheidungsfindung bei Wasserbauprojekten. Zurich: Eawag, WSL, LCH-EPFL, VAW-ETHZ.

Hrnčiarová, T., Mackovčin, P. & Zvara, I. (eds.) (2009). Landscape Atlas of the Czech Republic. Prague: Ministry of the Environment; The Silva Tarouca Research Institute for Landscape and Ornamental Gardening.

HROTE—Croatian Energy Market Operator (2015). Annual Report on the System of Encouraging the Production of Electricity from Renewable Energy Sources and Cogeneration in the Republic of Croatia for 2014' (online). Accessed 15 December 2017. Available at <http://files.hrote.hr/files/PDF/OIEIK/GI_2014_HROTE_OIEiK_web.pdf>.

Hughes, R. & Buchan, N. (1999). 'Chapter 1: The Landscape Character Assessment of Scotland'. In Usher, M. (ed.). Landscape Character: Perspectives on Management, 1–12. Edinburgh: The Stationery Office.

Hunziker, M., Michel, A., Buchecker, M. (2014). Landschaftsveränderungen durch erneuerbare Energien aus Sicht der Bevölkerung. WSL Bericht, 21: 43–49.

Hurtado, J. P., Fernandez, J., Parrondo, J. L. & Blanco, E. (2004). 'Spanish Method of Visual Impact Evaluation in Wind Farms'. Renewable Sustainable Energy Reviews 8: 483–91.

Hurter, S. & Haenel, R. (eds.) (2002). Atlas of Geothermal Resources in Europe. Luxembourg: European Communities.

Hydroenergy Association (2017). 'Market Analysis of Electric Power Produced by Renewable Energy in Bulgaria'. Available at <http://hidro-energia.org/wpcontent/uploads/

2016/01/AnalizNaVEIvBulgaria.pdf> (in Bulgarian).

Höppner, C., Whittle, R., Brundl, M. & Buchecker, M. (2012). 'Linking Social Capacities and Risk Communication in Europe: A Gap between Theory and Practice?' Natural Hazards 64(2):1753–78.

IEA (2013). Energy Policies of IEA Countries: Sweden, 2013 Review.

IEA (2016). Energy Policies of IEA Countries: Belgium. Accessed 23 December 2017. Available at <https://www.iea.org/publications/freepublications/publication/Energy_Policies_of_IEA_Countries_Belgium_2016_Review.pdf>.

IEA (International Energy Agency) (2015). 'IEA Energy Atlas'. Accessed 2 January 2018. Available at <http://energyatlas.iea.org/>.

IEA-RETD (2016). 'Re Transition: Transitioning to Policy Frameworks for Cost-Competitive Renewables' (Jacobs et al., IET—International Energy Transition GmbH), IEA Technology Collaboration Programme for Renewable Energy Technology Deployment (IEA-RETD), Utrecht. Accessed 30 March 2016.

INEGI/APREN (2018). e2p Endogenous Energies of Portugal, Database of electric power plants based on renewable energy sources. Available at <http://e2p.inegi.up.pt/?Lang=EN>.

Inger, R., Attrill, M. J., Bearhop, S., Broderick, A. C., Grecian, W. J., Hodgson, D. J., Mills, C., Sheehan, E., Votier, S. C., Witt, M. J. & Godley, B. J. (2009). 'Marine Renewable Energy: Potential Benefits to Biodiversity? An Urgent Call for Research'. Journal of Applied Ecology 46: 1145–1153.

Inglehart, R. & Welzel, C. (2010). 'Changing Mass Priorities: The Link between Modernization and Democracy'. Perspectives on Politics 8: 551–567. doi: 10.1017/S1537592710001258

Integrated Energy Consultants (2017). 'Glossary of Terms for Renewable Energy'. Accessed

10 October 2017. Available at <http://www.iecl-energy.co.uk/glossary.html>.

Intergovernmental Panel on Climate Change. Renewable energy sources and climate change mitigation. Accessed 15 July 2016. Available at <https://www.ipcc.ch/pdf/special-reports/srren/SRREN_FD_SPM_final.pdf> 2011.

International Association of Public Participation (2006). 'IAP2s Public Participation Spectrum'.

International Energy Agency (2017). 'Glossary'. Accessed 10 October 2017. Available at <http://www.iea.org/about/glossary/>.

International Energy Agency (2017). 'IEA Atlas of Energy, Country Profiles'. Available at <http://energyatlas.iea.org/#!/tellmap/-1076250891>.

International Finance Cooperation (2013). *Albania Renewable Energy: Removing Market Barriers to Support Clean Energy Development.*

International Hydropower Association (2017). 'The 2017 Hydropower Status Report: An Insight into Recent Hydropower Development and Sector Trends around The World'. Available at <https://www.hydropower.org/>.

IPCC (2013). 'Annex III: Glossary', Planton, S. (ed.). In Stocker, T. F., Qin, D., Plattner, G.-K., Tignor, M., Allen, S. K., Boschung, J., Nauels, A., Xia, Y., Bex, V. & Midgley, P. M. (eds.). *Climate Change 2013: The Physical Science Basis: Contribution of Working Group I to the Fifth Assessment Report of the Intergovernmental Panel on Climate Change.* Cambridge, Cambridge University Press, 1447–1466, doi: 10.1017/CBO9781107415324.031

IRENA (International Renewable Energy Agency) (2017). *Renewable Capacity Statistics 2017.*

Ironside Farrar (2014). 'Strategic Landscape Capacity Assessment for Wind Energy in Aberdeenshire. Final Report to Aberdeenshire Council and Scottish Natural Heritage'. Accessed 1 April 2018. Available at <https://www.aberdeenshire.gov.uk/media/11378/

section1introductionaslcassessment-march2014.pdf>.

Italian National Energy Balance (2013). Available at <http://dgsaie.mise.gov.it/dgerm/ben.asp>.

Jami, A. A. N. & P. R. Walsh (2014). 'The Role of Public Participation in Identifying Stakeholder Synergies in Wind Power Project Development: The Case Study of Ontario, Canada'. *Renewable Energy* 68:194–202.

Jančura, P., Bohálová, I. (2009). The Assessment of the Tvrdošín Medvedie Wind Park Location in Terms of Visual Landscape Characteristics. Pro Tempore, Kolégium, DKD; The Slovak Environmental Agency Banská Bystrica.

Jančura, P., Bohálová, I., Slámová, M. & Mišíková, P. (2010). 'Method of Identification and Assessment of Landscape's Characteristic Appearance' (Metodika identifikácie a hodnotenia charakteristického vzhľadu krajiny) (online). *Bulletin of the Ministry of the Environment of the Slovak Republic* 12(1b): 2–51.

Jauhiainen, J. S. (2014). 'New Spatial Patterns and Territorial–Administrative Structures in the European Union: Reflections on Eastern Europe'. *European Planning Studies* 22: 694–711. doi: 10.1080/09654313.2013.772732

Jenner, S., Groba, F. & Indvik, J. (2013). 'Assessing the Strength and Effectiveness of Renewable Electricity Feed-In Tariffs in European Union Countries'. *Energy Policy* 52: 385–401.

Jerpåsen, G. B. & Larsen, K. C. (2011). 'Visual Impact of Wind Farms on Cultural Heritage: A Norwegian Case Study'. *Environmental Impact Assessment Review* 31(3): 206–215.

Jiménez Herrero, L. (dir.) (2009). *Patrimonio Natural, Cultural y Paisajístico: Claves para la sostenibilidad territorial.* Madrid: Observatorio de la Sostenibilidad en España.

Jiricka, A., Salak, B., Eder, R., Arnberger, A. & Pröbstl, U. (2010). 'Energetic Tourism: Exploring the Experience Quality of Renewable Energies as a New Sustainable Tourism

Market'. In Brebbia, C. A. & Pineda, F. D. (eds.). *WIT Transactions on Ecology and the Environment* 139(4): 55–68.

Johansson, V., Felton, A. & Ranius, T. (2016). 'Long-Term Landscape Scale Effects of Bioenergy Extraction on Dead Wood-Dependent Species'. *Forest Ecology and Management* 371: 103–113.

Jombach, S., Drexler, D. & Sallay, Á. (2010). 'Using GIS for Visibility Assessment of a Wind Farm in Perenye, Hungary'. In Buhmann, Pietsch, Kretzler (seds.). *Peer Reviewed Proceedings of Digital Landscape Architecture*, 322–331. Berlin: Wichmann Verlag im Verlag VDE GmbH.

Jones, C. R. & Richard Eiser, J. (2010). 'Understanding "Local" Opposition to Wind Development in the UK: How Big Is a Backyard?' *Energy Policy* 38: 3106–3117. doi: 10.1016/j.enpol.2010.01.051

Kaldellis, J. K., Kapsali, M. & Katsanou, E. (2012). 'Renewable Energy Applications in Greece: What is the public attitude?' *Energy Policy* 42: 37–48.

Kil, J. (2011). 'Environmental and Spatial Effects of Constructing a Wind Power Plant'. *Contemporary Problems of Management and Environmental Protection* 7: 63–76.

Kluge, S. (2000). 'Empirically Grounded Construction of Types and Typologies in Qualitative Social Research'. *Forum: Qualitative Social Research* 1(1): 14.

Knieling, J. & Othengrafen, F. (2015). 'Planning Culture: A Concept to Explain the Evolution of Planning Policies and Processes in Europe?' *European Planning Studies* 23: 2133–2147. doi: 10.1080/09654313.2015.1018404

Koirala, B. P., Araghi, Y., Kroesen, M., Ghorbani, A., Hakvoort, R. A., & Herder, P. M. (2018). 'Trust, Awareness, and Independence: Insights from a Socio-Psychological Factor Analysis of Citizen Knowledge and Participation in Community Energy Systems'. *Energy Research & Social Science* 38: 33-40.

Kontogianni, A, Tourkolias, M. & Damigos, D. (2014). 'Planning Globally, Protesting Locally: Patterns in Community Perceptions towards the Installation of Wind Farms'. *Renewable Energy* 66: 170–177.

Koutsoyiannis, D. (2011). 'Scale of Water Resources Development and Sustainability: Small Is Beautiful, Large Is Great'. *Hydrological Sciences Journal* 55(4): 553–575.

Kristmannsdóttir, H. & Ármannsson, H. (2003). 'Environmental Aspects of Geothermal Energy Utilization'. *Geothermics* 32(4): 451–461.

Kruse, A. (ed.), Centeri, C., Renes, H., Roth, M., Printsmann, A., Palang, H., Benito Jorda, L., Velarde, M. D. & Kruckenberg, H. (2010). 'Glossary on Agricultural Landscapes'. *Hungarian Journal of Landscape Ecology* (special issue): 99–127.

Kruse, A. & Kruckenberg, H. (2010). 'VI.2 List of Terms and Travelling Exhibition'. In Pungetti, G., Kruse, A. (eds.). *European Culture Expressed in Agricultural Landscapes: Perspectives from the Eucaland Project*, 206–215. Rome, Palombi Editori.

Kruse, A., Marot, N. (eds.) (2018). 'Towards Common Terminology on Energy Landscape'. *Hungarian Journal of Landscape Ecology* (special issue forthcoming).

Kunc, J., Frantál, B. & Klusáček, P. (2011). 'Brownfields as Places for Renewable Sources Location?' In Klimová, V. & Zitek, V. (eds.). *14th International Colloquium on Regional Sciences* 132–140. Brno: Masaryk University.

Ladenburg, J. (2009). 'Visual Impact Assessment of Offshore Wind Farms and Prior Experience'. *Applied Energy* 86: 380–387.

Ladenburg, J. & Dubgaard, A. (2009). 'Preferences of Coastal Zone User Groups Regarding the Siting of Offshore Wind Farms'. *Ocean & Coastal Management* 52: 233–242.

LAEF (n.d.). 'Atjaunojamā enerģija un klimata pārmaiņas' (Renewable energy and climate change) (online). Latvian Renewable Energy Federation Website. Accessed 10 December 2017. Available at <https://www.laef.lv/files/LAEF_buklets_web_62cf2309.pdf>.

Land OÖ (2017). 'Masterplan for Wind Turbines 2017 and Forbidden Areas'. Accessed 11 December 2017. Available at <https://www.land-oberoesterreich.gv.at/110625.htm>.

Lange, E. (1994). 'Integration of Computerized Visual Simulation and Visual Assessment in Environmental Planning'. *Landscape and Urban Planning* 30(1): 99–112.

Langer, K., Decker, T. & Menrad, K. (2017). 'Public Participation in Wind Energy Projects Located in Germany: Which Form of Participation Is the Key to Acceptance?' *Renewable Energy* 112: 63–73.

Law 3851 (2010). 'Acceleration of the RES Development to Address the Climate Change, and Other Provisions on the Competence of the Ministry of Environment, Energy and Climate Change' (in Greek).

Leindecker, G. & Krstic-Furundic, A. (2017). 'New Architectural Design Options'. In Kalogirou, S. (ed.). *Building Integrated Solar Thermal Systems*. Limassol, Cost, 251–261.

Leung, D. Y. C. & Yang, Y. (2012). 'Wind Energy Development and Its Environmental Impact: A Review'. *Renewable and Sustainable Energy Reviews* 16: 1031–1039. doi: 10.1016/j.rser.2011.09.024

LIAA (n.d.). 'Environment and Renewable Energy Industry in LATVIA' (online). Investment and Development Agency in Latvia Website. Accessed 22 December 2017. Available at <http://www.liaa.gov.lv/files/liaa/attachments/liaa_environment_catalogue_m_4.pdf>.

Lietuvos rytas (2012). 'Lietuvoje įjungtos dvi didžiausios saulės jėgainės Baltijos šalyse' (Two largest solar power plants in the Baltic States switched on in Lithuania). 26 September 2012 (online). Accessed 30 December 2017. Available at <https://verslas.lrytas.lt/energetika/lietuvoje-ijungtos-dvi-didziausios-saules-jegaines-baltijos-salyse.htm>.

LI-IEMA, Landscape Institute—Institute of Environmental Management & Assessment (2002). *Guidelines for Landscape and Visual Impact Assessment*. 2nd edition. London/New York: Spon Press.

Lindeboom, H. J., Kouwenhoven, H. J., Bergman, M. J. N., Bouma, S., Brasseur, S., Daan, R., Fijn, R. C., de Haan, D., Dirksen, S., van Hal, R., Hille Ris Lambers, R., ter Hofstede, R., Krijgsveld, K. L., Leopold, M. & Scheidat, M. (2011). 'Short-Term Ecological Effects of an Offshore Wind Farm in the Dutch Coastal Zone: A Compilation'. *Environmental Research Letters* 6(3): 035101.

Llewellyn, D. H., Rohse, M., Day. R. & Fyfe, H. (2017). 'Evolving Energy Landscapes in the South Wales Valleys: Exploring Community Perception and Participation'. *Energy Policy* 108: 818–828, with a 'Corriegendum' in *Energy Policy* 110: 403.

López, E. C. (2016). *Sweden's Renewable Energy Policies towards 2020 and 2030*. Swedish Ministry of Environment and Energy.

Loupa Ramos, I., Bernardo, F., Carvalho Ribeiro, S. & Van Eetvelde, V. (2016). 'Landscape Identity: Implications for Policy Making'. *Land Use Policy* 53 (May 2016): 36–43. doi: 10.1016/j.landusepol.2015.01.030

LVK (2017). 'Myter om kraftkommuner'. The National Association of Hydropower Producing Municipalities. Available at <http://lvk.no/>.

Lüthi, S. & Wüstenhagen, R. (2012). *The Price of Policy Risk: Empirical Insights from Choice Experiments*. Oxford: Oxford University Press.

Majumdar, D. & Pasqualetti, M. J. (2018). 'Dual Use of Agricultural Land: Introducing 'Agrivoltaics' in Phoenix Metropolitan Statistical Area, USA'. *Landscape and Urban Planning* 170: 150–168.

Malu, P. R., Sharma, U. S. & Pearce, J. M. (2017). 'Agrivoltaic Potential on Grape Farms in India'. *Sustainable Energy Technologies and Assessments* 23: 104–110.

277

Mata Olmo, R. & Sanz Herraiz, R. (eds.) (2004). *Atlas de los Paisajes de España*. Madrid: Ministerio de Medio Ambiente, Medio Rural y Marino.

Mann, C. & Jeanneaux, P. (2009). 'Two Approaches for Understanding Land-Use Conflict to Improve Rural Planning and Management'. *Journal of Rural and Community Development* 4(1): 118–141.

Mann, J. & Teilmann, J. (2013). 'Environmental Impact of Wind Energy'. *Environmental Research Letters* 8(3): 035001. doi: 10.1088/1748-9326/8/3/035001

Manyoky, M., Wissen Hayek, U., Heutschi, K., Pieren, R., Grêt-Regamey, A. (2014). Developing a GIS-Based Visual-Acoustic 3D Simulation for Wind Farm Assessment. International Journal of Geo- Information 3(1): 29–48.

Marčiukaitis, M. (2017). 'Perspektyvinių VE plėtrai teritorijų ir prijungimo prie elektros tinklų Lietuvoje galimybių studija'. (Feasibility study on territories prospective for wind energy development and its connection to the power grid: project report on wind energy development and biodiversity-related areas) (online). Accessed 30 December 2017. Available at <http://corpi.lt/venbis/files/reports/VENBIS_LEI_galimybiu_studija.pdf>.

Mariestads Tiedningen (2017). 'Electrivillage, spektakulärt spännande project'. Accessed 20 February 2017. Available at <https://mariestadstidningen.se/asikter/fria-ord/2017/02/02/electrivillage-spektakulart>.

Marot, N. (ed.). Kruse, A. (ed.). Benediktsson, K., Bottarelli, M., Brito, P., Centeri, Cs., Dolacek Alduk, Z., Eiter, S., Frantal, B., Frolova, M., Gaillard, B., Háyrynen, M., Hernandez Jimenez, V., Hewitt, R., Hunziker, M., Kabai, R., Karan, I., Lachowska, M., Martinat, S., Martinopoulos, G., Mestre, N., Mickovski, S. B., Miller, D., Otte, P., Roehner, S., Munoz-Rojas, J. R. M., Schroth, O., Scognamiglio, A., Slupinski, M., Stremke, S. & Teschner, N. (2018). 'Glossary on Renewable Energy and Landscape Quality: The Glossary'. *Hungarian Journal of Landscape Ecology*, (special issue), in press.

Marrou, H., Guilioni, L., Dufour, L., Dupraz, C. & Wéry, J. (2013). 'Microclimate Under Agrivoltaic Systems: Is Crop Growth Rate Affected in the Partial Shade of Solar Panels?' *Agricultural and Forest Meteorology* 177: 117–132.

Martinez Alonso, P., Hewitt, R., Diaz Pacheco, J, Roman Bermejo, L., Hernandez Jimenez, V., Vicente Guillen, J. & de Boer, C. (2016). 'Losing the Roadmap: Renewable Energy Paralysis in Spain and Its Implications for the EU Low Carbon Economy'. *Renewable Energy* 89: 680–694.

Martín-Gamboa, M., Iribarren, D. & Dufour, J. (2015). On the Environmental Suitability of High and Low-Enthalpy Geothermal Systems'. *Geothermics* 53: 27–37.

Martinopoulos, G. (2016). 'Energy Efficiency and Environmental Impact of Solar Heating and Cooling Systems'. In Wang, R. Z. & Ge, T. S. (eds.). Advances in Solar Heating and Cooling, 43–59. Amsterdam: Elsevier.

Martinopoulos, G. & Tsalikis, G. (2018). 'Diffusion and Adoption of Solar Energy Conversion Systems: The Case of Greece'. *Energy* 144C: 800–807.

Marusic, J., Ogrin, D. & Jancic, M. (1998). Methodological Bases. Ljubljana, Ministry of Environment and Physical Planning, National Office for Physical Planning.

Masini, A. & Menichetti, E. (2012). 'The Impact of Behavioural Factors in the Renewable Energy Investment Decision Making Process: Conceptual Framework and Empirical Findings'. *Energy Policy* 40: 28–38.

Mátrai Erőmű ZRt. (n.d.). 'Mátrai erőmű: átadták magyarország legnagyobb naperőművét' (Mátra power plant: Hungary's largest solar thermal energy power plant delivered). Accessed 26 March 2018. Available at <http://www.mert.hu/atadtak-magyarorszag-legnagyobb-naperomuvet>.

'Áætlun um vernd og orkunýtingu landsvæða' (The Master Plan for Nature Protection and Energy Utilization) 2017. Accessed 2 January 2018. Available at <www.ramma.is>.

McKendry (2002). 'Energy Production from Biomass (Part 1): Overview of Biomass'. Bioresource Technology 83(1): 37–46.

Meallem, I., Garb, Y. & Cwikel, J. (2010). 'Environmental Hazards of Waste Disposal Patterns: A Multimethod Study in an Unrecognized Bedouin Village in the Negev Area of Israel'. *Archives of Environmental & Occupational Health* 65(4): 230–237.

MEEDDM. Ministère de l'Écologie, de l'Énergie, du Développement durable et de la Mer (2010). 'Le guide de l'étude d'impact des projets éoliens'. Availablet at <http://www.developpement-durable.gouv.fr/IMG/pdf/guide_eolien_15072010_complet.pdf>.

Mérida-Rodríguez, M., Lobón-Martín, R. & Perles-Roselló, M. J. (2015a). 'The Production of Solar Photovoltaic Power and Its Landscape Dimension in Andalusia (Spain)'. In Frolova, M., Prados, M.-J. & Nadaï, A. (eds.). *Renewable Energies and European Landscapes: Lessons from Southern European Cases*, 255–277. New York/London: Springer.

Mérida-Rodríguez, M., Reyes-Corredera, S., Pardo-García, S. & Zayas-Fernández, B. (2015b). 'Solar Photovoltaic Power in Spain: Expansion Factors and Emerging Landscapes'. In Frolova, M., Prados, M.-J. & Nadaï, A. (eds.). *Renewable Energies and European Landscapes: Lessons from Southern European Cases*, 63–80. New York/London: Springer.

Miles, I., Saritas, O. & Sokorov, A. (2016). 'Intelligence: Delphi'. In *Foresight for Science, Technology and Innovation*, 95–124. Cham (ZG): Springer International.

Miller, D. R., Bell, S., McKeen, M., Horne, P. L., Morrice, J. G. & Donnelly, D. (2010). 'Assessment of Landscape Sensitivity to Wind Turbine Development in Highland: Report for Highland Council, Macaulay Land Use Research Institute'. Accessed 1 April 2018.

Available at file:///D:/Downloads%20-C/ Assessment_of_Landscape_Sensitivity_to_ Wind_Turbine_Development_in_Highland_ Summary_Report%20(2).pdf.

Ministère de l'Environnement, de l'Energie et de la Mer (2017). *Évaluation environnementale: Guide de lecture de la nomenclature des études d'impact* (R.122-2). February 2017.

Ministère de l'Environnement, de l'Energie et de la Mer (2017). *Guide d'évaluation des impacts sur l'environnement des parcs éoliens en mer.* 2017 Edition.

Ministry for the Environment and Spatial Planning (2017). 'Glossary of the Natura2000 '(in Slovene). Accessed 20 September 2017. Available at <http://www.natura2000.si/index. php?id=46&no_cache=1&L>.

Ministry of Economy (2015). 'Energy in Croatia: Annual Energy Report; Entrepreneurship and Crafts, Energy Institute Hrvoje Pozar, Zagreb' (online). Accessed 20 November 2017. Available at <http://www.eihp.hr/ wp-content/uploads/2016/12/Energija2015. pdf>.

Ministry of Economy of the Republic of Macedonia (2017). 'Guidelines on the Procedures for the Development and Construction of Power Plants Utilising Renewable Energy Sources, Photovoltaic Plants' (online). Accessed December 2017. Available at <http://www.economy.gov.mk/Upload/Documents/Guidelines_on_photovoltaic_plants_ B5_f inal_3.pdf>.

Ministry of Economy of the Republic of Macedonia. 'Strategy for Utilization of Renewable Energy Sources in the Republic of Macedonia by 2020' (online). Accessed December 2017. Available at <http://iceor.manu.edu.mk/ Documents/ICEIM/Strategies/Strategy%20 for%20utilization%20 RES.pdf>.

Ministry of Energy and Industry (2015). *National Action Plan for Renewable Energy Resources in Albania: Albania, as a Contractual Party in the Treaty of Energy Community.* Tirana, Albania

Ministry of Environment (2007). 'St.meld. nr. 34 (2006–2007) Norsk klimapolitikk' (Governmental White Paper Norwegian Climate Policy). Oslo.

Ministry of Housing, Communities and Local Government (2014). 'Environmental Impact Assessment'. Accessed 1 December 2017. Available at <https://www.gov.uk/guidance/ environmental-impact-assessment>.

Mirasgedis, S., Tourkolias, C., Tzovla, E., Diakoulaki, D. (2014). 'Valuing the Visual Impact of Wind Farms: An Application in South Evia, Greece'. *Renewable and Sustainable Energy Reviews* 39: 296–311.

MoE (Ministry of Energy of the Republic of Lithuania) (2016). 'Renewable Energy Sources: Overview' (online). Accessed 30 December 2017. Available at <https://enmin. lrv.lt/en/sectoral-policy/renewable-energy-sources>.

MoE (Ministry of Energy of the Republic of Lithuania) (2017a). 'Nacionalinė energetinės nepriklausomybės strategija' (National strategy for energetic independence) (online). Accessed 30 December 2017. Available at <https://e-seimas.lrs.lt/portal/legalAct/lt/ TAD/TAIS.429490>.

MoE (Ministry of Energy of the Republic of Lithuania) (2017b). 'Nacionalinės energetinės nepriklausomybės strategijos SPAV ataskaita' (The report of the strategic environment impact assessment of the national strategy for energetic independence) (online). Accessed 30 December 2017. Available at <http:// enmin.lrv.lt/uploads/enmin/documents/files/ NENS_SPAV_enmin_03_rev_su_priedais_ compressed.pdf>.

MoEnv (Ministry of Environment of the Republic of Lithuania) (2015). 'Nacionalinis kraštovaizdžio tvarkymo planas' (National landscape management plan) (online). Accessed 30 December 2017. Available at <https://drive.google.com/file/d/0B0vZmxk-FijDnWGtyWjhzaTVZeWs/view>.

Mostert, E., Craps, M. & Pahl-Wostl, C. (2008). 'Social Learning: The Key to Integrated Water Resources Management?' *Water International* 33(3): 293–304.

Mullaj, A., Hoda, P., Shuka, L., Miho, A., Bego, F., & Qirjo, M. (2017). 'About Green Practices for Albania'. *Albanian Journal of Agricultural Sciences* (2017, special issue): 31–50.

Mumford, J. & Gray, D. (2010). 'Consumer Engagement in Alternative Energy: Can the Regulators and Suppliers Be Trusted?' *Energy Policy* 38(6): 2664–71.

Musall, F. D. & Kuik, O. (2011). 'Local Acceptance of Renewable Energy: A Case Study from Southeast Germany'. *Energy Policy* 39(6): 3252–3260.

Möller, B. (2010). 'Spatial Analyses of Emerging and Fading Wind Energy Landscapes in Denmark'. Land Use Policy 26(3): 233–241.

Müller, S., Buchecker, M., Backhaus, N. (2017). Participatory Mapping of Place Meanings in Community Wind Energy Planning: An Evaluation for Different Landscape Types in Switzerland. Nordic Geographers Meeting, Stockholm. Available at https://stockholmuniversity. app.box.com/s/xg3lptu8j11wyhsi9ethyu8bb-535pjfk.

Nadaï, A. & Prados, M. J. (2015). 'Landscapes of Energies: A Perspective on the Energy Transition'. In Frolova, M., Prados, M. J. & Nadaï, A. (eds.). *Renewable Energies and European Landscapes*, 25–42. Dordrecht/Heidelberg/ New York/London: Springer.

Nadaï, A. & Van der Horst, D. (2010). 'Introduction: Landscapes of Energies'. *Landscape Research* 35:2: 143–155. doi: 10.1080/01426390903557543

National Agency of Territorial Planning (AKPT) (2016). *National General Plan.*

National Energy Strategy (2017). Available at <http://www.sviluppoeconomico.gov.it/ images/stories/documenti/Testo-integrale-SEN-2017.pdf>.

Ndreu, A. (2016). 'Implementation of the Administrative-Territorial Reform for a Sustainable Development of the Economy in Albania'. Social and Natural Sciences 10(1): 1–4.

NEA 2015. *Albania Country Briefing: The European Environment; State and Outlook 2015*. Copenhagen: European Environment Agency.

Nerep, M. (2017). *Post-Soviet, Post-mining and Post-agriculture Landscape Wind Parks in Estonia: Displaying Regional Pride and Being Possible Touristic Attraction*. Master Thesis, Department of Landscape Architecture, Estonian University of Life Sciences. Available at <https://dspace.emu.ee/xmlui/bitstream/handle/10492/3309/Merlin%20Nerep_2017MA_AR_t%c3%a4istekst.pdf?sequence=1&isAllowed=y>.

Nersten, N. K., Puschmann, O., Hofsten, J., Elgersma, A., Stokstad, G. & Gudem, R. (1999). 'The Importance of Norwegian Agriculture for the Cultural Landscape'. *Notat* 11(1999). Norwegian Institute of Land Inventory (NIJOS) and Norwegian Agricultural Economics Research Institute (NILF), Ås/Oslo.

Neuman, S. & Hopkins, Ch. (2009). 'Renewable Energy Projects on Contaminated Property: Managing the Risks'. *Environmental Claims Journal* 21(4): 296–312.

Newig, J. & Fritsch, O. (2009). 'More Input—Better Output: Does Citizen Involvement Improve Environmental Governance?' In Blühdorn, Ingolfur (ed.). *In Search of Legitimacy: Policy Making in Europe and the Challenge of Complexity*, 205–224. Opladen: B. Budrich.

Nikodemus, O. (2018). 'Latvijas dabas un ainavu kartēšana un rajonēšana' (Mapping and typology of Latvia's nature and landscapes). In Nikodemus, O. et al. (eds.). Latvija. *Zeme, daba, tauta, valsts*, 571–574. Rīga: Latvijas Universitātes Akadēmiskais apgāds.

NMPE—Norwegian Ministry of Petroleum and Energy (2015). *Fakta. Energi- og vannressurser i Norge*. Oslo: Olje- og energidepartementet.

Northern Ireland Environment Agency (NIEA) (2014). 'Northern Ireland's Landscape Charter'. Available at <www.doeni.gov.uk/niea/landscape_charter_document.pdf>.

NREL (2010). *A Policymaker's Guide to Feed-in Tariff Policy Design: Technical Report*. 30 July 2010.

NVE (2016). 'Samlet plan for vassdrag er avviklet. 11.10.2016'. The Norwegian Water Resources and Energy Directorate. Accessed 18 December 2017. Available at <https://www.nve.no/nytt- fra-nve/nyheter-konsesjon/samlet-plan-for-vassdrag-er-avviklet/>.

NVE (2017). 'Energiforsyning og konsesjon'. The Norwegian Water Resources and Energy Directorate. Accessed 18 December 2017. Available at <https://www.nve.no/energi-forsyning-og- konsesjon/>.

Oddsdóttir, A. L. & Ketilsson, J. (2012). 'Vinnslusvæði hitaveitna' (Production areas of district heating utilities). Reykjavík: Orkustofnun (report OS-2012/07).

Olwig, K. (2007). The practice of Landscape 'Conventions' and the just Landscape. The case of the European Landscape Convention. Landscape Research 32 (5): 579–594.

Orkustofnun (National Energy Authority) (2017a). 'Energy Statistics in Iceland 2016'. Accessed 2 January 2018. Available at <www.orkustofnun.is/orkustofnun/utgafa/orku-tolur/>.

Orkustofnun (National Energy Authority) (2017b). 'Orkuvefsjá' (Iceland Energy Portal). Accessed 2 January 2018. Available at <www.orkuvefsjá.is/vefsja/orkuvefsja.html>.

Othengrafen, F. (2010). 'Spatial Planning as Expression of Culturised Planning Practices: The Examples of Helsinki, Finland and Athens, Greece'. *Town Planning Review* 81: 83–110. doi: 10.3828/tpr.2009.25

Otte, P., Rønningen, K. & Moe, E. (in preparation, forthcoming 2019). 'Contested Wind Energy: Investigating Discourses of Energy Impacts and Their Significance for Energy Justice in The Case of Europe's Largest Onshore Wind Power Project'. In Szolucha, A., Knudsen, S. & Haarstad, H. (eds.). *Energy Impacts and Contested Futures*. Abingdon: Routledge.

Pagan, H. C. & Vollmer, E. (eds.) (2017). 'Advocating for Sustainable Energy in Central and Eastern Europe'. Estonia: Ecoprint AS. Available at <http://dspace.emu.ee/xmlui/handle/10492/3755>.

Pahl-Wostl, C., Craps, M., Dewulf, A., Mostert, E., Tabara, D. & Taillieu, T. (2007). 'Social Learning and Water Resources Management'. *Ecology and Society* 12(2): 5.

Palang, H., Alumäe, H., Printsmann, A., Rehema, M., Sepp, K. & Sooväli-Sepping, H. (2011) 'Social Landscape: Ten years of Planning "Valuable Landscapes" in Estonia'. *Land Use Policy* 28: 19–25.

Palit, D. & Bandyopadhyay, K. R. (2016). 'Rural electricity Access in South Asia: Is grid Extension the Remedy? A Critical Review'. *Renewable and Sustainable Energy Reviews* 60: 1505–1515.

Palmas, C., Abis, E., von Haaren, C. & Lovett, A. (2012). 'Renewables in Residential Development: An Integrated GIS-Based Multicriteria Approach for Decentralized Micro-Renewable Energy Production in New Settlement Development: A Case Study of the Eastern Metropolitan Area of Cagliari, Sardinia, Italy'. Energy, *Sustainability and Society* 2(1): 10.

Palmer, J. R. (2014). 'Biofuels and the Politis of Land-Use Change: Tracing the Interactions of Discourse and Place in European Policy Making'. *Environment and Planning* A 46: 337–352.

Paloniemi, R., Apostolopoulou, E., Cent, J., Bormpoudakis, D., Scott, A., Grodzińska-Jurczak, M., Tzanopoulos, J., Koivulehto, M., Pietrzyk-Kaszyńska, A. & Pantis, J.D. (2015). 'Public Participation and Environmental Justice in Biodiversity Governance in Finland, Greece, Poland and the UK'. *Environmental Policy and Governance* 25: 330–342. doi: 10.1002/eet.1672

Papadopoulos, A. M., Glinou, G. L. & Papachristos, D. A (2008). 'Developments in the Utilisation of Wind Energy in Greece'. *Renewable Energy* 33(1): 105–110.

Papamanolis, N. (2015). 'The First Indications of The Effects of the New Legislation Concerning the Energy Performance of Buildings on Renewable Energy Applications in Buildings in Greece'. *International Journal of Sustainable Built Environment* 4(2): 391–399.

Pasqualetti, M. (2000). 'Morality, Space, and the Power of Wind'. *Geographical Review* 90: 381–394.

Pasqualetti, M. J. (2011). 'Opposing Wind Energy Landscapes: A Search for Common Cause'. *Annals of the Association of American Geographers* 101: 907–917. doi: 10.1080/000 45608.2011.568879

Pasqualetti, M. J. (2011). 'Social Barriers to Renewable Energy Landscapes'. *Geogr. Rev.* 101: 201–223.

Pasqualetti, M. & Stremke, S. (2018). 'Energy Landscapes in a Crowded World: A First Typology of Origins and Expressions'. *Energy Research & Social Science* 36: 94–105.

PCNPA (2013). 'Cumulative Impact of Wind Turbines on Landscape and Visual Amenity'. Accessed 1 December 2017. Available at <www.pembrokeshirecoast.org.uk/files/files/Dev%20Plans/ Cumulative%20Impact%20 SPG%20Final%20Jan2014.pdf>.

Pentru perioada 2007–2020. Accessed 4 December 2017. Available at <http://www.minind. ro/energie/STRATEGIA_energetica_actualizata.pdf>.

Piroozfar, P., Pomponi, F. & Farr, E. R. P. (2016). 'Life Cycle Assessment of Domestic Hot Water Systems: A Comparative Analysis'. *International Journal of Construction Management* 16(2): 109–125.

Plan for Management of Nature Park 'Vitosha' 2015–2024. Contract № ОПОС-03-145/10.12.2014. Available at <http://pu-vitosha.com/wp-content/uploads/2015/07/06_Analiz_DPPV.pdf> (in Bulgarian).

Plan for Management of Nature Park 'Pirin', 2014–2023 Г. Contract № УР-051/29. 01. 2014. Available at <http://www.pu-pirin.com/images/Sreshti/25- 26.02.2015_Proekt_PU_Pirin/4_Gorsko_25-26.02.15.pdf> (in Bulgarian).

Planning Authority (2015). 'Development Control Policy (2015)'. Accessed December 1, 2017. Available at <https://www.pa.org.mt/en/supplementary-guidance-details/development-control-design- policy-guidance-and-standards-2015-dc15->.

Polatidis, H. & Haralambopoulos, D. A. (2004). 'Local Renewable Energy Planning: A Participatory Multi-Criteria Approach'. *Energy Sources* 26(13): 1253–64.

Prato, T. (2017). 'Conceptual Framework for Adaptive Management of Coupled Human and Natural Systems with Respect to Climate Change Uncertainty'. *Australasian Journal of Environmental Management* 24(1): 47–63.

Prodanuks, T., Cimdiņa, G., Veidenbergs I. & Blumberga, D. (2016). 'Sustainable Development of Biomass CHP in Latvia'. *Energy Procedia* 95: 372–376.

Puschmann O. (1998). *The Norwegian Landscape Reference System: Use of Different Sources as a Base to Describe Landscape Regions*. NIJOS report 12/98. Ås: Norsk institutt for jord- og skogkartlegging/Norwegian Institute of Land Inventory (NIJOS).

PV portal (2017). 'Solar Power Plants in Slovenia' (online). Accessed December 2017. Available at <http://pv.fe.uni- lj.si/SEvSLO.aspx>.

Quadu, F. (2013). 'Impacts of Biomass Energy Development on the Walloon Territory'. *Territoire(s)* 2.

Ramans, K. (1994). 'Ainavrajonēšana' (Landscape typology and regions). In Kavacs, G.

(ed.). *Latvijas Daba I*, 22–24. Rīga: Latvijas Enciklopēdija.

Rathmann, R., Szklo, A. & Schaeffer, R. (2010). 'Land Use Competition for Production of Food and Liquid Biofuels: An Analysis of the Arguments in the Current Debate'. *Renewable Energy* 35: 14–22.

Raven, R. P., Heiskanen, E., Lovio, R., Hodson, M. & Brohmann, B. (2008). 'The Contribution of Local Experiments and Negotiation Processes to Field-Level Learning in Emerging (Niche) Technologies: Meta-Analysis of 27 New Energy Projects in Europe'. *Bulletin of Science, Technology & Society* 28(6): 464–477.

Raven, R. P., Mourik, R. M., Feenstra, C. F. J. & Heiskanen, E. (2009). 'Modulating Societal Acceptance in New Energy Projects: Towards a Toolkit Methodology for Project Managers'. *Energy* 34(5): 564–74.

Ravena, R. & Gregersen, K. (2007). 'Biogas Plants in Denmark: Successes and Setbacks'. *Renewable and Sustainable Energy Reviews* 11: 116–132

Real Estate Cadastre Agency, 'A Rulebook on Cadastral Grading' (online). Accessed December 2017. Available at <http://web01. katastar.gov.mk/userfiles/file/pravilnici/Pavilnik_za_katastarsko_klasiranje_i_ utvrduvanjeto_i_zapisuvanjeto_na_promenata_na_katastarskata_kultura_i_klasa.pdf>.

Reed, J., van Vianen, J., Barlow, J. & Sunderland, T. (2017). 'Have Integrated Landscape Approaches Reconciled Societal and Environmental Issues in the Tropics?' *Land Use Policy* 63: 481–492.

Reimer, M. & Blotevogel, H. H. (2012). 'Comparing Spatial Planning Practice in Europe: A Plea for Cultural Sensitization'. *Planning Practice and Research* 27(1): 7–24. doi: 10.1080/0 2697459.2012.659517

Renn, O. (1998). 'The Role of Risk Perception for Risk Management'. *Reliability Engineering & System Safety* 59(1): 49–62.

Rey, L., Hunziker, M., Stremlow, M., Arn, D., Rudaz, G. & Kienast, F. (2017). *Wandel der Landschaft: Erkenntnisse aus dem Monitoringprogramm Landschaftsbeobachtung Schweiz (LABES), Bern, Umwelt-Zustand Nr. 1641.* Bern: Bundesamt für Umwelt; Birmensdorf: Eidgenössische Forschungsanstalt für Wald, Schnee und Landschaft.

Rio, P. del, Peñasco, C. & Mir-Artigues, P. (2018). An Overview of Drivers and Barriers to Concentrated Solar Power in the European Union. Renew Sustain Energy Rev, 81: 1019–1029. doi:10.1016/J.RSER.2017.06.038.

RIPCHNH RS Republic Institute for the Protection of Cultural, Historical and Natural Heritage of the Republic of Srpska (2016). 'Protected Regions in Bosnia and Herzegovina' (online). Accessed 23 December 2017. Available at <http://nasljedje.org/prirodno-nasljedje/266>.

Rosenberg, D. M., Bodaly, R. A. & Usher, P. J. (1995). 'Environmental and Social Impacts of Large Scale Hydroelectric Development: Who Is Listening?' *Global Environmental Change* 5(2): 127–148.

Roth, M. & Bruns, E. (2016). 'Landschaftsbildbewertung in Deutschland'. BfN-Skripten 439. Available at <https://www.bfn.de/fileadmin/BfN/service/Dokumente/skripten/skript439.pdf>.

Roth, M. & Gruehn, D. (2014). 'Digital Participatory Landscape Planning for Renewable Energy–Interactive Visual Landscape Assessment as Basis for the Geodesign of Wind Parks in Germany'. *Peer Reviewed Proceedings of Digital Landscape Architecture*: 84–94.

Rousso, A. (n.d.). 'Le droit du paysage un nouveau droit pour une nouvelle politique'. *Courrier de l'environnement de l'INRA* 26.

Rugg, D. S. (1994). 'Communist Legacies in the Albanian Landscape'. *Geographical Review* 84(1): 59–73.

Ruggiero, F. & Scaletta, G. (2014). 'Environmental Integration of Wind Farms: The Territorial Governance'. *Energy and Power Engineering* 6(11): 386–400. doi: 10.4236/epe.2014.611033

Rusten, G. (2013). 'Chapter 7: The Hydropower Industry: Technology, Regulations and Regional Effects'. In Rusten, G., Potthoff, K. & Sangolt, L. (eds.). *Norway: Nature, Industry and Society*. Bergen: Fagbokforlaget.

Röhring, A. & Gailing, L. (2005). 'Institutional Problems and Management Aspects of Sharp Sultural Landscapes'. Working Paper No. 33. Erkner: Leibniz-Institute for Regional Development and Structural Planning.

Röpcke, I. (2007). 'Stille um die Lärmschutzwand'. *Sonne Wind und Wärme* 12: 100–105.

Sacchelli, S., Garegnani, G., Geri, F., Grilli, G., Paletto, A., Zambelli, P. & Vettorato, D. (2016). 'Trade-off between Photovoltaic Systems Installation and Agricultural Practices on Arable Lands: An Environmental and Socio-Economic Impact Analysis for Italy'. *Land Use Policy* 56: 90–99.

Saidur, R., Rahim, N. A., Islam, M. R. & Solangi, K. H. (2011). 'Environmental Impact of Wind Energy'. *Renewable and Sustainable Energy Reviews* 15: 2423–2430. doi: 10.1016/j.rser.2011.02.024

Schmeer, O. (2009). 'Windräder auf der Halde'. In WAZ. Accessed 10 April 2018. Available at <https://www.waz.de/staedte/gelsenkirchen/windraeder-auf-der-halde-id11839.html>.

Schmidt, C., von Gagern, M., Lachor, M., Hage, G., Schuster, L., Hoppenstedt, A., Kühne, O., Rossmeier, A., Weber, F., Bruns, D., Münderlein, D. & Bernstein, F (in press). 'Landschaftsbild und Energiewende'. In preparation on behalf of the Federal Agency for Nature Conservation (BfN). Available at <https://www.natur-und-erneuerbare.de/projektdatenbank/projekte/landschaftsbild-und-energiewende/>.

Schürmann, M. (2015). 'Energiewende auf alter Bergbauhalde in Gelsenkirchen'. In *Der Westen*. Accessed 10 April 2018. Available at <https://www.derwesten.de/wochenende/energiewende-auf-alter-bergbauhalde-in-gelsenkirchen-id11350874.html>.

Scognamiglio, A. (2016). 'Photovoltaic Landscapes: Design and Assessment; A Critical Review for a New Transdisciplinary Design Vision'. *Renewable and Sustainable Energy Reviews* 55: 629–661.

Scottish Government (2014). 'Agri-renewable Strategy'. Scottish Government.

Scottish Government (2016). 'Getting the Best from Our Land: A Land Use Strategy for Scotland 2016–2021'. Scottish Government.

Scottish National Heritage (2017). 'Landscape Glossary of Terms'. Accessed 20 September 2017. Available at <http://www.snh.gov.uk/protecting-scotlands-nature/looking-after-landscapes/landscape-resource-library/glossary-of-terms/>.

Scottish Natural Heritage (2013). 'Onshore Windfarm Proposals'. Accessed 1 April 2013. Available at <https://gateway.snh.gov.uk/natural-spaces/dataset.jsp?dsid=WINDFARM>.

Scottish Natural Heritage (2013). 'Onshore Wind Farm Proposals: Digital Datasets'. Scottish Natural Heritage. Accessed 1 September 2013. Available at <https://gateway.snh.gov.uk/natural-spaces/dataset.jsp?dsid=WINDFARM>.

Scottish Natural Heritage (2015). 'Landscape Character Assessment Digital Data'. Scottish Natural Heritage. Accessed 1 April 2013, last accessed 1 March 2018. Available at <https://gateway.snh.gov.uk/natural-spaces/dataset.jsp?dsid=LCA>.

Selman, P. (2004). 'Community Participation in the Planning and Management of Cultural Landscapes'. *Journal of Environmental Planning and Management* 47(3): 365-392.

Şenel, M. C. & Koç, E. (2015). 'The State of Wind Energy in Turkey and in the World: A General Evaluation' (in Turkish). *Engineer and Machine* 56(663): 46–56.

Service de l'Observation et des Statistiques (2017). 'Ministère de l'Environnement, de l'Energie et de la Mer en charge des relations internationales sur le climat' (online) Accessed 27 December 2017. Available at <http://www.statistiques.developpement-durable.gouv.fr/fileadmin/documents/Produits_editoriaux/Publications/Datalab/2017/Datalab-9-CC-de-l-environnement-edition-2016-fevrier2017.pdf>.

Shaaban, M. & Petinrin, J. O. (2014). 'Renewable Energy Potentials in Nigeria: Meeting Rural Energy Needs'. *Renewable and Sustainable Energy Reviews* 29: 72–84.

Shortall, R., Davidsdottir, B. & Axelsson, G. (2015). 'Geothermal Energy for Sustainable Development: A Review of Sustainability Impacts and Assessment Frameworks'. *Renewable and Sustainable Energy Reviews* 44: 391–406.

Silva, L. & Delicado, A. (2017). 'Wind Farms and Rural Tourism: A Portuguese Case Study of Residents' and Visitors' Attitudes and Perceptions'. *Moravian Geographical Reports* 25(4): 248–256.

Simcock, N. (2014). 'Exploring How Stakeholders in Two Community Wind Projects Use a "Those Affected" Principle to Evaluate Fairness of Each Project's Spatial Boundary'. *Local Environment* 19(3): 241–258.

Simcock, N. (2016). 'Procedural Justice and the Implementation of Community Wind Energy Projects: A Case Study from South Yorkshire, UK'. *Land Use Policy* 59: 467–77.

Sismani, G., Babarit, A. & Loukogeorgaki, E. (2017). 'Impact of Fixed Bottom Offshore Wind Farms on the Surrounding Wave Field'. *Proc. of 26th International Offshore and Polar Engineering Conference (ISOPE), Rhodes, Greece, June 26–July 1, 2016* 1: 511–518.

SISTAN—Terna (2017). Available at <http://download.terna.it/terna/0000/0994/85.PDF>.

SKDS (2008). 'Sabiedrības attieksme pret dažādiem enerģētikas jautājumiem' (Societal attitude towards various issues of energetics) (online). Latvenergo Website. Accessed 16 December 2017. Available at <http://www.latvenergo.lv/portal/page/portal/Latvian/files/aktualitates/atskaite_ene rgetika_032008.pdf>.

Slamova, M., Belacek, B. & Pous, R. (2012). 'Identification of Landscape's Values of the Turcianska Kotlina Basin'. *Landscape Architecture* (风景园林) 5:126–144.

SLU—Swedish University of Agricultural Science (2017). 'National Inventory of Landscapes in Sweden, NILS'. Accessed 29 March 2018. Available at <https://www.slu.se/en/Collaborative-Centres-and-Projects/nils/>.

Smardon, R. & Pasqualetti, M. (2017). 'Societal Acceptance of Renewable Energy Landscape'. In Apostol, D. et al. (eds.). *The Renewable Energy Landscape: Preserving Scenic Values in Our Sustainable Future*, 108–144. Abingdon: Routledge.

SNH—Scottish Natural Heritage (2009). 'Siting and Designing Windfarms in the Landscape'. Available at <http://www.snh.gov.uk/planning-and-development/renewable-energy/onshore-wind/landscape-impacts-guidance>.

Sodha, M. S. (2014). 'Renewable Energy Education and Awareness'. *J. Fundam Renew Energy Appl* 4: e101.

Sokka, L., Sinkko, T., Holma, A., Manninen, K., Pasanen, K., Rantala, M. & Leskinen, P. (2016). 'Environmental Impacts of the National Renewable Energy Targets: A Case Study from Finland'. *Renewable and Sustainable Energy Reviews* 59: 1599–1610.

Solli, J. (2010). 'Where the Eagles Dare? Enacting Resistance to Wind Farms Through Hybrid Collectives'. *Environmental Politics* 19(1): 45–60.

Solomon, B. D. & Barnett, J. B. (2017). 'The Changing Landscape of Biofuels: A Global Overview'. In Bouzarovski, S., Pasqualetti, M. J. & Castán Broto, V. *The Routledge Research Companion to Energy Geographies*, 61–78. London/New York: Routledge Publishing.

Sovacool, B. K. & Ratan, P. L. (2012). 'Conceptualizing the Acceptance of Wind and Solar Electricity'. *Renewable & Sustainable Energy Reviews* 16(7): 5268–79.

Sperling, K. (2017). 'How Does a Pioneer Community Energy Project Succeed in Practice? The Case of the Samso Renewable Energy Island'. *Renewable & Sustainable Energy Reviews* 71: 884–97.

SPF Economie (2010). 'Belgique: Plan d'action national en matière d'énergies renouvelables conformément à la Directive 2009/28/CE, 95p'. Accessed 5 December 2017. Available at <http://economie.fgov.be/fr/binaries/NREAP-BE-v25-FR_tcm326-112992.pdf>.

Sporrong, Ulf (1994). 'The Old Agrarian Landscape before 1750'. In Helmfrid, Staffan (1994). *Landscape and Settlements: National Atlas of Sweden*, 30ff.

Späth, L. & Scolobig, A. (2017). 'Stakeholder Empowerment through Participatory Processes: The Case of Electricity Transmission Lines in France and Norway'. *Energy Research and Social Science* 23: 189–198.

Späth, L., Scolobig, A., Patt, A., Hildebrand, J., Molinengo, V., Eversen, J. & Rudberg, B. (2014). 'Improved and Enhanced Stakeholder Participation in Reinforcement of Electricity Grid (Inspiregrid): D3.2 Establishing the Best Practices and Determining a Tool Box'. Zurich: ETH Zürich. Available at <www.inspire-grid.eu/index.php/inspire-grid-publications/>.

SSPA (State Service for Protected Areas under the Ministry of Environment of the Republic of Lithuania) (2017). 'Overview of the System of Protected Areas' (online). Accessed 30 December 2017. Available at <http://www.vstt.lt/en/VI/index.php#r/75>.

State Statistical Office of the Republic of Macedonia. Website. Accessed December 2017. Available at <http://www.stat.gov.mk/Default_en.aspx>.

Statistical Office of the Republic of Slovenia (2017). *Statopis 2016*. Ljubljana: Statistical Office of the Republic of Slovenia.

Statistical Office of the Slovak Republic (ŠÚ SR), Ministry of Economy of the Slovak Republic (MH SR) (2014). *Obnoviteľné zdroje energie* [Renewable energy resources]. Accessed 14 December 2017. Available at <https://www.enviroportal.sk/indicator/detail?id=2601&pdf=true>.

Statistics Norway (2017a). 'Electricity'. Accessed 18 December 2017. Available at <https://www.ssb.no/en/energi-og-industri/statistikker/elektrisitet/aar>.

Statkraft (2016). 'Europe's Largest Onshore Wind Power Project to Be Built in Central-Norway'. Accessed 18 December 2017. Available at <http://www.statkraft.com/IR/stock-exchange- notice/2016/europes-largest-onshore-wind-power-project--to-be-built-in-central-norway--/>.

Stead, D. (2013). 'Convergence, Divergence, or Constancy of Spatial Planning? Connecting Theoretical Concepts with Empirical Evidence from Europe'. *Journal of Planning Literature* 28:19–31. doi: 10.1177/0885412212471562

Steinhilber, B. U. (2010). 'Mit der Wernerschen Mühle haben Betzinger weiteres Kleinod'. In tagblatt.de. Accessed 10 April 2018. Available at <https://www.tagblatt.de/Nachrichten/Mit-der-Wernerschen-Muehle-haben-Betzinger-weiteres-Kleinod-218879.html>.

Steinhäußer, R., Siebert, R., Steinführer, A. & Hellmich, M. (2015). 'National and Regional Land-Use Conflicts in Germany from the Perspective of Stakeholders'. *Land Use Policy* 49: 183–194. doi: 10.1016/j.landusepol.2015.08.009

Stenström, M. & Laine, K. (eds.) (2006). *Towards Good Practices for Practice-Oriented Assessment in European Vocational Education*. Jyväskylä: Finish Institute for Educational Research, University of University of Jyväskylä.

Stevovic, S., Milovanovic, Z., & Stamatovic, M. (2015). Sustainable model of hydro power development—Drina river case study. Renewable and Sustainable Energy Reviews, 50, 363–371.

Stirling, A. (2006). 'Analysis, Participation and Power: Justification and Closure in Participatory Multi Criteria Analysis'. *Land Use Policy* 23(1): 95–107.

Strand, O., Colman, J. E., Eftestøl, S., Sandström, P., Skarin, A. & Thomassen, J. (2017). *Vindkraft og reinsdyr: en kunnskapssyntese*. NINA report. 1305: 62. Norwegian Institute for Nature Research, Trondheim.

Strazzari, F. (2009). 'Trans-Albanian vs. Pan-Albanian Spaces: The Urban Dimension of the "Albanian Question"'. *Southeastern Europe* 33: 77–99.

Stremke, S. (2015). 'Sustainable Energy Landscape: Implementing Energy Transition in the Physical Realm'. Jørgensen, S.V. (ed.). *Encyclopedia of Environmental Management*. Boca Raton, FL: CRC Press. doi: http://dx.doi.org/10.1081/E-EEM-120053717.

Stremke, S. & Dobbelsteen, A. (van den) (eds.) (2012). *Sustainable Energy Landscapes*. Boca Raton: CRC Press

Stremke, S. & Dobbelsten, A. (van den) (2013). 'Sustainable Energy Landscapes: An Introduction'. In Stremke, S. Dobbelsteen, A. (van den). *Sustainable Energy Landscapes: Designing, Planning and Development*. Boca Raton: CRC Press.

Strunz, S., Gawel, E. & Lehmann, P. (2016). 'The Political Economy of Renewable Energy Policies in Germany and the EU'. *Utilities Policy* 42: 33–41.

Sullivan, R. G., Kirchler, L. B., Cothren, J. & Winters, S. L. (2012). 'Offshore Wind Turbine Visibility and Visual Impact Threshold Distances'. *Science, Politics and Policy: Environmental Nexus; Conference Proceedings, National Association of Environmental Professionals, 37th Annual NAEP Conference (May 21–24, 2012, Portland Oregon)*, 943–969.

Swaffield, S. (2005). 'Landscape as a Way of Knowing the World'. In Harvey, S. et al. (eds.). *The Cultured Landscape: Designing the environment in the 21st century*, 3–24. Abingdon: Routledge.

Swanwick, C. (2004). 'The Assessment of Countryside and Landscape Character in England: An Overview'. In Philips, A. et al. (eds.). *Countryside Planning: New Approaches to Management and Conservation*, 109–124. London: Earthscan.

Swedish Energy Agency (2015). 'Sankey Diagram: Sweden´s Energy System 2014'.

Swedish Gas Centre (2007). 'Biogas: Basic Data on Biogas'.

Swedish Government (Regeringskansliet) (2010). 'The Swedish National Action Plan for the Promotion of the Use of Renewable Energy in Accordance with Directive 2009/28/EC and the Commission Decision of 30.06.2009'.

Swedish Institute (SI) (2015). 'Facts about Sweden: Energy'. Available at <https://sweden.se/wp-content/uploads/2013/09/Energy_high_resolution.pdf>.

Swiss Federal Office of Energy SFOE (2012). 'Energiestrategie 2050: Erstes Massnahmenpaket' (online). Accessed 25.06.2016. Available at <https://www.uvek.admin.ch/uvek/de/home/energie/energiestrategie-2050.html>.

Swiss Federal Office of Energy SFOE (2015a). 'Schweizerische Statistik der Erneuerbaren Energien: Ausgabe 2014' (online). Accessed 24 December 2017. Available at <http://www.SFOE.admin.ch/themen/00526/00541/00543/index.html?lang=de&dossier_id=00772>.

Swiss Federal Office of Energy SFOE (2015b). 'Schweizerische Elektrizitätsstatistik' (online). Accessed 14 March 2018. Available at <http://www.SFOE.admin.ch/themen/00526/00541/00542/00630/index.html?lang=de&doss ier_id=00700>.

Swiss Federal Office of Energy SFOE (2016). 'Geothermal Energy' (online). Accessed 14 March 2018. Available at <http://www.SFOE. admin.ch/themen/00490/00501/index.html?lang=en>.

Swiss Federal Office of Energy SFOE (2017a). 'Small-Scale Hydropower' (online). Accessed 14 March 2018. Available at <http://www.SFOE.admin.ch/themen/00490/00491/00493/index.html?lang=en>.

Swiss Federal Office of Energy SFOE (2017b). 'Wind Energy' (online). Accessed 14 March 2018. Available at <http://www.SFOE. admin.ch/themen/00490/00500/index.html?lang=en>.

Swiss Spatial Planning Association VLP-ASPAN (2014). *Einführung in die Raumplanung*. Bern.

Tauš, P., Rybár, R., Kudelas, D., Domaracký D. &Kuzevič, Š. (2005). 'Potention of Renewable Energy Sources in Slovakia in Term of Production of Electricity'. *Acta Montanistica Slovaca* 10(3): 317–326.

Terkenli, T. S. (2011). In Search of the Greek Landscape: A Cultural Geography'. In *The European Landscape Convention*, 121–141. Dordrecht: Springer.

Teschner, N. & Alterman, R. (2018). 'Preparing the Ground for Renewable Energy: Regulatory Challenges in Siting Small-Scale Wind Turbines in Urban Areas'. *Renewable & Sustainable Energy Reviews* 81(2): 1660–1668.

Teschner, N. & Paavola, J. (2013). 'Discourses of Abundance: Transitions in Israel's Energy Regime'. *Journal of Environmental Policy & Planning* 15(3): 447–466.

The Countryside Agency and Scottish Natural Heritage (2002). 'Landscape Character Assessment Guidance for England and Scotland'.

The Energy Regulatory Commission of the Republic of Macedonia. Website. Accessed December 2017. Available at <http://www.erc.org.mk/Default_en.aspx>.

Tobias, S., Buser, T. & Buchecker, M. (2016). 'Does Real-Time Visualization Support Local Stakeholders in Developing Landscape Visions?' *Environment and Planning B: Planning and Design* 43(1): 184–197.

Torres-Sibille, A., Cloquell-Ballester, V., Cloquell-Ballester, V. & Darton, R. (2009). 'Development of a Multicriteria Indicator for the Assessment of Objective Aesthetic Impact of Wind Farms'. *Renewable Sustainable Energy Review* 13: 40–55.

Tóth, A., Štěpánková, R. & Feriancová, Ľ. (2016). *Landscape Architecture and Green Infrastructure in the Slovak Countryside*. Prague: Czech University of Life Sciences/ Powerprint Prague.

Tougaard, J., Madsen, P. T. & Wahlberg, M. (2008). 'Underwater Noise from Construction and Operation of Offshore Wind Farms'. *Bioacoustics* 17 (1–3): 143–146.

Trafikverket—Swedish Transport Administration (2015). 'Integrated Landscape Character Assessment'. Accessed 29 March 2018. Available at <https://www.trafikverket.se/en/startpage/planning/Landscape-planning/>.

Transparency International B&H 2014 Percepcija javne uprave, Bosna i Hercegovina (2014) (online). Transparency International Bosnia and Herzegovina website. Accessed 24 April 2018. Available at <http://ti-Bosnia-Herzegovina.org/wp-content/uploads/2015/03/TI-BOSNIA-HERZEGOVINA-Percepcija-Javne-Uprave-Bosnia-Herzegovina-2014.pdf>.

Tsikalakis, A., Tomtsi, T., Hatziargyriou, N., Poullikkas, A., Malamatenios, C., Giakoumelos, E., Cherkaoui, J., Chenak, A., Fayek, A., Matarg, T. & Yasin, A. (2011). 'Review of Best Practices of Solar Electricity Resources Applications in Selected Middle East and North Africa (MENA) Countries'. *Renewable and Sustainable Energy Reviews* 15(6): 2838–2849.

Tsilingiridis, G., Martinopoulos, G. & Kyriakis, N. (2004). 'Life Cycle Environmental Impact of a Thermosyphonic Domestic Solar Hot Water System in Comparison with Electrical and Gas Water Heating'. *Renewable Energy* 29(8): 1277–1288.

Tsoutsos, T., Frantzeskaki, N. & Gekas, V. (2005). 'Environmental Impacts from the Solar Energy Technologies'. *Energy Policy* 33(3): 289–296.

Tulcea County Council. Website: Accessed 29 December 2017. Available at <https://www.cjtulcea.ro>.

Tveit, M., Ode, A., Fry, G. (2007). 'Key Concepts in a Framework for Analysing Visual Landscape Character'. *Landscape Research* 31(3): 229–255.

U.S. Department of Energy (2013). 'Glossary of Energy Related Terms'. Accessed 10 October 2017. Available at <https://energy.gov/eere/energybasics/articles/glossary-energy-related-terms>.

U.S. Green Building Council (2017). 'Glossary on LEED buildings'. Accessed 10 October 2017. Available at <https://www.usgbc.org/glossary/>.

UNESCO (2015). *UNESCO Country Programming Document for Albania* 2014–2017. Venice: UNESCO. Available at <http://unesdoc.unesco.org/images/0023/002330/233036E.pdf>.

United Nations (2000). 'Convention on Biological Diversity'. United Nations.

United Nations (2015). 'Paris Agreement: United Nations Framework Convention on Climate Change on Climate Change'. pp 27. Available at <http://unfccc.int/paris_agreement/items/9485.php>.

United Nations Economic Commission for Europe (1998). 'Convention on Access to Information, Public Participation in Decision-Making and Access to Justice in Environmental Matters'. Aarhus Convention (June).

United Nations Economic Commission for Europe (2000). 'Convention on Access to Information, Public Participation in Decision-Making and Access to Justice in Environmental Matters'. Accessed 1 April 2018. Available at <www.unece.org/fileadmin/DAM/env/pp/documents/cep43e.pdf>.

Upham, P., Oltra, C. & Boso, À. (2015) 'Towards a Cross-Paradigmatic Framework of the Social Acceptance of Energy Systems'. *Energy Research & Social Science* 8: 100–112. doi: 10.1016/j.erss.2015.05.003

Upreti, B. R. & Van Der Horst, D. (2004). 'National renewable Energy Policy and Local Opposition in the UK: The Failed Development of a Biomass Electricity Plant'. *Biomass and Bioenergy* 26(1): 61–69.

Vaab, T., Keerberg, L. & Vaarmari, K. (2010). 'Tuulikud ja tuulepargid Eestis. Senine planeerimine. Probleemid. Ettepanekud lahendusteks. Eesti Keskkonnaühenduste Koda, Keskkonnaõiguse Keskus, Eesti Roheline Liikumine'. Available at <http://www.eko.org.ee/wp-content/uploads/2010/06/Tuulikud-ja-tuulepargid-Eestis.pdf> (in Estonian).

Van der Horst, D. (2007). 'NIMBY or Not? Exploring the Relevance of Location and the Politics of Voiced Opinions in Renewable Energy Siting Controversies'. *Energy Policy* 35(5): 2705–2714.

Van der Horst, D. (2009). 'Spatial Planning of Wind Turbines and the Limits of "Objective" Science'. *Moravian Geographical Reports* 17(2): 46–51.

Van der Horst D. & Vermeylen, S. (2012). 'Ownership Claims, Valuation Practices and the Unpacking of Energy-Landscape Conflicts'. *International Review of Sociology* 22(3): 421–437.

Van der Molen, J., Smith, H. C. M., Lepper, P., Limpenny, S. & Rees, J. (2014). 'Predicting the Large-Scale Consequences of Offshore Wind Turbine Array Development on a North Sea Ecosystem'. *Continental Shelf Research* 85: 60–72.

Van Hecke, E., Antrop, M., Schmitz, S., Van Eetvelde, V. & Sevenant, M. (2010). *Land-schap, platteland en landbouw, Atlas van Belgïe*, vol. 2. Gent: Academia Press.

Van Rijnsoever, F. J., van Mossel, A. & Broecks, K. P. F. (2015). 'Public Acceptance of Energy Technologies: The Effects of Labeling, Time, and Heterogeneity in a Discrete Choice Experiment'. *Renewable and Sustainable Energy Reviews* 45: 817–829. doi: 10.1016/j.rser.2015.02.040

Vaughan, A. (2017). 'Time to Shine: Solar Power Is Fastest-Growing Source of New Energy'. Accessed 5 October 2017. Available at <https://www.theguardian.com/environment/2017/oct/04/solar-power-renewables-international-energy-agency>.

Vaughan, A. & Hopkins, N. (2017). 'Renewable Power Critic Is Chosen to Head Energy Price Review'. Accessed 5 October 2017. Available at <https://www.theguardian.com/environment/2017/jul/12/renewable-power-energy-costs-review-dieter-helm>.

Ven, H. (van de) (2014). 'Cultural Heritage Agency, Sun Energy in A Historical Environment'. Accessed 12 March 2018. Available at <https://cultureelerfgoed.nl/publicaties/zonne-energie-in-de-historische-omgeving-groene-gids>.

Vermeylen, V. & Walker, G. (2011). 'Environmental Justice, Values, and Biological Diversity: The San and Hoodia Benefit-Sharing-Agreement'. In *Environmental inequalities beyond borders: local perspectives on global injustices*, 105–128. Cambridge, MA: MIT Press.

Veselý, A. (2011). 'Theory and Methodology of Best Practice Research: A Critical Review of the Current State'. *Central European Journal of Public Policy* 5(2): 98–117.

Vlada RS (2016). 'Energetska bilanca RS' (Government of the Republic of Slovenia, Energy balance sheet of the Republic of Slovenia) (online). Accessed December 2017. Available at <http://www.energetika-portal.si/fileadmin/dokumenti/publikacije/energetska_bilanca/ebrs_2016.pdf>.

Von der Dunk, A., Grot-Regamey, A., Dalang, T. & Hersperger, A. M. (2011). 'Defining a Typology of Peri-Urban Land-Use Conflicts: A Case Study from Switzerland'. *Landscape and Urban Planning* 101(2): 149–156.

Vorel, I., Bukáček, R., Matějka, P., Culek, M. & Sklenička, P. (2006). *A Method for Assessing the Visual Impact on Landscape Character of Proposed Construction, Activities or Changes in Land Use: A Method for Spatial and Character Differentiation of an Area*. Prague: Centre for Landscape.

Waage, E. R. H. (2013). *The Concept of Landslag: Meanings and Value for Nature Conservation. University of Iceland, PhD dissertation in Geography*. Available at http://hdl.handle.net/1946/25744.

Waage, E. R. H. & Benediktsson, K. (2010). 'Performing Expertise: Landscape, Governmentality and Conservation Planning in Iceland'. *Journal of Environmental Policy and Planning* 12(1): 1–22; 21 December 2016, Code of Practice for Wind Energy Development in Ireland Guidelines for Community Engagement1 Department of Communications, Climate Action and Environment

Walbiom (2017). *Panorama de la filière biométhanisation en Wallonie*. Namur: SPW.

Walker G. (2010). 'Environmental Justice, Impact Assessment and the Politics of Knowledge: The Implications of Assessing the Social Distribution of Environmental Outcomes'. *Environ Impact Assess Rev* 30(5): 312–8.

Walker, G. & Devine-Wright, P. (2008). 'Community Renewable Energy: What Should It Mean?' *Energy Policy* 36(2): 497–500.

Walker, G., Devine-Wright, P., Hunter, S., High, H. & Evans, B. (2010). 'Trust and Community: Exploring the Meanings, Contexts and Dynamics of Community Renewable Energy'. *Energy Policy* 38(6): 2655–63.

Wang, C., Miller. D., Brown, I., Jiang, Y. & Castellazzi, M. (2016). 'Visualisation Techniques to Support Public Interpretation of Future Climate Change and Land-Use Choices: A

Case Study from N-E Scotland'. *International Journal of Digital Earth* 9(6): 586–605.

Warren, C. R. (2014). 'Scales of disconnection: Mismatches Shaping the Geographies of Emerging Energy Landscapes'. *Moravian Geographical Reports* 22(2): 7–14.

Warren, C. R. & McFadyen, M. (2010). 'Does Community Ownership Affect Public Attitudes to Wind Energy? A Case Study from South-West Scotland'. *Land Use Policy* 27(2): 204–213.

Warren, Ch., Cowell, R., Ellis, G., Strachan, P. A. & Szarka, J. (2012). 'Wind Power: Towards a Sustainable Energy Future?' In Szarka, J., Cowell, R., Ellis, G., Strachan, P. & Warren, Ch. (eds.) (2012). *Learning from Wind Power: Governance, Societal and Policy Perspectives on Sustainable Energy*, 1–14. Basingstoke: Palgrave MacMillan.

WEDG, Wind Energy Development Guidelines (2006). Department of the Environmental, Heritage and Local Government (Ireland). Available at: http://www.environ.ie/en/.

Weiland, U., Wüstneck, T., Lichte V. & Scholles, F. (2016). 'Zur Bündelung von Stromtrassen mit anderen linearen Infrastrukturen: ein strittiges Thema'. *UVP-report* 30(3): 159–170.

Wesselink, A., Paavola, J, Fritsch, O. & Renn, O. (2011). 'Rationales for Public Participation in Environmental Policy and Governance: Practitioners' Perspectives'. *Environment and Planning A* 43(11): 2688–2704.

Whipple, T. (2011). 'The Peak Oil Crisis: The 3rd Transition'. *Falls Church News-Press*, 17 February 2011 (online). Accessed 20 March 2018. Available at <http://www.fcnp.com/commentary/national/8548-the-peak-oil-crisis-the-3rd-transition.html>.

Wicklow County Council (2016). *Wicklow County Development Plan 2016–2022*, Appendix 6, Wicklow Wind Energy Strategy.

Wikipedia (2017a). 'Definition of Energy Landscape'. Accessed 5 October 2017. Available at <https://en.wikipedia.org/wiki/Energy_landscape>.

Wikipedia (2017b). 'Definition of the Glossary'. Accessed 5 October 2017. Available at <https://en.wikipedia.org/wiki/Glossary>.

Wilson, J. C. & Elliott, M. (2009). 'The Potential for Habitat Creation Produced by Offshore Wind Farms'. *Wind Energy* 12, 2: 203–212.

Wind Europe (2017). 'Wind Energy in Europe: Scenarios for 2030'. Available at <https://windeurope.org/about-wind/reports/wind-energy-in-europe-scenarios-for-2030/>.

Wolsink, M. (2006). 'Invalid Theory Impedes Our Understanding: A Critique on the Persistence of the Language of NIMBY'. *Transactions of the Institute of British Geographers* NS31: 85–91.

Wolsink, M. (2007). 'Planning of Renewables Schemes: Deliberative and Fair Decision-Making on Landscape Issues instead of Reproachful Accusations of Non-Cooperation'. Energy policy 35(5), 2692–2704. Oslo: Governmental White Paper Norwegian Climate Policy. doi: 10.1016/j.enpol.2006.12.002

Wolsink, M. (2010). 'Contested Environmental Policy Infrastructure: Socio-Political Acceptance of Renewable Energy, Water, and Waste Facilities'. *Environmental Impact Assessment Review* 30: 302–311. doi: 10.1016/j.eiar.2010.01.001

Wolsink, M. & Breukers, S. (2010). 'Contrasting the Core Beliefs Regarding the Effective Implementation of Wind Power: An International Study of Stakeholder Perspectives'. *Journal of Environmental Planning and Management* 53(5): 535–58.

World Bank (2014). 'Access to Electricity, Rural (% of Rural Population)' (online). Accessed 20 March 2018. Available at <https://data.worldbank.org/indicator/EG.ELC.ACCS.RU.ZS>.

World Commission on Dams (2000). *Dams and Development: A New Framework for Decision-making*. London and Sterling, VA: Earthscan Publications Ltd.

World Data. Accessed 12 January 2018. Available at: <https://www.worlddata.info/europe/sweden/energy-consumption.php>.

Worldenergy (2016). Accessed 29 March 2018. Available at <https://www.worldenergy.org/data/resources/country/sweden/>.

Wüstenhagen, R., Wolsink, M. & Burer, M.J. (2007). 'Social Acceptance of Renewable Energy Innovation: An Introduction to the Concept'. *Energy Policy* 35(5): 2683–91. doi: 10.1016/j.enpol.2006.12.001

Zanon, B. & Verones, S. (2013). 'Climate Change, Urban Energy and Planning Practices: Italian Experiences of Innovation in Land Management Tools'. *Land Use Policy* 32: 343–355.

Zoellner, J., Schweizer-Ries, P. & Wemheuer, C. (2008). 'Public Acceptance of Renewable Energies: Results from Case Studies in Germany'. *Energy Policy* 36: 4136–4141. doi: 10.1016/j.enpol.2008.06.026

Editor Profiles

alphabetical order

Buchecker, Matthias
Swiss Federal Research Institute WSL, Research Unit Economics and Social Sciences, Zürcherstrasse 111, CH-8903 Birmensdorf
Phone: +41 44 739 23 60
Email: matthias.buchecker@wsl.ch
Homepage: https://www.wsl.ch/en/employees/buchecke.html

Matthias Buchecker is senior researcher at the research unit Economics and Social Sciences of the Swiss Federal Institute WSL. He studied geography at the university of Berne and completed a doctoral thesis on participatory landscape development. Since 1999, he has directed projects at WSL on people-environment-interactions in the research fields landscape development, recreation, river management, natural hazards, and most recently also renewable energies.

2009–2012: Leader of WP 'Risk communication' in the EU-7th Framework Project „Social Capacity building for Natural Hazards" (Cap-Haz Net).
2011-2013 Leader of WSL project within the EU-7th Framework Project 'Building a Culture of risk prevention in Europe' (KULTURisk).
2012–2013: Leader of the project 'Social Acceptance of photovoltaic panel sites in a Alpine tourism region' in collaboration with the Energy Region Obergoms.
2013: Leader of the project 'Wahrnehmung des Thema Flussrevitalisierungen in den lokalen und regionalen Medien im Berner Oberland. Medienanalyse der Periode von 1999–2012.' Renaturierungsfonds des Kantons Bern, BAFU.
2013–2018: Leader of the PhD-project 'Analyzing the contribution of deliberative planning in risk management to social learning.' Swiss National Found
2016–2019: Leader of the PhD project 'Mapping meaningful places as a tool for participatory planning of renewable energy projects'. SBFI, COST Action TU1401

Centeri, Csaba
Szent István University, Inst. of Nature Conservation and Landscape Management, 2100-Gödöllő, Páter K. u. 1., Hungary
Email: Centeri.Csaba@mkk.szie.hu
Phone: +36 302027336

Csaba Centeri is an associate professor at the Szent István University. His main research interest is nature conservation and land use change related issues, especially in connection with soil and soil erosion, and wildlife impact. He also started working with agricultural landscapes, landscape protection, and renewable energy production, where soil and nature conservation play an important role.

Fekete István Programme (mapping the possible areas for producing renewable energy from biomass in a Trans-Danubian region, Hungary)
EUCALAND Network
COST-RELY
Bio-Bio, Indicators for biodiversity in organic and low-input farming systems (FP7)
EU-FP7 ENVIEVAL, Development and application of new methodological frameworks for the evaluation of environmental impacts of rural development programmes in the EU (Grant Agreement No. 312071)

Eiter, Sebastian
NIBIO—
Norwegian Institute of Bioeconomy Research
PO Box 115
1431 Ås
NORWAY
Phone: +47 6497 1044
Email: sebastian.eiter@nibio.no

Dr Sebastian Eiter is a geographer and landscape ecologist. He is employed as a research scientist in the Department of Landscape Monitoring at NIBIO, the Norwegian Institute of Bioeconomy Research. His recent research topics include causes and consequences of agricultural landscape change, cultural heritage, biodiversity, public participation, and urban agriculture.

Edible Cities Network: Integrating Edible City Solutions for Socially Resilient and Sustainably Productive Cities (EU H2020) 2018–2023.
Agricultural Landscapes in Norway: Occurrence, Sustainability, Characteristics, and Local Variations and Values (Research Council of Norway) 2015–2017.
Urban Agriculture Europe (EU COST) 2012–2016.
Monitoring cultural heritage environments protected by law—development of a method (Norwegian Directorate for Cultural Heritage) 2011–2016.
Monitoring the Norwegian mountain dairy farm landscape (Norwegian Forest and Landscape Institute) 2009–2014.
Landscape change (Research Council of Norway) 2008–2014.
Agricultural buildings and the cultural landscape (Norwegian Ministry of Agriculture and Food) 2009–2012.
Indicators for biodiversity in organic and low-input farming systems (EU FP7) 2009–2012
Land use changes in urban pressure areas—threats to food production and landscape qualities (Research Council of Norway) 2009–2012.

Frantál, Bohumil
Department of Environmental Geography, Institute of Geonics, The Czech Academy of Sciences, Drobneho 28, 602OO Brno, Czech Republic
Email: frantal@geonika.cz
Phone: +420 545 422 720

RNDr. Bohumil Frantál, PhD is a leading senior scientist at the Department of Environmental Geography, Institute of Geonics, Czech Academy of Sciences. In his research he focuses on social-spatial contexts of ongoing energy transition, particularly renewable energy development and related land use conflicts, environmental risk perceptions, urban renewal, and spatial models of behaviour. His principal publications address the issues of social acceptance of renewables, nuclear power, and coal mining. He also works as a lecturer at Palacký University in Olomouc.

Recent research projects:
2016–2018: 'Exploring social-spatial diffusion of renewable energy projects in the Czech Republic: lessons for adaptive governance of energy transition' (Czech Science Foundation No. 16-04483S)—Co-investigator
2014–2018: Integrated Spatial Planning, Land Use and Soil Management Research Action (INSPIRATION), (Horizon 2020, Grant No. 681256)—Member of research team
2011–2014: 'Energy Landscapes: innovation, development and internationalization of research' (ENGELA) (CZ.1.07/2.3.00/20.0025)—Project leader
2010–2013: 'Tailored Improvement of Brownfield Regeneration in Europe (TIMBRE)' (EU´s 7th Framework Programme, No. ENV.2010.3.1.5-2)—Member of research team
2009–2011: 'Spatial models of behavior in the changing urban space from the point of view of time geography' (Czech Science Foundation No. 403/09/0885)—Co-investigator
2008–2010: 'The use of wind energy: evaluation of spatial relations, environmental aspects and social context by the means of GIS' (Grant Agency of the Czech Academy of Sciences No. KJB700860801)—Member of research team

Frolova, Marina
Institute for Regional Development, University of Granada, c/Rector López Argueta s/n, 18071 Granada, Spain
Email: mfrolova@ugr.es
Phone: +34 958 24 10 00 Ext. 20218

Marina Frolova is professor at the Department for Regional and Physical Geography, and the Institute for Regional Development, University of Granada. She is an experienced contributor to and leader of multi-disciplinary research projects. Her research interests include landscape analysis, landscape and renewable energy policies, and their relevance for the social acceptance of energy technologies. She has published and edited several books, book chapters and articles on landscape—especially energy-related—issues.

Title: ADAPTation to sustainable energy transition in Europe: environmental, socio-economic and cultural aspects CSO2017-86975-R, funded by Spain's Ministry of Economy, Industry and Competitivity, 2018–2021 (chair);
Title: Energía eólica y paisaje: Evaluación del paisaje terrestre y maritimo para una ordenación sostenible, CSO2011-23670, funded by Spain's Ministry of Science and Innovation, 2012-2015 (chair);
Title: Ressources paysagères et ressources énergétiques dans les montagnes sud-européennes: Histoire, comparaison, expérimentation, fonded by Ministère de la Culture et de la Communication de France, Direction générale des patrimoines de France, Bureau de la recherche architecturale, urbaine et paysagère de France, Ministère de l'Écologie, du Développement durable, des Transports et du Logement de France, Direction de la recherche et de l'innovation de France y Atelier international du Grand Paris (AIGP) (Chair of WG)

Karan, Isidora
Center for Spatial Research
Slobodana Dubocanina 1, 78 000 Banjaluka, Bosnia and Herzegovina
Email: isidora_karan@yahoo.com
Phone: +387 65819623

Isidora Karan is an architect and urban designer. She was awarded a PhD in architecture (2015) from the University of Granada and pursued postdoctoral research at the National Autonomous University of Mexico in (2016). She has worked in practice on the projects of urban planning and design (2009–2011, 2016). She is a partner in the Center for Spatial Research (NGO) and teaching assistant (University of Banjaluka).

Projects:
Plan of detailed regulation for the area under the influence of Highway M-16.1., Bosnia and Herzegovina, 2009
General Urban Plan for Vlasenica Municipality, Bosnia and Herzegovina, 2016 (First prize at 26th International Urban Planners Exhibition in Serbia)
Small-scale urban intervention in Banjaluka, Bosnia and Herzegovina (2017–ongoing)

International competitions:

Recognition for project for The Sustainable Theatre, World Stage Design, Cardiff, 2013 (exposition and award)
First prize for the Urban Design of Memorial Zone DONJA GRADINA in Bosanska Dubica, Banja Luka, Bosnia and Herzegovina, 2009 (exposition and award)
First prize for the Urban Design of Old Market Veselin Maslesa in Banja Luka, Bosnia and Herzegovina, 2009

Kruse, Alexandra
Institute for Research on European Agricultural Landscapes e.V. (EUCALAND) institute World Heritage Consulting
10bis, rue du Haras – 78530 Buc/France –
Email: kruse@eucalandnetwork.eu
Phone: +33 6 28138569

Dr. Alexandra Kruse is an expert in UNESCO World Heritage and cultural landscapes. She is director of insitu World Heritage (whconsult. eu). In addition, she is an expert in natural heritage and nature sites, and project management. She has long experience in documentation. She is initiator and secretary general of the international network 'Institute for

Research on European Agricultural Landscapes (EUCALAND) e.V.' (eucaland.net). Since 2006 this network has been working on classifying and raising the awareness of the importance of European agricultural landscapes, their values for cultural heritage, and biodiversity. Projects:" Erasmus - FEAL: Multifunctional farming and agricultural landscapes (feal-future.org), UNESCO nomination dossier 'Großglockner High Alpine Road', Republic of Austria, Altes Land, Germany: Holler Route (cycling route within the framework of EC program of European Cultural Routes) and preparation of a UNESCO World Heritage nomination, COMUS—Community-Led Urban Strategies in Historic Towns—by the Council of Europe, assigned as expert for Armenia including Eco-tourism concepts (http://pjp-eu.coe.int/en/web/comus)

Röhner, Sina
Nürtingen-Geislingen University
Institute for Landscape and Environment
Schelmenwasen 4–8
72622 Nürtingen
Phone.: +49(0)7022 / 201-192
Email: sina.roehner@hfwu.de

Sina Röhner studied landscape planning at Nürtingen-Geislingen University and geographical information science at Salzburg University. Since 2015, she has been employed as research associate at the Institute of Landscape and Environment of Nürtingen-Geislingen University where she is involved in research projects on landscape quality, renewable energy, and grid expansion and in lectures on GIS and landscape related topics.

Projects: 2015–2018: COST Action TU1401 'Renewable Energy and Landscape QualitY (RELY)'
2015–2017: Nationwide scenic quality assessment for Germany as a basis for high voltage grid expansion

2015: Dezent Zivil—Decisions on decentral energy production in civil society
2015: Development and application of methods for interactive visualisation of different building alternatives for the 'Hölderlinhaus' in Nürtingen
2015: Visual landscape assessment in Germany—State of Research and Practice

Roth, Michael
Nürtingen-Geislingen University
School of Landscape Architecture,
Environmental and Urban Planning
Schelmenwasen 4-8
72622 Nürtingen
Germany
Tel: +49 7022 201-181
Fax: +49 7022 201-166
E-Mail: michael.roth@hfwu.de
WWW: http://www.hfwu.de/michaelroth

Prof. Dr. Michael Roth studied landscape architecture and landscape planning in Dresden. He obtained his PhD at the School of Spatial Planning, Dortmund University of Technology. He held research and teaching positions at TU Berlin (2002–2006), TU Dortmund (2006–2013), Michigan State University (2011–2012), University of British Columbia (2013), and University of Natural Resources and Life Sciences in Vienna (2016). Since 2013 he has been professor for landscape planning and landscape informatics at Nürtingen-Geislingen University (Germany).

Projects: 'Visual landscape assessment as a basis for the impact mitigation regulation' funded by the Thuringian Agency for Environment and Geology (2018)
COST Action TU1401 RELY 'Renewable Energy and Landscape Quality' funded by the EU COST framework with funds from the EU framework programme Horizon 2020 (2014–2018)
'Nationwide scenic quality assessment for Germany as a basis for high voltage grid

expansion' funded by the Federal Agency for Nature Conservation (2015–2017)
'Decisions on decentral energy production in civil society' funded by the Federal Ministry of Education and Research (2015)
'Visual landscape assessment in Germany: State of Research and Practice' funded by the Federal Agency for Nature Conservation (2015)

Schmitz, Serge
University of Liege, Clos Mercator, 3, 4000 Liege (Belgium)
Email: S.Schmitz@uliege.be
Phone: +32 43665629

Serge Schmitz is professor of human geography at the University of Liege where he teaches courses in rural geography, tourism, and landscape analysis. In 2007, he created the Laboratory for the Analysis of Places, Landscape, and European Countryside (www.laplec.ulg.ac.be). Since 2016, he has co-chaired the Commission on the Sustainability of Rural Systems of the International Geographical Union.

Selected publications:
Dubois C., Firmino A., Kim D. C., Schmitz S. (eds.), 2017. Balancing Heritage and Innovation: Pathways towards the Sustainability of Rural Systems. BSGLg, 69, 94 p.
Ciervo M., Schmitz S., 2017. Sustainable biofuel: a question of scale and aims, Moravian Geographical Reports, 25/4, 220–233.
Schmitz S., Vanderheyden V., 2016, Reflexive Loops on Scaling issues in Landscape Quality Assessment, Land Use Policy, 53, 3–7.
Schmitz S., Vanderheyden V., Vanden Broucke S., Loopmans M., 2012, The Shaping of Social Attitudes toward Energy-Parks in the Belgian Countryside, Horizons in geography, 81, 83–93.
Van Hecke E., Antrop M., Schmitz S., Van Eetvelde V., Sevenant M., 2010, Paysages, Monde rural et agriculture, Atlas de Belgique, vol. 2, Academia Press, Gent, 74

Stober, Dina

Faculty of Civil Engineering Osijek
Josip Juraj Strossmayer University of Osijek
Vladimira Preloga 3
HR-31000 Osijek
Croatia
dstober@gfos.hr

Dina Stober has studied architecture and
urban planning at the Faculty of Architecture,
University of Zagreb, Croatia. She was awarded
a PhD in spatial and urban planning (2013)
from the University of Ljubljana, Slovenia. She
is Assistant Professor at the Faculty of Civil Engineering Osijek, University of Osijek where she
teaches courses in rural planning, urban planning and design of industrial buildings. Her
interest is in digital heritage. She is active me-
meber of Association od Architects Osijek. She
has published papers on rural settlement, urban
public spaces and place attachment topics.

COST Action TU1401 RELY 'Renewable Energy and Landscape Quality'
funded by the EU COST framework
with funds from the EU framework programme Horizon 2020 (2014–2018)
UNIREG IMPULSE Regional Universities as initiators of transnational region of
knowledge, funded by IPA Cross-border
Co-operation Programme 2011 (2011–2013)
CARDS 2003 Sustainable development of
small family farms in Baranja (2005–2006)

Van der Horst, Dan

Geography Building
University of Edinburgh,
Drummond Street,
EH 89XP, Edinburgh
United Kingdom
Email: Dan.vanderHorst@ed.ac.uk

Dan van der Horst is a reader in energy,
environment, and society at the School
of Geosciences, University of Edinburgh.
His research queries in particular the role
of geography and the potential fate of our
landscapes in the aspired transition towards a
society that is less environmentally unsustain-
able and less socially unfair. His publications
examine amongst others the spatially unequal
impacts of energy policy, the characteristics
and politics of ('local') resistance to renewable energy projects and the ethics of visual
aesthetics in relation to landscape change.

Projects: 'Landscapes of Energies', funded by
the European Science Foundation.
'Energyscapes and Ecosystem Services',
funded by the UK research councils.
'Seanergy 2020'; best practice in marine spatial
planning for renewables, funded by the EU.
'Energy Landscapes: Innovation, development
& internationalization of research',
with Geonika (Czech Republic),
funded by the European Social Fund.
'TEDDINET; Transforming Energy
Demand in Buildings through Digital
Innovation', research network funded
by the UK research councils.
'Review of biomass energy technologies
for sub Saharan Africa', funded by the UK
Department for International Development.
'Renewable energy & adaptive governance
in rural policy', funded by
the Czech Council of Research.
'Forest 2020'; Political economy of
forest monitoring & protection, funded
by the UK Space Agency.

List of Authors

First name	Last name	Institution	Country
Rachelle	Alterman	Technion - Israel Institute of Technology, Center for Urban and Regional Planning, Haifa	Israel
Gisele	Alves	Glasgow Caledonian University (GCU), School of Engineering and Built Environment	United Kingdom
Kathrin	Ammermann	Federal Agency for Nature Conservation (BfN), Leipzig	Germany
Dragi	Antonijevic	Innovation Center, Faculty of Mechanical Engineering, University of Belgrade, Belgrade	Serbia
Brian	Azzopardi	Institute of Electrical and Electronic Engineering - Malta College of Arts, Science and Technology (MCAST), Paola	Malta
Henk	Baas	Cultural Heritage Agency of the Netherlands (RCE)	The Netherlands
Emel	Baylan	Department of Landscape Architecture, Faculty of Architecture and Design, University of Yuzuncu Yil, Van	Turkey
Karl	Benediktsson	Faculty of Life and Environmental Sciences, University of Iceland - School of Engineering and Natural Sciences, Reykjavik	Iceland
Tadej	Bevk	Department of Landscape Architecture, Biotechnical Faculty, University of Ljubljana	Slovenia
Maria	Boştenaru Dan	Department of Urban and Landscape Design, Faculty of Urban Planning, Ion Mincu University of Architecture and Urban Planning, Bucharest	Romania
Michele	Bottarelli	University of Ferrara	Italy
Ken	Boyle	Department of Environment and Planning, Dublin Institute of Technology - School of Planning and Transport Engineering	Ireland
Pat	Brereton	Dublin City University - School of Communications	Ireland
Paulo	Brito	Instituto Politécnico de Portalegre	Portugal
Matthias	Buchecker	Swiss Federa | Research Institute WSL, Economics and Social Sciences, Birmensdorf	Switzerland
Csaba	Centeri	Department of Nature Conservation and Landscape Ecology, Szent István University, Gödöllö	Hungary
Csaba	Csontos	Eötvös Loránd University, Budapest	Hungary
Cheryl	De Boer	University of Twente	The Netherlands
Aleksandra	Dedinec	Faculty of Computer Science and Engineering, "Ss. Cyril and Methodius" University, Skopje	Republic of Macedonia
Blanca	Del Espino	Department of Urban and Land Planning, University of Seville	Spain
Sokol	Dervishi	Department of Architecture, Epoka University, Tirana	Albania
Zlata	Dolacek-Alduk	University of Josip Juraj Strossmayer Osijek, Faculty of Civil Engineering, Department for Organisation, Technology and Management	Croatia
Sebastian	Eiter	Department of Landscape Monitoring, NIBIO – Norwegian Institute of Bioeconomy Research, Ås	Norway
Wendy	Fjellstad	Department of Landscape Monitoring, NIBIO – Norwegian Institute of Bioeconomy Research, Ås	Norway

Bohumil	Frantál	Department of Environmental Geography, Institute of Geonics, Czech Academy of Sciences, Brno	Czech Republic
Marina	Frolova	Institute of Regional Development, University of Granada	Spain
Bénédicte	Gaillard	Consulting in World Heritage Conflict Management, Les Issambres	France
Giedrė	Godienė	Institute of Geosciences, Vilnius University	Lithuania
Mojca	Golobič	Department of Landscape Architecture, Biotechnical Faculty, University of Ljubljana	Slovenia
Mihaela	Hărmănescu	Faculty of Architecture, Ion Mincu University of Architecture and Urbanism, Bucharest	Romania
Marton	Havas	Eotvos Lorand University, Budapest	Hungary
Benjamin	Hennig	Faculty of Life and Environmental Sciences, University of Iceland	Iceland
Verónica	Hernán-dez-Jiménez	Research Group of Ecology and Landscape in the Polytechnis Univesity of Madrid, Observatory for a culture of the territory	Spain
Daniel	Herrero	Dpto. Geografía, Facultad Filosofia y Letras, Universidad de Valladolid	Spain
Richard	Hewitt	Research Group of Ecology and Landscape in the Polytechnis Univesity of Madrid, Observatory for a culture of the territory	Spain
Marcel	Hunziker	Swiss Federal Institute for Forest, Snow and Landscape Research WSL, Birmensdorf	Switzerland
Artan	Hysa	Department of Architecture, Epoka University, Tirana	Albania
Nela	Jantol	OIKON Ltd. – Institute of Applied Ecology, Zagreb	Croatia
Berthe	Jongejan	Cultural Heritage Agency of the Netherlands (RCE)	The Netherlands
Jovana	Jovanovic	Faculty of Managenemt Herceg Novi	Montenegro
Robert	Kabai	Hunscapes Co. Ltd., Budapest	Hungary
Vania	Kachova	Forest Research Institute, Bulgarian academy of Sciences, Sofia	Bulgaria
Matea	Kalcicek	Ekoinvest – Consulting Ltd., Zagreb	Croatia
Isidora	Karan	Center for Spatial Research, Banjaluka	Bosnia and Herzegovina
Mirko	Komatina	Faculty of Mechanical Engineering, University of Belgrade	Serbia
Alexandra	Kruse	Institute for Research on European Agricultural Landscapes (EUCALAND) e.V., Buc	France
Ain	Kull	Department of Geography, University of Tartu	Estonia
Mart	Külvik	Institute of Agricultural and Environmental Sciences, Estonian University of Life Sciences, Tartu	Estonia
Igor	Kuvac	Center for Spatial Research, Banjaluka	Bosnia and Herzegovina
Marija	Lalosevic	Urban Planning Institute of Belgrade	Serbia
Raffaella	Laviscio	Department of architecture, built environment and construction engineering A.B.C., Polytecnico die Milano	Italy
Gerald	Leindecker	Hochschule Oberösterreich, Linz	Austria
Naja	Marot	Department of Landscape Architecture, Biotechnical Faculty, University of Ljubljana	Slovenia
Stanislav	Martinát	Department of Environmental Geography, Institute of Geonics, Czech Academy of Sciences, Brno	Czech Republic
Georgios	Martinopoulos	International Hellenic University - School of Science and Technology, Thessaloniki	Greece
Sennan	Mattar	Glasgow Caledonian University (GCU), School of Engineering and Built Environment	United Kingdom
Adolfo	Mejia Montero	University of Edinburgh	United Kingdom
Daniel	Micallef	Faculty of the Built Environment, Department of Environmental Design, University of Malta, Valletta	Malta
Renata	Mikalauskiene	Malta College of Arts, Science and Technology (MCAST), Paola	Malta
David	Miller	James Hutton Institute, Aberdeen	United Kingdom
Stefanie	Müller	Swiss Federal Institute of Forest, Snow and Landscape (WSL), Birmensdorf	Switzerland

Bela Attila	Munkácsy	Environmental Education Network, Szigetszentmiklós	Hungary
Emilio	Munoz-Ceron	Engineering Projects Area, University of Jaén	Spain
Symi	Nyns	Department of Geography, University of Liege	Belgium
Pia	Otte	RURALIS – Institute of Rural and Regional Research, Trondheim	Norway
Hannes	Palang	Centre for Landscape and Culture, Tallinn University, Tallinn	Estonia
Nikos	Papamanolis	Technical University of Crete, Chania	Greece
Paolo	Picchi	University of Trento	Italy
Christiane	Plum	Swiss Federal Institute for Forest, Snow and Landscape Research WSL, Birmensdorf	Switzerland
María-José	Prados	Dpto. Geografia Humana, Universidad de Sevilla, Sevilla	Spain
Sina	Röhner	Institute for Landscape and Environment, Nürtingen-Geislingen University	Germany
Katrina	Rønningen	RURALIS – Institute of Rural and Regional Research, Trondheim	Norway
MIchael	Roth	Nürtingen-Geislingen University, School of Landscape Architecture, Environmental and Urban Planning, Nürtingen	Germany
Mădălina	Sbarcea	Danube Delta National Institute for Research and Development, Tulcea	Romania
Serge	Schmitz	Department of Geography, University of Liege	Belgium
Kim Philip	Schumacher	Institute for Spatial Analysis and Planning in Areas of Intensive Agriculture (ISPA), University of Vechta	Germany
Alessandra	Scognamiglio	Dipartimento Tecnologie Energetiche; Divisione Fotovoltaico e Smart Networks, ENEA, Portici	Italy
Luis	Silva	Faculdade de Ciências Sociais e Humanas, Centre for Research in Anthropology, Universidade Nova de Lisboa	Portugal
Georgia	Sismani	School of Civil Engineering, Aristotle University of Thessaloniki	Greece
Martina	Slámová	Department of Landscape Planning and Design, Technical University in Zvolen	Slovakia
Mateusz	Slupinski	Department of Environmental Engineering, Wroclaw University of Technology	Poland
Tamás	Soha	Eötvös Loránd University, Budapest	Hungary
Svetlana	Stevovic	Faculty of Managenemt Herceg Novi	Montenegro
Ivan	Stevovic	Faculty of Managenemt Herceg Novi	Montenegro
Dina	Stober	Faculty of Civil Engineering, Josip Juraj Strossmayer University, Osijek	Croatia
Biljana Risteska	Stojkoska	Faculty of Computer Science and Engineering, "Ss. Cyril and Methodius" University, Skopje	Republic of Macedonia
Sven	Stremke	Amsterdam University of Arts, Academy of Architecture	The Netherlands
Monika	Suškevičs	Institute of Agricultural and Environmental Sciences, Estonian University of Life Sciences, Tartu	Estonia
Na'ama	Teschner	Technion - Israel Institute of Technology	Israel
Attila	Tóth	Department of Garden and Landscape Architecture, Slovak University of Agriculture in Nitra	Slovakia
Dan	Van der Horst	Department of Environmental Geography, Institute of Geonics, Czech Academy of Sciences, Brno and School of Geosciences, University of Edinburgh	United Kingdom
Anneloes	Van Noordt	Spatial Development Department Flanders, Ruimte Vlaanderen, Afdeling Onderzoek, Ondersteuning en Monitoring, Brussels	Belgium
Darijus	Veteikis	Institute of Geosciences, Vilnius University	Lithuania
Elis	Vollmer	Institute of Agricultural and Environmental Sciences, Estonian University of Life Sciences, Tartu	Estonia
Margarita	Vološina	Department of Geography, University of Latvia, Riga	Latvia
Edda R.H.	Waage	University of Iceland, Reykjavik	Iceland
Bruno	Zanon	Department of Civil, Environmental and Mechanical Engineering, University of Trento	Italy
Anita	Zariņa	Department of Geography, University of Latvia, Riga	Latvia

Imprint

© 2018 by jovis Verlag GmbH
Texts by kind permission of the authors.
Pictures by kind permission of the photographers/holders of the picture rights.

Language Check: David Miller, Pat Brereton
Proofreading: Michael Taylor
Design and setting: Susanne Rösler, jovis
Lithography: Bild1Druck, Berlin
Printed in the European Union

Bibliographic information published by the Deutsche Nationalbibliothek.
The Deutsche Nationalbibliothek lists this publication in the Deutsche Nationalbibliografie; detailed bibliographic data are available on the Internet at http://dnb.d-nb.de.

jovis Verlag GmbH
Kurfürstenstraße 15/16
10785 Berlin

www.jovis.de

jovis books are available worldwide in selected bookstores. Please contact your nearest bookseller or visit www.jovis.de for information concerning your local distribution.

ISBN 978-3-86859-524-6